THE SOCIETY FOR APPLIED BACTERIOLOGY
SYMPOSIUM SERIES NO. 23

Fundamental and Applied Aspects of Bacterial Spores

Edited by G.W. Gould,
A.D. Russell
and
D.E.S. Stewart-Tull

BLACKWELL SCIENTIFIC PUBLICATIONS
OXFORD LONDON EDINBURGH BOSTON
MELBOURNE PARIS BERLIN VIENNA

ISBN 978-0-86542-897-3

Contents

vS Contributors

viiS Preface

1S Classification and identification of endospore-forming bacteria
R.C.W. BERKELEY & N. ALI

9S The genetic analysis of bacterial spore germination
ANNE MOIR, E. HELEN KEMP, C. ROBINSON & B.M. CORFE

17S The trigger mechanism of spore germination: current concepts
K. JOHNSTONE

25S The role and regulation of cell wall structural dynamics during differentiation of endospore-forming bacteria
S.J. FOSTER

40S Molecular mechanisms of resistance to heat and oxidative damage
R.E. MARQUIS, J. SIM & S.Y. SHIN

49S Mechanisms which contribute to the long-term survival of spores of *Bacillus* species
P. SETLOW

61S *Bacillus cereus* and its toxins
P.E. GRANUM

67S Spore resistance and ultra heat treatment processes
K.L. BROWN

81S Tolerance of spores to ionizing radiation: mechanisms of inactivation, injury and repair
J. FARKAS

91S Mechanisms of inactivation and resistance of spores to chemical biocides
SALLY F. BLOOMFIELD & M. ARTHUR

105S Effects of water activity and pH on growth of *Clostridium botulinum*
P.J. MCCLURE, M.B. COLE & J.P.P.M. SMELT

115S Heat resistance and recovery of spores of non-proteolytic *Clostridium botulinum* in relation to refrigerated, processed foods with an extended shelf-life
BARBARA M. LUND & M.W. PECK

129S Bioluminescence and spores as biological indicators of inimical processes
P.J. HILL. L. HALL, D.A. VINICOMBE, C.J. SOPER, P. SETLOW, W.M. WAITES, S. DENYER & G.S.A.B. STEWART

135S Subject index

Contributors

N. ALI
Department of Pathology and Microbiology, School of Medical Sciences, University Walk, Bristol BS8 1TD, UK

M. ARTHUR
Department of Pharmacy, King's College London, Manresa Road, London SW3 6LX, UK

R.C.W. BERKELEY
Department of Pathology and Microbiology, School of Medical Sciences, University Walk, Bristol BS8 1TD, UK

SALLY F. BLOOMFIELD
Department of Pharmacy, King's College London, Manresa Road, London SW3 6LX, UK

K.L. BROWN
Food Hygiene Department, Campden Food and Drink Research Association, Chipping Campden, Glos GL55 6LD, UK

M.B. COLE
Unilever Research, Colworth Laboratory, Sharnbrook, Bedford MK44 1LQ, UK

B.M. CORFE
Krebs Institute for Biomolecular Research, Department of Molecular Biology and Biotechnology, University of Sheffield, PO Box 594, Sheffield S10 2UH, UK

S. DENYER
Department of Pharmacy, University of Brighton, Cockcroft Building, Moulsecoomb, Brighton BN2 4GJ, UK

J. FARKAS
Department of Refrigeration and Livestock Products Technology, University of Horticulture and Food Industry, Menesi ut 45, 1118 Budapest, Hungary

S.J. FOSTER
Department of Molecular Biology and Biotechnology, University of Sheffield, PO Box 594, Sheffield S10 2UH, UK

P.E. GRANUM
Department of Pharmacology, Microbiology and Food Hygiene, Norwegian College of Veterinary Medicine, PO Box 8146, 0033 Oslo, Norway

L. HALL
Department of Applied Biochemistry and Food Science, University of Nottingham, Sutton Bonington Campus, Sutton Bonington, Leicestershire LE12 5RD, UK

P. J. HILL
Department of Applied Biochemistry and Food Science, University of Nottingham, Sutton Bonington Campus, Sutton Bonington, Leicestershire LE12 5RD, UK

K. JOHNSTONE
Department of Plant Sciences, University of Cambridge, Downing Street, Cambridge CB2 3EA, UK

E. HELEN KEMP
Krebs Institute for Biomolecular Research, Department of Molecular Biology and Biotechnology, University of Sheffield, PO Box 594, Sheffield S10 2UH, UK

BARBARA M. LUND
8 The Walnuts, Branksome Road, Norwich NR4 6SR, UK

R.E. MARQUIS
Department of Microbiology and Immunology, Box 672, University of Rochester Medical Center, Rochester, NY 14642-8672, USA

P.J. McCLURE
Unilever Research, Colworth Laboratory, Sharnbrook, Bedford MK44 1LQ, UK

ANNE MOIR
Krebs Institute for Biomolecular Research, Department of Molecular Biology and Biotechnology, University of Sheffield, PO Box 594, Sheffield S10 2UH, UK

M.W. PECK
Institute of Food Research, Norwich Laboratory, Norwich Research Park, Colney, Norwich NR4 7UA, UK

C. ROBINSON
Krebs Institute for Biomolecular Research, Department of Molecular Biology and Biotechnology, University of Sheffield, PO Box 594, Sheffield S10 2UH, UK

P. SETLOW
Department of Biochemistry, University of Connecticut Health Center, Farmington, CT 06030-3305, USA

S.Y. SHIN
Department of Microbiology and Immunology, Box 672, University of Rochester Medical Center, Rochester, NY 14642-8672, USA

J. SIM
Department of Microbiology and Immunology, Box 672, University of Rochester Medical Center, Rochester, NY 14642-8672, USA

J.P.P.M. SMELT
Unilever Research, Vlaardingen Laboratory, Olivier van Noortlaan 120, Vlaardingen, The Netherlands

C.J. SOPER
School of Pharmacy and Pharmacology, University of Bath, Bath BA2 7AY, UK

G.S.A.B. STEWART
Department of Applied Biochemistry and Food Science, University of Nottingham, Sutton Bonington Campus, Sutton Bonington, Leicestershire LE12 5RD, UK

D.A. VINICOMBE
School of Pharmacy and Pharmacology, University of Bath, Bath BA2 7AY, UK

W.M. WAITES
Department of Applied Biochemistry and Food Science, University of Nottingham, Sutton Bonington Campus, Sutton Bonington, Leicestershire LE12 5RD, UK

Preface

Because no major Symposium on bacterial endospores had been held in the UK for several years, we deemed it apposite and timely to bring together experts from around the world to present new data and to review recent advances, targeting both the fundamental and the applied aspects of this topic. The deliberations of that meeting, held at the University of Nottingham on the 14th and 15th July 1993, form the basis of this volume. They provide a detailed and up-to-date account of the unique and fascinating form of life that is exemplified by the bacterial spore.

The contributions are broad-ranging. They deal with the basics of taxonomy and the classification and identification of sporeformers, the genetics of sporulation and germination, the 'trigger' mechanism of germination, the regulation of cell-wall structural dynamics and the toxins of *Bacillus cereus*. Recent advances in the understanding of mechanisms of resistance are covered in papers on the roles of the small acid-soluble proteins that are characteristic components of spores, on heat resistance, on ultraheat treatment processing, on biocides and oxidizing agents and on ionizing radiation. The growth of *Clostridium botulinum*, and its prevention and control in foods are considered, and a final paper describes the novel application of genetically-modified spores, employing bioluminescence, as indicators of inimical processes.

It is hoped that this Symposium volume will be of value not only to research microbiologists, geneticists and biochemists who are interested in the fundamentals of spore science, but also to those scientists who are concerned with the thoroughly practical problems of spore destruction and control in the food and pharmaceutical industries and in public health.

The Editors are grateful to all those who gave generously of their time, at the Symposium itself, and in the preparation of these proceedings.

G.W. Gould
*A.D. Russell**
D.E.S. Stewart-Tull†
Unilever Research Laboratory
Colworth House, Sharnbrook
Bedford MK44 1LQ, UK
**Welsh School of Pharmacy*
University of Wales College of Cardiff
Cardiff CF1 3XF, UK, and
†Department of Microbiology
Glasgow University
Glasgow G12 8QQ, UK

Journal of Applied Bacteriology Symposium Supplement 1994, **76**, 1S–8S

Classification and identification of endospore-forming bacteria

R.C.W. Berkeley and N. Ali
Department of Pathology and Microbiology, School of Medical Sciences, Bristol, UK

1. Introduction, 1S
2. Classification within the endospore-forming genera
   2.1 *Alicyclobacillus*, 2S
   2.2 *Amphibacillus*, 2S
   2.3 *Bacillus*, 2S
   2.4 *Clostridium*, 3S
   2.5 *Desulfotomaculum*, 3S
   2.6 *Oscillospira*, 3S
   2.7 *Pasteuria*, 3S
   2.8 *Sporohalobacter*, 3S
   2.9 *Sporolactobacillus*, 3S
   2.10 *Sporosarcina*, 4S
   2.11 *Sulfobacillus*, 4S
   2.12 *Syntrophospora*, 4S
   2.13 *Thermoactinomyces*, 4S
   2.14 Genera *incertae sedis*, 4S
   2.15 Overall taxonomic arrangement, 4S
3. Identification
   3.1 Identification to the generic level, 5S
   3.2 Identification of species, 5S
   3.3 Non-classical approaches to identification, 6S
4. Concluding remarks, 6S
5. References, 6S

1. INTRODUCTION

At the end of 1992 the endospore-formers were allocated to 13 validly published genera which, taken together, form a group of some morphological, physiological and genetic diversity. Applying the generally accepted rule that bacteria with DNA base compositions differing by more than 10 mol% G + C (%GC) should not be regarded as members of the same genus (see Bull *et al.* 1992) to the values shown in Table 1, it can be seen that several of these genera are heterogeneous.

Also, within some of the genera, there are species with a range of %GC values well in excess of the 5% span widely recognized as the maximum difference permissible between organisms belonging to the same species (see Bull *et al.* 1992).

To complicate an unsatisfactory situation still further, there are a number of endospore-formers which have been described on the basis of their microscopical appearance but not isolated, or been cultured, characterized and the original culture lost, or been isolated but insufficiently well studied for their taxonomic relationships to be reasonably clear.

Taxonomists will, therefore, see room for improvement in the current arrangement of the endospore-forming organisms.

Before considering possible approaches to achieving this improvement, the present classification in each genus will be described briefly. Approaches to the identification of endospore-forming isolates will then be summarized.

Table 1 Validly described genera of endospore-forming bacteria and their DNA base composition

Genus	Mol% GC
Alicyclobacillus	52–60
Amphibacillus	36–38
Bacillus	32–69
Clostridium	22–54
Desulfotomaculum	38–52
Oscillospira	—
Pasteuria	—
Sporohalobacter	30–32
Sporolactobacillus	38–40
Sporosarcina	40–42
Sulfobacillus	54
Syntrophospora	38
Thermoactinomyces	52–55

— = No information available.

Note—the review on which this article is based was completed at the end of December 1992. Exceptionally, the ninth edition of *Bergey's Manual of Determinative Bacteriology* (Holt *et al.* 1994), which was published as this article went to press, is referred to as it contains information on the identification of many, but not all, the organisms mentioned in this article.

Correspondence to: R.C.W. Berkeley, Department of Pathology and Microbiology, School of Medical Sciences, University Walk, Bristol BS8 1TD, UK.

2 CLASSIFICATION WITHIN THE ENDOSPORE-FORMING GENERA

2.1 *Alicyclobacillus*

As is noted below (Section 2.3), the phenotypic and genetic heterogeneity of the genus *Bacillus* is great and it has been evident for some considerable time that division of the genus is warranted. One step in this direction is the creation of the genus *Alicyclobacillus* for the thermoacidophilic species peviously known as *Bacillus acidocaldarius*, *Bacillus acidoterrestris* and *Bacillus cycloheptanicus*. This new taxon is based on 16S rRNA catalogues of five strains, including three of *B. acidocaldarius* (Wisotzkey *et al.* 1992). The similarity of *B. acidocaldarius* to *B. acidoterrestris* is extremely high (98·8%) and invites the question of whether these two organisms really do belong to separate species. The similarities of *B. acidocaldarius* and *B. acidoterrestris* to *B. cycloheptanicus* are respectively 93·2 and 92·7% (Wisotzkey *et al.* 1992). In keeping with this rearrangement, Ash and her colleagues (1991) found that *B. cycloheptanicus* had a very low similarity with all the other strains of *Bacillus* examined but *B. acidoterrestris* was found to lie in group 1, in an entirely different part of the phylogenetic tree to *B. cycloheptanicus*. It has subsequently emerged that the strain of *B. acidoterrestris* studied in this work was different from that used by Wisotzkey *et al.* (1992).

2.2 *Amphibacillus*

Organisms newly isolated by Niimura and his colleagues (1990) form the basis of this new genus. These grow and form spores under both aerobic and anaerobic conditions. Spore formation in the presence and absence of air distinguishes them from the genera *Bacillus*, *Clostridium* and *Sporolactobacillus*. The low DNA homology between three strains of *Amphibacillus xylanus*, the only species, which have a %GC range of 36–38, and a number of *Bacillus* and *Clostridium* species with similar DNA contents and *Sporolactobacillus inulinus*, confirms that these organisms represent a new taxon.

2.3 *Bacillus*

The work which underlies the current classification of this genus was on 621 mesophilic *Bacillus* strains which were allocated to 15 species. The study was extended by the inclusion of 513 more strains and the addition of four more species (Smith *et al.* 1952). This extremely careful and thorough investigation led to the arrangement of the 19 species in three groups based on morphology and physiology, giving some indication of the supposed relationships of the species within those groups to each other.

Several numerical studies, by Sneath (1962), Bonde (1975), Logan and Berkeley (1981), Priest *et al.* (1981) and Priest *et al.* (1988), as well as early molecular work (e.g. Doi and Igarashi 1965; Fox *et al.* 1977) provided a good measure of support for the arrangement of Smith and his colleagues (1952) although the taxometric studies all indicated the probable existence of more than three groups of species. In some instances the groups were regarded as of generic rank. No attempt was made, however, to create new genera pending the establishment of a firm basis on which to do this.

The molecular investigations also indicated close, and sometimes unexpected, relationships with species other than *Bacillus* species. *Sporosarcina urea*, *B. megaterium*, *B. pasteurii* and *B. subtilis* were shown to have DNA : rRNA homologies in excess of 73% (Herndon and Bott 1969) and *Sporosarcina urea* was demonstrated to be more closely related to *B. pasteurii* than the latter is to *B. subtilis* or to *B. stearothermophilus* (Pechman *et al.* 1976). Further confirmation of the close connection between *Bacillus* species with round spores, such as *B. pasteurii*, and *Sporosarcina urea* was provided by Stakebrandt and his collaborators (1987) who also demonstrated the affinity between another round-spore-former, *B. sphaericus*, and non-spore-forming organisms such as *Caryophanon latum*, *Filibacter limicola* and *Planococcus citreus*.

Meanwhile, early observations indicating lack of homogeneity of some species in the genus, for example *B. circulans* (Gibson and Topping 1938), and of the genus itself (Marmur *et al.* 1963), were confirmed by very many workers (see Jones and Sneath 1970; Claus and Berkeley 1986).

The unsatisfactory nature of the classification of this genus arising from its demonstrated close relationships with non-spore-formers and the heterogeneity of many of its species as well as of the genus itself, which now has a %GC span from 33 to 69, has not always been improved by the valid description of the many new species since the publication of the approved list in 1980 (Skerman *et al.*). At that time it contained 31 species; by the end of 1992 the number had risen to 67. Some of the additions are based on only very few strains and it may be that further study, including more isolates, will reveal that the creation of some of these taxa was unwise.

Some of the earliest molecular systematics studies included *Bacillus* strains but until 1991 only nine species had been investigated by 16S rRNA cataloguing. In that year Ash and her colleagues examined the sequences of the small subunit rRNA of 51 species represented by a single strain (in all but one instance the type strain). The kind of problem that may occur as a consequence of the lack of the internal experimental evidence provided by the use of more than one strain has been outlined in Section 2.1.

The results of this work indicate the existence of five phylogenetically distinct clusters. There is some measure of agreement between these and the groupings which emerged from earlier phenotypic work (Smith *et al.* 1952; Logan and Berkeley 1981; Priest *et al.* 1988), often in areas of the genus for which large numbers of strains are available and which have been extensively studied. There are, however, areas where the grouping based on molecular evidence is not consistent with that based on existing phenotypic data.

2.4 *Clostridium*

Clostridium species have %GC values of 22–55 with a majority having values in the range 25–30%GC (Cato *et al.* 1986) and fall clearly into four groups on the basis of 16S rRNA catalogues. As with the genus *Bacillus*, the molecular evidence indicates relationships with non-spore-forming genera (Fox *et al.* 1980; Woese *et al.* 1980) and again, as for *Bacillus*, the present classification of *Clostridium* is recognized to be unsatisfactory (Williams *et al.* 1992). The number of species in the approved lists (Skerman *et al.* 1980) had, by the end of 1992, increased by more than 50% to well over 100 with many represented by only one strain whereas for others hundreds are available and their classification is not based on sound, modern taxonomic procedures (Williams *et al.* 1992).

Two steps towards the rearrangement of this taxon have been made. One new taxon, *Sporohalobacter* (see section 2.8), based on an organism originally described as *Cl. lortetii* has been created for two strains of a halophilic organism showing structural and physiological differences from *Clostridium* (Oren 1983). Another new group, *Syntrophospora* (see Section 2.12), has been erected for *Cl. bryantii*, which also differs from *Clostridium* structurally and physiologically and has been shown to be phylogenetically distinct (Zhao *et al.* 1990).

2.5 *Desulfotomaculum*

Once included with the genus *Clostridium*, the members of the ten validly described species of the genus *Desulfotomaculum*, a large number of which are represented by only one strain, have in common the ability to reduce sulphate to hydrogen sulphide. Four species, *D. geothermicum*, *D. kuznetsovii*, *D. nigrificans* and *D. thermobenzoicum*, are thermophilic and six are mesophilic. *Desulfotomaculum antarcticum* and *D. guttoideum* have growth temperature optima between 20 and 32°C whereas those for *D. acetoxidans*, *D. orientis*, *D. ruminis* and *D. sapomandens* are from 36 to 38°C (Gogotova and Vainstein 1983; Hobbs and Cross 1984; Cord-Ruwisch and Garcia 1985; Campbell and Singleton 1986; Daumas *et al.* 1988; Nazina *et al.* 1988; Tasaki *et al.* 1993).

This genus, too, is heterogeneous. The %GC values for its members range from 38 to 52 (Campbell and Singleton 1986). Diversity is also indicated by 16S rRNA cataloguing (Fowler *et al.* 1986), with the 16S rRNA sequences of *D. orientis* and *D. ruminis* being only 83% similar (Devereux *et al.* 1989). Devereux and his colleagues infer independent rather than common origins for these organisms.

2.6 *Oscillospira*

One of the endosporing organisms apparently not yet grown in pure culture is the genus *Oscillospira*. It is, however, validly described and included in *Bergey's Manual* (Gibson 1986). The genus has one species, *O. guillermondii*.

2.7 *Pasteuria*

These organisms have not been grown in pure culture; they are cultured with a nematode host. There has been considerable taxonomic confusion about this endospore-former, involving the designation of a non-sporing budding bacterium as the type culture. This has been clearly summarized by Sayre and Starr (1989). *Pasteuria* species have branching hyphae, the terminal parts of which enlarge to form sporangia in which ellipsoidal to nearly spherical endospores, of major dimension 1·0–3·0 μm, are formed. Four species are recognized, three described by Sayre and Starr (1989) and another by Sayre and colleagues (1991). Some molecular evidence indicates that *P. penetrans* is a deeply rooted member of the *Clostridium*/*Bacillus* line of descent, neither related to the actinomycetes nor closely to the true endospore-formers (Stakebrandt, personal communication).

2.8 *Sporohalobacter*

Studies with 16S rRNA show that this genus is related to *Halobacteroides* and *Haloanaerobium*. It contains two species of halophilic anaerobic organisms, one of which, *Sporohalobacter lortetii*, has gas vacuoles associated with its spore. Both species are represented by only one strain (Oren *et al.* 1987).

2.9 *Sporolactobacillus*

Sporolactobacillus inulinus is the single species in this genus. Although it differs from the majority of *Bacillus* species, and resembles members of the genus *Lactobacillus*, in carrying out a homolactic fermentation, not producing catalase, not having cytochromes and not reducing nitrate or forming indole, *Sporolactobacillus inulinus* does produce tyical endospores and has the MK-7 menaquinone usually found in the genus *Bacillus*. Furthermore 16S rRNA studies have confirmed that it is correctly regarded as a close relative of these aerobic endospore-forming organisms (Fox *et al.* 1977; Stakebrandt *et al.* 1981).

2.10 *Sporosarcina*

Although the spore-forming ability of the organisms in this genus was first noted and their possible close relationship with the genus *Bacillus* suggested by Beijerinck (1901), when he first described these organisms, microbiologists have been uncomfortable with the idea of spore-forming cocci (Norris 1981) until quite recently. Molecular evidence of the close relationship of *Sporosarcina* with *B. pasteurii* has, however, led to the acceptance of *Sporosarcina* as a close relative of the aerobic endospore-forming rods (see Section 2.2). There are two species—*S. ureae* and *S. halophila* (Claus and Fahmy 1986) each represented by a few strains.

2.11 *Sulfobacillus*

This genus contains one species, based on a single strain, which resembles *Thiobacillus ferroxidans* in its substrates for energy metabolism and in being acidophilic, but differs from it in producing endospores and being a thermophile (Golovacheva and Karavaiko 1978).

2.12 *Syntrophospora*

Once classified as *Clostridium bryantii* and only grown in syntrophic coculture with *Methanospirillum hungatei*, the one species in this genus has now been grown in pure culture and its 16S rRNA sequence analysed. This shows that it is not closely related to any *Clostridium* species and it has been renamed *Syntrophospora bryantii*. It is closely related to *Syntrophomonas wolfei* which does not form spores but is not classified with this organism because of morphological differences (Zhao *et al.* 1990).

2.13 *Thermoactinomyces*

The mycelial character of the growth on agar of organisms in this genus led to the inclusion at one time of *Thermoactinomyces* with the actinomycetes. Members of this genus, however, produce single, phase bright and refractile, spores which are not stained by the Gram reaction but take up hot malachite green and contain dipicolinic acid. The production of endospores is consistent with their taxonomic position close to the genus *Bacillus* indicated by evidence from 16S rRNA studies (Stakebrandt and Woese 1981). The genus contains seven species (Lacey and Cross 1989). Some of these are represented by a few strains; others have only one.

2.14 Genera *incertae sedis*

Several endosporers of uncertain position have been described but, like the genera *Oscillospira* and *Pasteuria*, not isolated in pure culture. References to the description of these organisms can be obtained from Claus and Berkeley (1986).

2.15 Overall taxonomic arrangement

It is clear from the above review that the taxonomy of the endospore-forming organisms leaves a lot to be desired. This is not surprising given that no representative of at least two genera, and almost certainly several more, as yet to be validly described, has been obtained in pure culture. Another factor contributing to the unsatisfactory taxonomy is the description of some genera and several species on the basis of single isolates.

Molecular approaches to establishing phylogenetic relationships will point to ways for some reordering but there are problems involved in so doing. These are discussed in the final section of this article. Thus an unsatisfactory classification of these organisms must be assumed to be with us for some time to come.

3. IDENTIFICATION

Given an unsatisfactory classification, identification may not be easy. Practitioners, however, using classical procedures, rapid techniques, miniatuarized methods, chemotaxonomic approaches or specific probes need to, and generally do, achieve identification, sometimes very quickly, but often rather slowly. Intellectually, the need for a rational classification is compelling; practically, the greater need is for a universally applicable method allowing identification within minutes of the arrival of a specimen or sample in the laboratory.

Using classical approaches, identification to the generic and species levels can be achieved as summarized below.

3.1 Identification to the generic level

(I) Cells rod shaped.
(1) Aerobic or facultative and non-sporing under anaerobic conditions, catalase usually produced.
(i) Growth at 45°C and pH 3·0.
(a) Ferrous iron, sulphide and sulphur not used as energy sources.

Alicyclobacillus

(b) Ferrous iron, sulphide and sulphur used as energy sources.

Sulfobacillus

(ii) No growth at 45°C and pH 3·0.

Bacillus

(2) Facultative, produce spores anaerobically, catalase not produced.

Amphibacillus

(3) Microaerophilic, catalase not produced.

Sporolactobacillus

(4) Anaerobic.
(i) Sulphate not reduced to sulphide.
(a) Gram-positive type cell walls, do not require 0·5 mol l^{-1} NaCl for growth.
(α) Grow axenically.

Clostridium

(β) Isolated in syntrophic culture.

Syntrophospora

(b) Gram-negative type cell wall, require 0·5 mol l NaCl for growth.

Sporohalobacter

(ii) Sulphate reduced to sulphide.

Desulfotomaculum

(II) Cells spherical, in packets, aerobic.

Sporosarcina

(III) Mycelium formed.
(1) Grow axenically.

Thermoactinomyces

(2) Require an invertebrate host for growth.

Pasteuria

(IV) Rods or filaments more than 3·0 μm diameter.

Oscillospira

3.2 Identification of species

Information for the identification of the then described species of the genera then validly published are given by Hobbs and Cross (1984), in *Bergey's Manual of Systematic Bacteriology*, volumes 2 (Sneath *et al.* 1986) and 4 (Williams *et al.* 1989) and in the ninth edition of *Bergey's Manual of Determinative Bacteriology* (Holt *et al.* 1994).

Sources of information for the identification of other species are as follows: the separation of the three species of the genus *Alicyclobacillus* may be achieved with the data given by Wisotzkey *et al.* (1992).

Amphibacillus has only one species (Niimura *et al.* 1990).

Reviews of approaches to the identification of *Bacillus* species by Berkeley and his colleagues (1984) and by Priest (1989) exist. The former emphasized the use of the API system and instrumented methods, particularly pyrolysis mass spectrometry; the latter has given weight to traditional approaches. Both consider computer-assisted identification methodology. More recently Kämper (1991) has applied miniaturized tests to the identification of *Bacillus* species. The number of validly published species is now nearly double the number dealt with in these articles and information about newer species has to be sought in the original literature through the validation lists in the *International Journal of Systematic Bacteriology*.

Methods for the identification of *Clostridium* species are those developed for pathogens. They depend largely on the toxins produced and are frequently found wanting when applied to clostridia from animal sources not recognized as pathogens (Williams *et al.* 1992) or to strains isolated from the environment (e.g. Timmis *et al.* 1974). Just over 80 species were described by Cato and his colleagues (1986). Since then the number of validly described species has increased by about 50% and information about newly described strains has to be sought in the same way as for *Bacillus* species (see above).

References to information for the identification of the ten species of *Desulfotomaculum* are given in Holt *et al.* (1994) and Tasaki *et al.* (1991). The one species of *Oscillospira* is described by Gibson (1986). The morphological and structural features and the host which are the distinguishing features of *Pasteuria* species are described by Sayre *et al.* (1991).

Information for the differentiation of *Sporohalobacter lortetii* from *Sporohalobacter marismortui* is given by Oren and his colleagues (1987). Data for the identification of the one species of *Sporolactobacillus* is in Kandler and Weiss (1986), that for the two species of *Sporosarcina* in Claus and Fahmy (1986), that for the one species of *Syntrophospora* in Zhao *et al.* (1990) and for the only species of *Sulfobacillus* in Golovacheva and Karavaiko (1978).

The seven species of *Thermoactinomyces*, as well as a yet unnamed related phenon, can be recognized using the descriptions in Lacey and Cross (1989).

3.3 Non-classical approaches to identification

These have been reviewed by Berkeley and his colleagues (1984) and by Priest (1989). The merits of pyrolysis mass spectrometry (see Berkeley *et al.* 1992) as judged by the criteria of accuracy, rapidity, transferability, universal applicability and cost effectiveness are considerable. The development of laser pyrolysis techniques enabling identification to be based on single cells may allow the achievement of the goal of identification within minutes of the arrival of a new isolate in the diagnostic laboratory.

4. CONCLUDING REMARKS

There has probably not been a more difficult time since the first sensible attempts to classify the endosporers, some fifty years ago, to write the comprehensive treatment requested by the editors of the classification and identification of the endospore-formers.

There are several reasons for this, amongst which are: many of these organisms are ill-studied; following the lull after the rationalization of taxa in the approved lists (Skerman *et al.* 1980) there has been an explosive description of new taxa, sometimes based on too few strains; phylogenetic evidence points to close relationships between structurally and/or physiologically very different organisms (e.g. *Syntrophospora bryantii* and *Syntrophomonas wolfei*—see Section 2.12); confusion due to the use in different studies of different strains with the same label, echoing some of the problems faced in early taxonomic studies with this group of organisms.

Of these the most fundamental is that of the disparities between phylogenetic and phenotypic groupings. The view has been expressed that it is unlikely that it will always be possible to define phylogenetic groups in phenotypic terms (Ash *et al.* 1991). Evidence that this is probably so is provided by the *Syntrophospora bryantii*/*Syntrophomonas wolfei* example (Zhao *et al.* 1990). Inability to achieve this will lead to difficulties for diagnostic bacteriologists unless the example is followed of these authors, who placed in separate species the two phylogenetically very similar organisms 'to keep the classification practical and functional'. In doing this they followed well established precedent. That *B. cereus* and *B. anthracis* belong to the same species has been recognized since well before any molecular evidence to this effect was available, yet because of the different practical importance of the two organisms they have been kept separate. Gordon (1981) has pointed out that taxonomists are responsible for the names and descriptions that give meaning to the names of strains used by other bacteriologists. If organisms with different phenotypic properties are given the same name there will be a schism between taxonomists and users of their taxonomic arrangements. Clearly, this is to be avoided.

5. REFERENCES

Ash, C., Farrow, J.A.E., Wallbanks, S. and Collins, M.D. (1991) Phylogenetic heterogeneity of the genus *Bacillus* revealed by comparative analysis of small-subunit-ribosomal RNA sequences. *Letters in Applied Microbiology* **13**, 202–206.

Beijerinck, M.W. (1901) Anhäufungsversuche mit Ureumbakterien. Ureumspaltung durch Urease und durch Katabolismus. *Zentralblatt für Bakteriologie Parasitenkunde, Infectionskrankheiten und Hygiene. Abteilung II* **7**, 33–61.

Berkeley, R.C.W., Logan, N.A., Shute, L.A. and Capey, A.G. (1984) Identification of *Bacillus* species. In *Methods in Microbiology*, Vol. 16. ed. Bergan, T. pp. 292–328. London: Academic Press.

Berkeley, R.C.W., Goodacre, R., Helyer, R.J. and Kelley, T. (1992) Pyrolysis mass spectrometry in the identification of *Bacillus* species. In *Identification Methods in Applied and Environmental Microbiology* ed. Board, R.G., Jones, D. and Skinner, F.A. pp. 257–262. London: Blackwell.

Bonde, G.J. (1975) The genus *Bacillus*. *Danish Medical Bulletin* **23**, 41–61.

Bull, A.T., Goodfellow, M. and Slater, J.H. (1992) Biodiversity as a source of innovation in biotechnology. *Annual Reviews of Microbiology* **46**, 219–252.

Campbell, L.L. and Singleton, R., Jr (1986) Genus *Desulfotomaculum* Campbell and Postgate 1965, 362^{AL}. In *Bergey's Manual of Systematic Bacteriology*, Vol. 2. ed. Sneath, P.H.A., Mair, N.S. and Sharpe, M.E. pp. 1200–1202. Baltimore: Williams and Wilkins.

Cato, E.P., George, W.I. and Finegold, S.M. (1986) Genus *Clostridium* Prazmowski 1880, 23^{AL}. In *Bergey's Manual of Systematic Bacteriology*, Vol. 2. ed. Sneath, P.H.A., Mair, N.S. and Sharpe, M.E. pp. 1141–1200. Baltimore: Williams and Wilkins.

Claus, D. and Berkeley, R.C.W. (1986) Genus *Bacillus* Cohn 1872, 174^{AL}. In *Bergey's Manual of Systematic Bacteriology*, Vol. 2. ed. Sneath, P.H.A., Mair, N.S. and Sharpe, M.E. pp. 1105–1139. Baltimore: Williams and Wilkins.

Clause, D. and Fahmy, F. (1986) Genus *Sporosarcina* Kluyver and van Niel 1936, 401^{AL}. In *Bergey's Manual of Systematic Bacteriology*, Vol. 2. ed. Sneath, P.H.A., Mair, N.S. and Sharpe, M.E. pp. 1105–1139. Baltimore: Williams and Wilkins.

Cord-Ruswich, R. and Garcia, J.L. (1985) Isolation and characterization of an anaerobic benzoate-degrading spore-forming sulfate-reducing bacterium, *Desulfotomaculum sapomandens* sp. nov. *FEMS Microbiology Letters* **29**, 325–330.

Daumas, S., Cord-Ruwisch, R. and Garcia, J.L. (1988) *Desulfotomaculum geothermicum* sp. nov., a thermophilic, fatty acid-degrading, sulfate-reducing bacterium isolated with H_2 from geothermal ground water. *Antonie van Leuwenhoek* **54**, 165–178.

Devereux, R., Delaney, M., Widdel, F. and Stahl, D.A. (1989) Natural relationships among sulfate-reducing eubacteria. *Journal of Bacteriology* **171**, 6689–6695.

Doi, R.H. and Igarashi, R.T. (1965) Conservation of ribosomal and messenger ribonucleic acid cistrons in *Bacillus* species. *Journal of Bacteriology* **90**, 384–390.

Fowler, V.J.F., Widdel, F., Pfennig, N., Woese, C.R. and Stakebrandt, E. (1986) Phylogenetic relationships of sulfate- and sulfur-reducing eubacteria. *Systematic and Applied Microbiology* **8**, 32–41.

Fox, G.E., Pechman, K.R. and Woese, C.R. (1977) Comparative cataloging of 16S ribosomal ribonucleic acid: molecular approach to procaryotic systematics. *International Journal of Systematic Bacteriology* **27**, 44–57.

Fox, G.E., Stakebrandt, E., Hespell, R.B., Gibson, J., Maniloff, J., Dyer, T.A. *et al.* (1980) The phylogeny of prokaryotes. *Science* **209**, 457–463.

Gibson, T. (1986) Genus *Oscillospira* Chatton and Pérard 1913, 1159^{AL}. In *Bergey's Manual of Systematic Bacteriology*, Vol. 2. ed. Sneath, P.H.A., Mair, N.S. and Sharpe, M.E. pp. 1207. Baltimore: Williams and Wilkins.

Gibson, T. and Topping, L.E. (1938) Further studies of the aerobic spore-forming bacteria. *Society for Agriculture, Bacteriological Proceedings, Abstracts* 43–44.

Gogotova, G.T. and Vainstein, M.B. (1983) The sporogenous sulfate reducing bacterium *Desulfotomaculum guttoideum* sp. nov. *Mikrobiologia* **52**, 789–793.

Golovacheva, R.S. and Karavaiko, G.I. (1978) *Sulfobacillus*, a new genus of thermophilic sporeforming bacteria. *Mikrobiologia* **47**, 815–822.

Gordon, R.E. (1981) One hundred and seven years of the genus *Bacillus*. In *The Aerobic Endospore-forming Bacteria* ed. Berkeley, R.C.W. and Goodfellow, M. pp. 1–15. London and New York: Academic Press.

Herndon, S.E. and Bott, K.F. (1969) Genetic relationship between *Sporosarcina urea* and members of the genus *Bacillus*. *Journal of Bacteriology* **97**, 6–12.

Hobbs, G. and Cross, T. (1984) Identification of endospore-forming bacteria. In *The Bacterial Spore*, Vol. 2. ed. Hurst, A. and Gould, G.W. pp. 50–78. London: Academic Press.

Holt, J.G., Krieg, N.R., Sneath, P.H.A., Staley, J.T. and Williams, S.T. (1994) *Bergey's Manual of Determinative Bacteriology*, 9th edn. Baltimore: Williams and Wilkins.

Jones, D. and Sneath, P.H.A. (1970) Genetic transfer and bacterial taxonomy. *Bacteriological Reviews* **34**, 40–81.

Kandler, O. and Weiss, N. (1986) Genus *Sporolactobacillus* Kitahara and Suzuki 1963, 69^{AL}. In *Bergey's Manual of Systematic Bacteriology*, Vol. 2. ed. Sneath, P.H.A., Mair, N.S. and Sharpe, M.E. pp. 1139–1141. Baltimore: Williams and Wilkins.

Kämper, P. (1991) Application of miniaturized physiological tests in numerical classification and identification of some bacilli. *Journal of General and Applied Microbiology* **37**, 225–247.

Lacey, J. and Cross, T. (1989) Genus *Thermoactinomyces* Tslinsky 1899, 501^{AL} In *Bergey's Manual of Systematic Bacteriology*, Vol. 4. ed. Williams, S.T., Sharpe, M.E. and Holt, J.G. pp. 2574–2585. Baltimore: Williams and Wilkins.

Logan, N.A. and Berkeley, R.C.W. (1981) Classification and identification of members of the genus *Bacillus* using API tests. In *The Aerobic Endospore-forming Bacteria* ed. Berkeley, R.C.W. and Goodfellow, M. pp. 105–140. London and New York: Academic Press.

Marmur, J., Seaman, E. and Levine, J. (1963) Interspecific transformation in *Bacillus*. *Journal of Bacteriology* **85**, 461–467.

Nazina, T.N., Ivanova, A.E., Kanchaveii, L.P. and Rozanova, E.P. (1988) *Desulfotomaculum kuznetsovii* sp. nov., a new spore-forming thermophilic methylotrophic sulfate-reducing bacterium. *Mikrobiologia* **57**, 823–827.

Niimura, Y., Koh, E., Yanagida, F., Suzuki, K.-I., Komagata, K. and Kozaki, M. (1990) *Amphibacillus xylanus* gen. nov., sp. nov., a facultatively anaerobic sporeforming xylan-digesting bacterium which lacks cytochrome, quinone and catalase. *International Journal of Systematic Bacteriology* **40**, 297–301.

Norris, J.R. (1981) *Sporosarcina* and *Sporolactobacillus*. In *The Aerobic Endospore-forming Bacteria*, ed. Berkeley, R.C.W. and Goodfellow, M. pp. 105–140. London and New York: Academic Press.

Oren, A. (1983) *Clostridium lortetii* sp. nov., a halophilic obligatory anaerobic bacterium producing endospores with attached gas vacuoles. *Archives of Microbiology* **136**, 42–48.

Oren, A., Pohla, H. and Stakebrandt, E. (1987) Transfer of *Clostridium lortetii* to a new genus *Sporohalobacter* gen. nov. as *Sporohalobacter lortetii* comb. nov., and a description of *Sporohalobacter marismortui* sp. nov. *Systematic and Applied Microbiology* **9**, 239–246.

Pechman, K.J., Lewis, B.J. and Woese, C.R. (1976) Phylogenetic status of *Sporosarcina urea*. *International Journal of Systematic Bacteriology* **26**, 305–310.

Priest, F.G., Goodfellow, M. and Todd, C. (1988) A numerical classification of the genus *Bacillus*. *Journal of General Microbiology* **134**, 1847–1882.

Priest, F.G., Goodfellow, M. and Todd, C. (1981) The genus *Bacillus*: a numerical analysis. In *The Aerobic Endospore-forming Bacteria* ed. Berkeley, R.C.W. and Goodfellow, M. pp. 91–103. London and New York: Academic Press.

Priest, F.G., Goodfellow, M. and Todd, C. (1988) A numerical classification of the genus *Bacillus*. *Journal of General Microbiology* **134**, 11847–1882.

Sayre, R.M. and Starr, M.P. (1989) Genus *Pasteuria* Metchnikoff 1888, 166^{AL}. In *Bergey's Manual of Systematic Bacteriology*, Vol. 4. ed. Williams, S.T., Sharpe, M.E. and Holt, J.G. pp. 2601–2614. Baltimore: Williams and Wilkins.

Sayre, R.M., Wergin, W.P., Schmidt, J.M. and Starr, M.P. (1991) *Pasteuria nishizawae* sp. nov., a mycelial and endospore-forming bacterium parasitic on cyst nematodes of genera *Heterodera* and *Globodera*. *Research in Microbiology (Paris)* **147**, 551–564.

Skerman, V.B.D., McGowen, V. and Sneath, P.H.A. (1980) Approved lists of bacterial names. *International Journal of Systematic Bacteriology* **30**, 225–420.

Smith, N.R., Gordon, R.E. and Clark, F.E. (1952) *Aerobic Spore-forming Bacteria*. Agriculture Monograph No 16. Washington, DC: US Department of Agriculture.

Sneath, P.H.A. (1962) The construction of taxonomic groups. In *Microbial Classification*, ed. Ainsworth, G.C. and Sneath, P.H.A. pp. 289–322. Cambridge: Cambridge University Press.

Sneath, P.H.A., Mair, N.S. and Sharpe, M.E. (1986) *Bergey's Manual of Systematic Bacteriology*, Vol. 2. Baltimore: Williams and Wilkins.

Stakebrandt, E. and Woese, C.R. (1981) The evolution of prokaryotes. In *Molecular and Cellular Aspects of Microbial Evolution* ed. Carlile, M.J., Collins, J.F. and Moseley, B.E.B. pp. 1–31. Cambridge: Cambridge University Press.

Stakebrandt, E., Ludwig, W., Weizenegger, M., Dorn, S., McGill, T.J., Fox, G.E. *et al.* (1987) Comparative 16S RNA oligonucleotide analyses and murein types of round-spore-forming bacilli and non-spore-forming relatives. *Journal of General Microbiology* **133**, 2523–2529.

Tasaki, M., Kamagata, Y., Nakamura, K. and Mikami, E. (1993) Isolation and characterization of a thermophilic benzoate-degrading, sulfate-reducing bacterium, *Desulfotomaculum thermobenzoicum* sp. nov. *Archives of Microbiology* **155**, 348–352.

Timmis, K.N., Hobbs, G. and Berkeley, R.C.W. (1974) Chitinolytic clostridia isolated from marine mud. *Canadian Journal of Microbiology* **20**, 1284–1285.

Williams, K., Silley, P. and Hobbs, G. (1992) Identification methods for pathogenic *Clostridium* species. In *Identification Methods in Applied and Environmental Microbiology* ed. Board, R.G., Jones, D. and Skinner, F.A. pp. 263–281. London: Blackwell Scientific Publications.

Williams, S.T., Sharpe, M.E. and Holt, J.G. (1989) *Bergey's Manual of Systematic Bacteriology*, Vol. 4. Baltimore: Williams and Wilkins.

Wisotzkey, J.D., Jurtshuk, P., Jr, Fox, G.E., Deinhard, G. and Poralla, K. (1992) Comparative, sequence analysis on the 16S (rDNA) of *Bacillus acidocaldarius*, *Bacillus acidoterrestris* and *Bacillus cycloheptanicus* and a proposal for creation of a new genus, *Alicyclobacillus* gen. nov. *International Journal of Systematic Bacteriology* **42**, 263–269.

Woese, C.R., Maniloff, J. and Zablen, L.B. (1980) Phylogenetic analysis of the mycoplasmas. *Proceedings of the National Academy of Sciences* **77**, 494–498.

Zhao, H., Yang, D., Woese, C.R. and Bryant, M.R. (1990) Assignment of *Clostridium bryantii* to *Syntrophospora bryantii* gen. nov., comb. nov. on the basis of 16S rRNA sequence analysis of its crotonate-grown pure culture. *International Journal of Systematic Bacteriology* **40**, 40–44.

Journal of Applied Bacteriology Symposium Supplement 1994, **76**, 9S–16S

The genetic analysis of bacterial spore germination

Anne Moir, E. Helen Kemp, C. Robinson and B.M. Corfe
Krebs Institute for Biomolecular Research, Department of Molecular Biology and Biotechnology, University of Sheffield, Sheffield, UK

1. Introduction, 9S
2. How might germinants act? 9S
3. The genetic approach
   3.1 Germination mutants, 10S
   3.2 Germination genes, 11S
   3.3 Germination proteins—their predicted properties, 12S
4. Back to biochemistry: evidence for localization, 13S
5. Conclusions, 14S
6. References, 15S

1. INTRODUCTION

The formation of a bacterial endospore is a complex and sophisticatedly-regulated process of structural differentiation, responsible for the resistance and dormancy properties of the spore (Errington 1993). However, the success of this strategy for survival is dependent on the presence in the spore of an efficient mechanism for returning the organism to the vegetative state, allowing growth and multiplication when nutrients are available.

Despite the spore being insensitive to environmental insult, it must be able to respond to *particular* external chemical stimuli by germinating, losing the spore structural properties that confer dormancy and resistance (Gould 1969; Setlow 1981; Moir 1992). Germination, which may be defined as the loss of spore resistance properties, is followed by a period of outgrowth, when biosynthetic activity is resumed and an actively dividing rod-shaped cell is regenerated (Setlow 1984).

There is a general assumption that the molecular events in germination, as in sporulation, will be similar in nature across the range of endospore-formers. Certainly the gross morphological and biochemical changes and changes in spore structure at germination are common to all species, although the natures of the chemicals that are effective as germinants differ. Although the particularly thick coat and less synchronous germination response of spores of *Bacillus subtilis* 168 has meant that it is not the organism of choice for biochemical studies, it is the only spore former in which sophisticated genetic analysis is possible.

Correspondence to: Dr A. Moir, Krebs Institute for Biomolecular Research, Department of Molecular Biology and Biotechnology, University of Sheffield, PO Box 594, Sheffield S10 2UH, UK.

2. HOW MIGHT GERMINANTS ACT?

There is an extensive literature describing the response of spores of a variety of species to particular germinants (Gould 1969; Smoot and Pierson 1982; Moir 1992) and the structural and biochemical changes occurring during germination (Levinson and Hyatt 1966; Gould and Dring 1972; Scott *et al.* 1978; Foster and Johnstone 1989; Venkatasubramanian and Johnstone 1989). We have as yet no precise molecular description of how a germinant molecule initiates the series of physical, chemical and morphological changes which result in the breakage of dormancy. In general, theories have invoked either allosteric or metabolic roles for germinants (as in Halvorson *et al.* 1966 and Prasad *et al.* 1972). Sensitive experiments designed to detect any metabolism in the germinating spore population suggested that commitment to germinate precedes any significant metabolism (Scott and Ellar 1978a,b).

Germination is insensitive to inhibitors of RNA or protein synthesis, and thus involves proteins already present in the mature spore. It is presumed that the germinant interacts with a specific site in the spore, which we may call a germination receptor; the assumption is that this generates some sort of allosteric alteration in the structure and properties of the receptor protein. If the receptor is located in a membrane, as discussed later, consequent changes in the membrane might alter its permeability properties, leading to a redistribution of ions and water in the spore and to activation of specific degradative processes (Keynan 1978). An alternative postulate is that the interaction of germinant with receptor leads to the activation by specific proteolysis of a cortex lytic enzyme (Foster and Johnstone 1988). These models need not be mutually exclusive. Models of the germination process are discussed at more length in a separate article (see Johnstone, this Symposium, pp. 17S–24S).

3. THE GENETIC APPROACH

Biochemical approaches have not pin-pointed any specific protein that is yet proven to be necessary for germination. The plethora of models for germination derived from physiological studies, and the relative difficulty of studying biochemical events during the rapid and relatively asynchronous germination of a spore population, suggested that a genetic approach might provide useful insights.

The logic of the genetic approach is that a mutant unable to germinate in the normal manner contains a mutation in a gene whose product is required for germination (either directly or indirectly). This approach defines a gene, and therefore a gene product, that is required (either directly or indirectly) for germination. The first steps, transfer of the mutation into an unmutagenized background then classification of the mutants and genetic mapping, have been invaluable in defining germination *ger* genes—and therefore the encoded Ger proteins.

Germination mutants were enriched in a population of spores by incubating them in a germinant, then challenging with heat or chloroform to kill any germinated spores. Because the population of wild-type spores germinates asynchronously, and may include some 'superdormant' spores, the procedure was repeated. Spores that had still not germinated, but that retained the ability to germinate under different conditions or after a longer lag, were recovered by plating on rich medium. Individual colonies were then purified, the putative mutants allowed to sporulate and the germination behaviour of washed spore suspensions tested (Trowsdale and Smith 1975). A plate test for scoring the germination phenotype of a colony, measuring the resumption of respiratory metabolism by the reduction of a tetrazolium salt, served as a quick reporter of germination phenotype, invaluable for genetic mapping and for the transfer of mutations between strains in genetic crosses.

3.1 Germination mutants

Spores of *B. subtilis* respond to at least two different types of germinative stimulus: they will germinate in alanine (ALA) or some analogues of this compound (valine or cycloleucine, for example), or in a combination of asparagine, which is not a germinant on its own, along with glucose, fructose and a potassium salt (AGFK: Wax and Freese 1968).

Varying the germinant included in the enrichment procedure generated different types of conditionally defective mutants; these were classified by phenotype and by map location, defining a number of *ger* genes (Moir *et al.* 1979). Each group of mutants that mapped to a different location on the genetic map, and therefore represented a different gene from the others, was given a separate genetic designation; they are discussed in more detail in Moir and Smith (1990), and are summarized in Fig. 1.

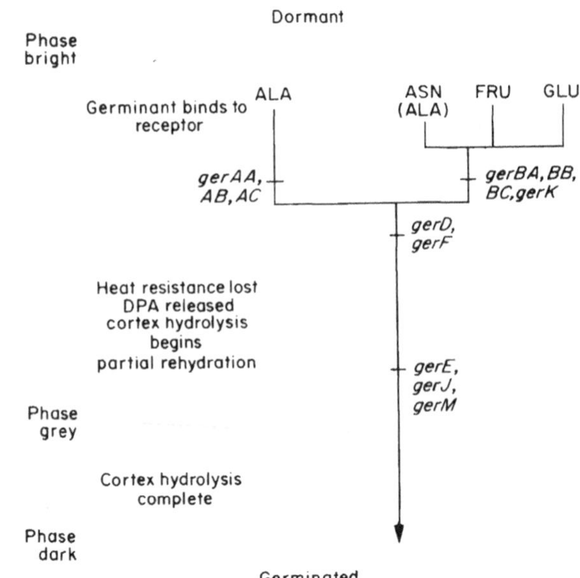

Fig. 1 A schematic representation of the mutational blocks in *Bacillus subtilis* spore germination mutants. DPA, dipicolinic acid

Like the *spo* loci, the *ger* loci are scattered around the chromosome of *B. subtilis*, rather than being clustered in one place. The classical genetic approaches available could not distinguish whether a single locus contained a cluster of *ger* genes or only one: such an analysis had to await cloning of the *ger* loci.

Most of the mutants obtained were blocked before the loss of heat resistance (*gerA*, *B*, *D*, *F* and *K*); of these, *gerA* mutants are defective in alanine-stimulated germination (Sammons *et al.* 1981) but germinate normally in AGFK, whereas *gerB* mutants (Moir *et al.* 1979) and *gerK* mutants (Irie *et al.* 1982) fail to germinate in the latter mixture, but germinate normally in alanine. As the genes are required for the germination response to particular germinants, the simplest interpretation would be that the *gerA* and the *gerB*/*K* genes encode different germination receptors in the spore, responding to different stimuli. The suggestion that *gerA* encoded a receptor for alanine was reinforced by the isolation of *gerA38* and *gerA44* mutants, which have an increased concentration requirement for alanine and its analogues in ALA germination (Sammons *et al.* 1981). As some of the *gerA* mutations were responsible for temperature-sensitive germination, we can be confident that the GerA proteins act during germination, and are not regulators of the expression of the germination apparatus.

Two classes of mutant, those with *gerD* or *gerF* mutations, are affected in their response to germinants of both the ALA and AGFK type (Moir *et al.* 1979; Warburg *et al.*

1985; Irie *et al.* 1986). They were slow to germinate in ALA, and did not germinate at all in AGFK. This suggests that each of the gene products has a role that is essential for AGFK germination, and that although both genes' products are involved in ALA germination, they are not essential for it to take place.

As these mutations are not germinant-specific, the gene products are not likely to be involved in germinant recognition; they have therefore been tentatively placed a little later in the series of germination events in Fig. 1; perhaps they are involved in transducing the initial germination stimulus?

Another group of mutants should never, logically, have been obtained by the enrichment procedures used; these are blocked at a later stage of germination, *after* the loss of heat resistance. The initial recognition of germinant by the spore is therefore still intact in these mutants. All three groups of mutants (*gerE*, *J* and *M*) have spore structural defects; these are examples of germination genes where the gene product influences spore germination without being directly involved. These germinate part-way, losing heat resistance, and starting but not completing the process of cortex hydrolysis (Moir 1981; Warburg and Moir 1981; Sammons *et al.* 1987).

Although not strictly germination genes, these represent interesting genes with a role in spore formation. The GerE protein is a DNA-binding protein that regulates expression of a number of spore coat genes (Errington 1993); the *gerE* mutant may lack a protein required for late stages of germination, either because its expression is dependent on GerE, or because the defective coats, which are permeable to lysozyme, allow the protein to leak from the spore.

Less is known about the function of GerM and GerJ proteins, although both types of mutant are known to be defective in sporulation. The altered heat-resistance properties and the late synthesis of spore-specific penicillin binding proteins in *gerJ* mutants (Warburg *et al.* 1986), and the multiple abnormalities in septum formation, cell division and cortex structure of *gerM* mutants, suggest that these, too, may be regulators of spore morphogenesis.

There are reports in the literature of metabolic mutants with defective germination, but these reports are of uncertain value. Unless the mutations are transferred into an unmutagenized background, it is always possible that the isolate carries two entirely separate mutations responsible for the two phenotypes. The genetic work reported in this section supports the hypothesis that metabolism of the germinant is not required. None of the mutants obtained are affected in metabolism. Germination mutants that lose the ability to germinate in alanine also lose the ability to germinate in non-metabolizable alanine analogues (Sammons *et al.* 1981), suggesting that their defect does not concern even a minor or alternative metabolic route. Glucose dehydrogenase null mutants germinate normally in germinant mixtures containing glucose (Irie, personal communication); this enzyme is therefore not concerned with the triggering of the germination response. Currently, there is no proven case where a metabolic defect prevents spore germination.

3.2 Germination genes

Most of the *ger* genes listed in Fig. 1 have been cloned and sequenced, defining at least some of the proteins involved in the germination response. The discussion that follows concentrates on the cloned genes whose products are required for loss of heat resistance in response to germinant, i.e. *gerA*, *gerB*, *gerD* and *gerK* genes; the *gerF* gene has not yet been cloned.

Because *ger* genes do not confer any selectable characteristic on a host cell, cloning strategies have often been indirect. Some genes have been cloned in phage λ vectors. The *gerA* genes were obtained along with the adjacent *citG* gene, which was directly selectable in *E. coli* (Moir 1983). The *gerB* clone was obtained by chromosome walking from a nearby cloned locus (Corfe *et al.* 1994). The *gerK* gene was obtained by screening a large number of λ clones for the ability of their DNA to transform a *gerK* mutant to *ger*$^+$ (Irie, personal communication).

An alternative general approach is based on Tn917 transposon technology (Youngman 1990); *ger* mutants are generated by an interruption of the gene by transposon Tn917. Modified versions of this transposon allow recovery from the chromosome of a section of DNA that includes one end of the transposon and the adjacent part of the interrupted *ger* gene. Once part of the gene has been cloned, it is possible to use this as a probe to screen a chromosomal gene library, and then recover the intact wild-type germination gene. This approach was used to clone the *gerD* gene (Yon *et al.* 1989).

Once cloned, information on the gene organization and regulation of expression of *ger* genes can be obtained by combinations of molecular and classical genetic techniques. The amino acid sequence of Ger proteins can be predicted, and strategies can be designed to overexpress the cloned gene products under the control of a foreign promoter.

The *gerA* locus contains three genes (*gerAA*, *AB* and *AC*), arranged in an operon (Feavers *et al.* 1985; Zuberi *et al.* 1985; Zuberi *et al.* 1987). The collection of known *gerA* mutants included mutations in each gene, indicating that all the GerA protein products are required for ALA germination. The *gerA38* and *A44* mutations that require higher concentrations of alanine for germination are both located in the middle gene, *gerAB*, suggesting that the GerAB protein is likely to bind alanine (Zuberi *et al.* 1985).

The regulation of expression of the *gerA* operon has been studied, by *lac* fusion analysis, promoter mapping and *in vitro* transcription (Feavers *et al.* 1990). As might have been predicted, the genes are subject to developmental control; they are not expressed in vegetative growth, but are switched on during sporulation in the forespore compartment, in response to the activity of the forespore-specific sigmaG-containing RNA polymerase. The level of expression of *gerA* genes appears to be very low, as judged by the very low level of expression of β-galactosidase from fusions to the *gerA* promoter (300-fold lower than expression from the fumarase (*citG*) promoter, for example), suggesting that the spore contains only low quantities of the GerA proteins.

The *gerB* locus contains three genes, *gerBA*, *BB* and *BC*, encoding three homologues of the GerA proteins, organized in the same order as the three genes of the *gerA* operon (Corfe *et al.* 1994; Fig 2). These too are dependent on the forespore-specific sigmaG-containing RNA polymerase for expression (Corfe and Moir, unpublished); the homology thus extends to their regulation. One difference is the level of expression: the *gerB* genes are expressed at an even lower level, about one-tenth of that of the *gerA* operon.

The *gerK* locus is at an earlier stage of analysis, but is already known to encode at least a homologue of GerBC and GerAC (Irie, personal communication; Fig. 2).

The *gerD* locus, in contrast, contains a single gene; it is expressed at a higher level (about 15-fold higher than *gerA*), but still under the control of sigmaG, in the forespore compartment of the sporulating cell (Kemp *et al.* 1991).

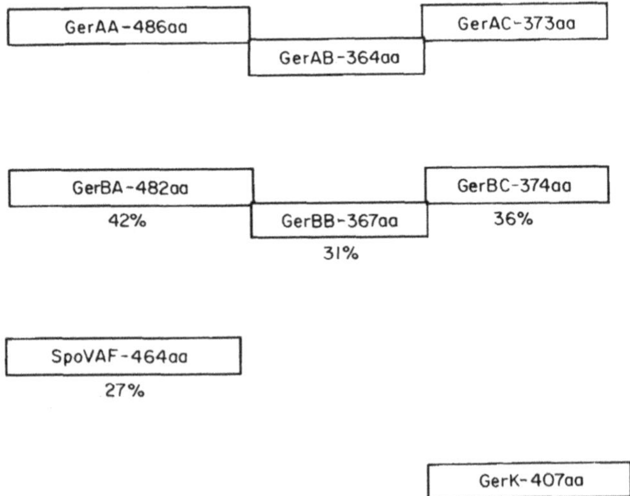

Fig. 2 GerA proteins and their homologues. Related proteins are aligned vertically. The percentage identity of amino acid residues with the corresponding GerA sequence is shown below the homologue

All the germination-specific *ger* genes so far analysed in detail (*gerA*, *B* and *D*) are expressed only during spore formation. The genes are all expressed in the forespore, which goes on to form the cellular compartment of the mature spore. As described in more detail below, the proteins encoded are predicted to be either membrane-associated or have N-terminal signal sequences, suggesting that they are variously located in, or transported across, the forespore membrane.

3.3 Germination proteins—their predicted properties

The 480 amino acid GerAA protein predicted from the DNA sequence would be organized in at least two, and probably three, distinct domains (Feavers *et al.* 1985; Fig. 3). Of predicted molecular weight 53 506 Da, it possesses a large hydrophilic (and therefore potentially cytoplasmic) N-terminal domain, followed by a membrane-associated domain of around 200 amino acids composed of five hydrophobic, potentially membrane-spanning helices, interspersed with charged regions. At the C terminus, there is another hydrophilic domain, of 50 amino acids—this is likely to be located on the opposite side of the membrane to the N-terminal one, if our prediction of five membrane-spanning helices (Fig. 3) is correct. The GerAB protein (41 257 Da) has the hydrophobicity profile characteristic of an integral membrane protein, with ten likely membrane-spanning helices (Zuberi *et al.* 1987). The GerAC protein (42 363 Da) is, in contrast, hydrophilic throughout, with the notable exception of a pre-lipoprotein signal sequence at the N-terminus (Fig. 3). The signal sequence suggests that this protein is transferred across the forespore membrane, and is predicted to be attached to the membrane via a cysteine–glyceride link (Zuberi *et al.* 1987).

Comparisons of these proteins with protein sequence databases have not identified similarities with other types of protein, with the possible exception of the GerAB protein. Apart from GerBB, of course, the proteins that score highest against GerAB in FASTA alignment searches are an *Escherichia coli* tryptophan transport protein, TnaB (Sarsero *et al.* 1991), sharing 24% identity over 345 amino acids, a *Pseudomonas* arginine/ornithine transporter ArcD (Luthi *et al.* 1990) and Gram-positive tetracycline resistance proteins (Noguchi *et al.* 1986). There may be a distant evolutionary relationship between GerAB and the extended family of single component membrane transport proteins (Griffith *et al.* 1992).

The importance of the GerA group of proteins to germination has been underlined by the discovery that the GerBA, BB and BC proteins are 42%, 31% and 35% identical to their respective GerA homologues, suggesting that a basic mechanism has been conserved in receptors responding to different germinants, and that these receptors

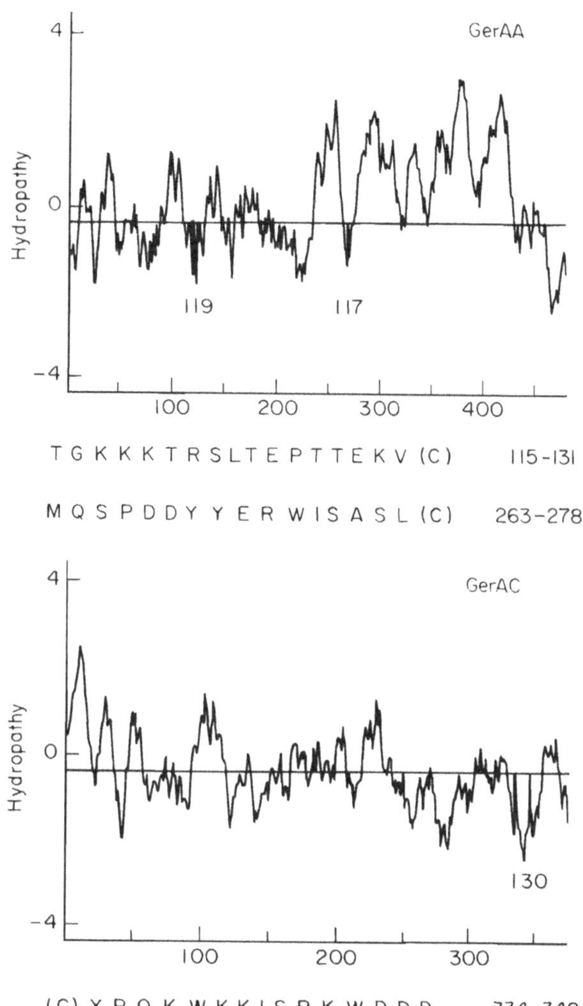

Fig. 3 The hydrophobicity profiles of GerAA and GerAC proteins. Also indicated are the synthetic peptides used for immunological studies. Peptides 119 and 117 were derived from hydrophilic sequences in GerAA, and peptide 130 from GerAC

have evolved by gene duplication and subsequent divergence. Figure 4 shows the alignment between GerAA and GerBA primary sequences. Most striking is the higher local homology in the GerAA/BA proteins in the 200 amino acid hydrophobic domain—especially the hydrophobic region of residues 346–391, where 36 out of 46 residues are identical. This degree of conservation, in proteins that otherwise have diverged considerably, suggests that these helices have a crucial functional importance, rather than being merely membrane anchors.

There are two more known members of this family (Fig. 2). The predicted GerK protein is a more distant homologue of GerAC and BC proteins (Irie, personal communication). In addition, the sixth protein encoded by the spoVA operon, SpoVAF, is a GerAA homologue, although there is no ALA or AGFK germination defect

```
GerAA   1 MEQTEFKEYIHDNLALVLPKLKENDDLVKNKKMLANG.LVFYYLYFSEMT 49
               ::.. :::||. : ..| :|||::  .  ::::     :.||:.::|
GerBA   1 ...MQIDSDLQNNLDTLKKTLGQNDDMMFYTFAFGDSRQKACLLYIDGLT 47

       50 DENKVSEAIKTLIKDEETL....TLDQVKKRLDQLDARPVETAKKTIESI 95
          ::..:.: : . :..|.     .::::.  : .:. ..|.| |.. : :
       48 ENKMLAQYVISPLQKEALAHKECSIEDLSAFFFGFHHSVVSTMKEIEQLV 97

       96 LNGNCAVFINGLDKAYILTTGKKKTRSLTEPTTEKVVRGPKVAFVEDIDT 145
          :.|.. :: :|.  :. :.| .   ||||.||..| | ||||::|:|.: |
       98 FSGQAILLADGYRGGLAFDTKSVATRSLDEPSSEVVERGPKIGFIEKLRT 147

      146 NLALIRQRTSHPKLITKKIMIGENKLKPAAIMYIEGKAKKSVIKEVKARL 195
          | ||:|:|||.|.|:.|.: :|...  |. |: ||:: |.. |:|||  ||
      148 NTALLRERTSDPNLVIKEMTLGKRTKKKIAVAYIQDIAPDYVVKEVFKRL 197

      196 KNIQLEDIQDSGTLEELIEDNKYSPFPQIQNTERPDKVSSALFNGRVAIL 245
          |.:.::::.:|||||:||||:.:| || | .|||||:|.|.|::|||.||
      198 KSVNIDNLPESGTLEQLIEDEPFSIFPTILSTERPDRVESSLLEGRVSIL 247

      246 VDSSPFVLLVPVSLGILMQSPDDYYERWISASLIRSLRFASIFITLFLSS 295
          ||:.||.|:||..:: :::|||||| :|||. ||:| ||:.||:||::|.:
      248 VDGTPFALIVFATVDEFIHSPDDYSQRWIPMSLVRLLRYSSILITIYLPG 297
                 I
      296 IYIALVSFHQGLLPTALAVTISANRENVPFPPIFEALLMEVTIELLREAG 345
          :||.||||| ||||| :|:.|.:.| |||||||:.||::| .||||:||||
      298 LYISLVSFHTGLLPTRMAISIAGSRLNVPFPPFVEAFIMIFTIELIREAG 347
            II
      346 ..LPNPLGQTIGLVGGVVIGQAAVEANLVSSILVIVVSVIALASFTVPQY 393
            ||.|:|||||||:|||||||||||:|.:||.::|||||||.||||||||| |
      348 LRLPKPIGQTIGLIGGVVIGQAAVQAQIVSALMVIVVSVTALASFTVPSY 397
                    III                      IV
      394 GMGLSFRVLRFISMFSAAILGLYGIILFMLVVYTHLTRQTSFGSPYFSPN 443
          : .:.:|::|:   |:||. ||:||:|:. |.|..|| | .|::.  :|.:
      398 AYNFPLRIIRIGVMISATALGMYGVIMVYLFVIGHLMRLKSLARITLSDH 447
                          V
      444 GFFS..LKNTDDSIIRLPIKNKPKEVNNPNEPKTDSTET 480
          :  :   ||:|   .|  : :|.:|.  |:|::    :.
      448 AQPGQDLKDTVIRIPTMFLKRRPTR.NDPEDNIRQR... 482
```

Fig. 4 A comparison of GerAA and GerBA proteins. The GAP program of the UWGCG sequence analysis package was used to generate the alignment. Identical residues are indicated by a vertical bar. The five likely membrane-spanning regions have been underlined and indicated by Roman numerals

associated with the inactivation of this gene (Kemp, unpublished).

The GerD protein (21 117 Da) is probably not receptor related, and has no homology with the other Ger proteins, or with any other protein in the database. It is hydrophilic, but does have an N-terminal signal sequence, suggesting that it is transferred from the forespore across the inner membrane of the spore. It could be anchored in the membrane or released into the peptidoglycan layer.

4. BACK TO BIOCHEMISTRY: EVIDENCE FOR LOCALIZATION

The proof of these sequence-derived predictions requires evidence that the genes do actally encode proteins of the expected size, and that these proteins are present in spores. This has been demonstrated for the GerA proteins. The *gerA* genes were cloned behind a regulatable, highly-expressed, *E. coli* promoter and expressed in an *in vitro* transcription/translation system; the labelled products from individual ger genes were assigned by expressing versions of *gerA* deleted for individual genes (Kemp *et al.* submitted), and they correspond approximately to the sizes predicted from the DNA sequence.

GerAA-derived and GerAC-derived synthetic peptides, chosen from the predicted amino acid sequence (Fig. 4), were conjugated to a carrier protein (ovalbumin) through added cysteine residues. These antisera were used to probe Western blots of SDS-PAGE separated proteins from spore fractions. They identified a protein band, of the same size as *in vitro* expressed GerAA protein, in the spore membrane fraction (Fig. 5). This pattern was obtained with both of the two available anti-peptide probes. In contrast, the GerAC protein was found predominantly in the integument (coats and cortex) fraction (Kemp *et al.* submitted). The prediction had been that GerAC would be a lipoprotein attached to the outer surface of the forespore membrane; either this association is temporary, or not stable to the isolation procedure. More work will be needed to confirm this localization in the integument, using more gentle spore breakage procedures.

Interpretations have also to be qualified at the moment by the observation that there is some residual cross-reacting material of the same size as the GerA proteins in a GerA deletion strain: at the time that the peptide antibodies were raised, the authors were not aware of the GerA homologues. Deletion strains are being used to clarify the situation.

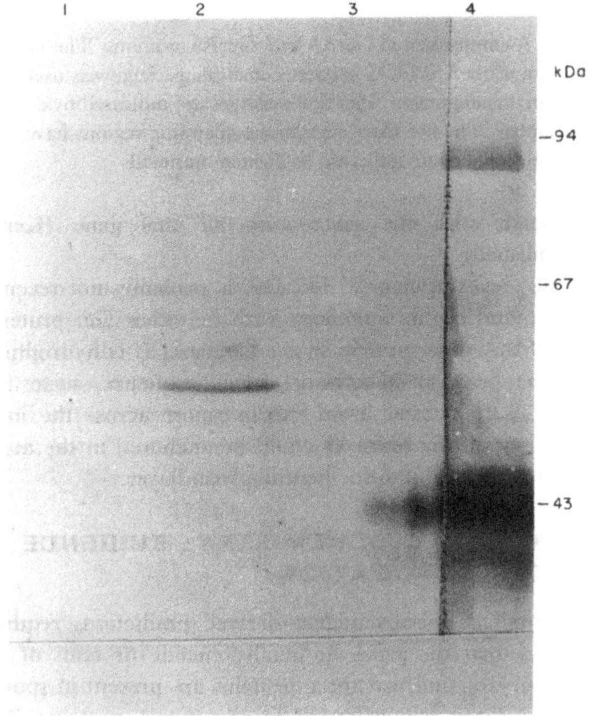

Fig. 5 The detection of GerAA protein in spore membranes. Spores were broken by homogenization with glass beads; the integument and membrane fractions were isolated in successive centrifugation steps. Proteins from each fraction were solubilized by boiling in SDS buffer and separated by SDS-PAGE. (100 μg per lane). Lane 1, soluble fraction. Lane 2, membrane fraction. Lane 3, integument fraction. Lane 4, molecular weight standards (including ovalbumin). Antiserum raised against peptide 119 (GerAA) coupled to ovalbumin was used to detect cross-reacting proteins. A band of the size of GerAA was detected, primarily in the membrane fraction

The GerAB protein has not been studied in this way, but as a membrane protein synthesized in the forespore, it is unlikely that its location could be anywhere other than in the membrane bounding this compartment, i.e. the inner spore membrane, in the same location as GerAA. Whether the proteins are present in association is not known.

A fusion of part of the GerD protein to glutathione *S*-transferase has been overexpressed, affinity purified, and the GerD moiety cleaved from the fusion protein and used to raise polyclonal antisera in rabbits; the GerD protein is detected at the predicted size, in the integument fraction of the spore, and is absent in a *gerD* null mutant (Robinson, unpublished).

This work represents the first physical demonstration of germination proteins in the spore. Models of the mechanism of germination must take these data into account.

5. CONCLUSIONS

Bacillus subtilis contains genes that have evolved significantly from each other and that form families of proteins serving as the germination apparatus in the spore for different germinants. These are spore-specific proteins, that are not present in vegetative cells; there is no suggestion that they have a metabolic role in the spore.

We can now confirm that proteins that were predicted to form a germinant receptor in the spore are found in the integument layers (GerAC) and inner spore membrane (GerAA and AB). These proteins are the first recognized members of a wider family, and conserved sequences between different members of the family of homologues may point to particular regions of functional importance. It is not yet known whether their functions encompass functions other than spore germination—one member, SpoVAF, is not required for either of the known germination systems in *B. subtilis*.

The spore germination receptor and associated proteins represent a new class of sensory transducer, whose precise mode of action is unknown. It is not clear how individual germinants and groups of germinants act on particular receptors to initiate germination, or how the local interaction is transduced throughout the spore. Returning to the models of germination discussed earlier, it is possible that some of the Ger proteins are concerned with initiating ion fluxes: the membrane-associated proteins GerAA/BA and GerAB/BB are obvious candidates. Other proteins may be concerned with transducing the signal to the spore—could GerAC or GerD protein, for example, be a cortex lytic enzyme, or an activator of such hydrolytic activity? What would be responsible for their activation during germination? We now have a detailed molecular description of some of the components of the germination apparatus, but we do not yet have direct proof of the interaction with the germinant, nor do we know what happens when the germinant interacts with this receptor.

Although some progress has been made in identifying parts of the germination apparatus, it would be naïve to think that genetic analysis alone will be sufficient to generate a complete molecular explanation of the process. For example, procedures to isolate germination mutants would not allow detection of mutations in any gene whose protein product is a member of a functionally equivalent family, so that one of several proteins could substitute for the defective one. We are now entering a phase of research in which genetic and biochemical approaches and information need to be integrated; any generalizations drawn from a study of germination receptor proteins in *B. subtilis* need also to be tested in other bacilli. It would not be possible to adopt such a rigorous genetic analysis in other bacilli, but neither would it be necessary, as the tools for detecting homologous Ger proteins have been developed through the study of *B. subtilis*, and reverse genetics could be applied to test the relevance to germination of any cloned genes.

6. REFERENCES

Corfe, B.M., Sammons, R.L., Smith, D.A. and Maüel, C. (1994) The *gerB* region of the *Bacillus subtilis* 168 chromosome encodes a homologue of the *gerA* spore germination operon. *Microbiology* **140**, 471–478.

Errington, J. (1993) *Bacillus subtilis* sporulation: regulation of gene expression and control of morphogenesis. *Microbiological Reviews* **57**, 1–33.

Feavers, I.M., Miles, J.S. and Moir, A. (1985) The nucleotide sequence of a spore germination gene (*gerA*) of *Bacillus subtilis*. *Gene* **38**, 95–102.

Feavers, I.M., Foulkes, J., Setlow, B., Sun, D., Nicholson, W., Setlow, P. and Moir, A. (1990) The regulation of transcription of the *gerA* spore germination operon of *Bacillus subtilis*. *Molecular Microbiology* **4**, 275–282.

Foster, S.J. and Johnstone, K. (1988) Germination-specific cortex lytic enzyme is activated during triggering of *Bacillus megaterium* KM spore germination. *Molecular Microbiology* **2**, 727–733.

Foster, S.J. and Johnstone, K. (1989) The trigger mechanism of bacterial spore germination. In *Regulation of Procaryotic Development* ed. Smith, I., Slepecky, R. and Setlow, P. pp. 89–108. Washington, DC: American Society for Microbiology.

Gould, G.W. (1969) Germination. In *The Bacterial Spore* ed. Gould, G.W. and Hurst, A. pp. 397–444. London: Academic Press.

Gould, G.W. and Dring, G.J. (1972) Biochemical mechanisms of spore germination. In *Spores V* ed. Hoch, J.A. and Setlow, P. pp. 401–408. Washington, DC: American Society for Microbiology.

Griffith, J.K., Baker, M.E., Rouch, D.A., Page, M.G.P., Skurray, R.A., Paulsen, I.T., Chater, K.C., Baldwin, S.A and Henderson, P.J.F. (1992) Membrane transport proteins: implications of sequence comparisons. *Current Opinion in Cell Biology* **4**, 684–695.

Halvorson, H.O., Vary, J.C. and Steinberg, W. (1966) Developmental changes during the formation and breaking of the dormant state in bacteria. *Annual Review of Microbiology* **20**, 169–186.

Irie, R. Okamoto, T. and Fujita, Y. (1982) A germination mutant of *Bacillus subtilis* deficient in response to glucose. *Journal of General and Applied Microbiology* **28**, 345–354.

Irie, R. Okamoto, T. and Fujita, Y. (1986) Characterisation and mapping of *Bacillus subtilis gerD* mutants. *Journal of General and Applied Microbiology* **32**, 303–315.

Kemp, E.H., Sammons, R.L., Moir, A., Sun, D. and Setlow, P. (1991) Analysis of transcriptional control of the *gerD* spore germination gene of *Bacillus subtilis* 168. *Journal of Bacteriology*, **173**, 4646–4652.

Keynan, A. (1978) Spore structure and its relations to resistance, dormancy and germination. In *Spores VII* ed Chambliss, G and Vary, J.C. pp. 43–53. Washington, DC: American Society for Microbiology.

Levinson, H.S. and Hyatt, M.T. (1966) Sequences of events during *Bacillus megaterium* spore germination. *Journal of Bacteriology* **91**, 1811–1818.

Luthi, E., Baur, H., Gamper, M, Brunner, F., Villeval, D., Mercenier, A. and Haas, D. (1990) The *arc* operon for anaerobic arginine catabolism in *Pseudomonas aeruginosa* contains an additional gene, *arcD*, encoding a membrane protein. *Gene* **87**, 37–43.

Moir, A. (1981) Germination properties of a spore-coat defective mutant of *Bacillus subtilis*. *Journal of Bacteriology* **146**, 1106–1116.

Moir, A. (1983) The isolation of λ transducing phages carrying the *citG* and *gerA* genes of *Bacillus subtilis*. *Journal of General Microbiology* **129**, 303–310.

Moir, A. (1992) Spore germination. In *Biology of Bacilli—Applications to Industry* ed. Doi, R. pp. 23–38. New York: Butterworth.

Moir, A. and Smith, D.A. (1990) The genetics of bacterial spore germination. *Annual Review of Microbiology* **44**, 531–553.

Moir, A., Lafferty, E. and Smith, D.A. (1979) Genetic analysis of spore germination mutants of *Bacillus subtilis* 168: the correlation of phenotype with map location. *Journal of General Microbiology* **111**, 165–180.

Noguchi, N., Aoki, T., Sasatu, M., Kono, M., Shishido, K. and Ando, T. (1986) Determination of the complete nucleotide sequence of pNS1, a staphylococcal tetracycline resistance plasmid propagated in *Bacillus subtilis*. *FEMS Microbiology Letters* **37**, 283–288.

Prasad, C., Diesterhaft, M. and Freese, E. (1972) Initiation of spore germination in glycolytic mutants of *Bacillus subtilis*. *Journal of Bacteriology* **110**, 321–328.

Sammons, R.L., Moir, A. and Smith, D.A. (1981) Isolation and properties of spore germination mutants of *Bacillus subtilis* 168 defective in the initiation of germination. *Journal of General Microbiology* **124**, 229–241.

Sammons, R.L., Slynn, G.M. and Smith, D.A. (1987) Genetical and molecular studies on *gerM*, a new developmental locus of *Bacillus subtilis*. *Journal of General Microbiology* **133**, 3299–3312.

Sarsero, J.P., Wookey, P.J., Gollnick, P., Yanofsky, C and Pittard, A.J. (1991) A new family of integral membrane proteins involved in transport of aromatic amino acids in *Escherichia coli*. *Journal of Bacteriology* **173**, 3231–3234.

Scott, I.R. and Ellar, D.J. (1978a) Metabolism and the triggering of germination of *Bacillus megaterium*: concentrations of amino acids, adenine nucleotides, and nicotinamide nucleotides during germination. *Biochemical Journal* **174**, 627–634.

Scott, I.R. and Ellar, D.J. (1978b) Metabolism and the triggering of germination in *Bacillus megaterium*: use of L-[^{3}H] alanine and tritiated water to detect metabolism. *Biochemical Journal* **174**, 635–640.

Scott, I.R., Stewart, G.S.A.B., Koncewicz, M.A., Ellar, D.J. and Crafts-Lighty, A. (1978) Sequence of biochemical events during germination of *Bacillus megaterium* spores. In *Spores VII* ed. Chambliss, G. and Vary, J.C. pp. 95–103. Washington, DC: American Society for Microbiology.

Setlow, P. (1981) Biochemistry of forespore development and spore germination. In *Sporulation and Germination* ed. Levinson, H.S., Sonenshein, A.L. and Tipper, D.J. pp. 13–28. Washington, DC: American Society for Microbiology.

Setlow, P. (1984) Germination and outgrowth. In *The Bacterial Spore*, Vol. 2. ed. Hurst, A. and Gould, G.W. pp. 211–254. London: Academic Press.

Smoot, L.A. and Pierson, M.D. (1982) Inhibition and control of bacterial spore germination. *Journal of Food Protection* **45**, 84–92.

Trowsdale, J. and Smith, D.A. (1975) Isolation, characterisation and mapping of *Bacillus subtilis* 168 spore germination mutants. *Journal of Bacteriology* **123**, 83–95.

Venkatasubramanian, P. and Johnstone, K. (1989) Biochemical analysis of the *Bacillus subtilis* 1604 spore germination response. *Journal of General Microbiology* **135**, 2723–2733.

Warburg, R.J. and Moir, A. (1981) Properties of a mutant of *Bacillus subtilis* in which spore germination is blocked at a late stage. *Journal of General Microbiology* **124**, 243–253.

Warburg, R.J., Moir, A. and Smith, D.A. (1985) Influence of alkali metal cations on the germination of spores of wild type and *gerD* mutants of *Bacillus subtilis*. *Journal of General Microbiology* **131**, 221–230.

Warburg, R.J., Buchanan, C.E., Parent, K. and Halvorson, H.O. (1986) A detailed study of *gerJ* mutants of *Bacillus subtilis*. *Journal of General Microbiology* **132**, 2309–2319.

Wax, R. and Freese, E. (1968) Initiation of the germination of *Bacillus subtilis* spores by a combination of compounds in place of L-alanine. *Journal of Bacteriology* **95**, 433–438.

Yon, J.R., Sammons, R.L. and Smith, D.A. (1989) Cloning and sequencing of the *gerD* gene of *Bacillus subtilis*. *Journal of General Microbiology* **135**, 3431–3445.

Youngman, P. (1990) Use of transposons and integrational vectors for mutagenesis and construction of gene fusions in *Bacillus* species. In *Molecular Biological Methods for Bacillus* ed. Harwood, C. and Cutting, S.M. pp 221–266. Chichester: Wiley.

Zuberi, A.R., Feavers, I.M. and Moir, A. (1985) Identification of three complementation units in the *gerA* spore germination locus of *Bacillus subtilis*. *Journal of Bacteriology* **162**, 756–762.

Zuberi, A.R., Moir, A. and Feavers, I.M. (1987) The nucleotide sequence and gene organisation of the *gerA* spore germination operon of *Bacillus subtilis* 168. *Gene* **51**, 1–11.

Journal of Applied Bacteriology Symposium Supplement 1994, **76**, 17S–24S

The trigger mechanism of spore germination: current concepts

K. Johnstone
Department of Plant Sciences, University of Cambridge, Cambridge, UK

1. Introduction, 17S
2. Germination and the germinant receptor, 17S
3. The commitment reaction, 18S
4. Biochemical characterization of the commitment reaction
   4.1 Germinant metabolism, 19S
   4.2 Identification of metabolic events during germination triggering, 19S
   4.3 Inhibitor studies, 19S
5. Peptidoglycan hydrolysis during germination triggering, 20S
6. A model for germination triggering, 20S
7. Conclusions, 21S
8. References, 22S

1. INTRODUCTION

The metabolic dormancy and the resistance properties of bacterial spores are both crucial to the ecological role of spores as survival structures. In order to complete this role it is also essential that spores are able to monitor their external environment so as to trigger germination in suitable environmental conditions. Thus paradoxically, in the absence of metabolic processes, the spore must retain an alert sensory mechanism which is able to initiate the germination process (Gould 1983). Metabolic dormancy and heat resistance are imposed on the spore core by a number of mechanisms (Gerhardt and Marquis 1989). These include immobilization of core macromolecules, enzymes and metabolites in a dehydrated calcium dipicolinate gel (Stewart *et al.* 1979; Stewart *et al.* 1980; Johnstone *et al.* 1980, 1982a). In addition the inner spore membrane is present in a semi-crystalline state (Stewart *et al.* 1979). The germination sensing mechanism must be able to function in the absence of general metabolic processes and thus must escape the mechanisms of dormancy and resistance generally imposed on spore constituents. This mechanism may therefore be located outside the spore core and inner membranes and yet must intrinsically possess the resistance properties of the intact spore.

This article reviews our current knowledge of the spore germinant sensing mechanism by building a conceptual framework from selected research articles to produce a working model of this mechanism. It does not attempt to review the vast wealth of spore germination literature—the reader is referred to reviews concerning spore structure (Ellar 1978; Keynan 1978; Warth 1978; Russell 1982), the genetics of germination (Moir *et al.*, this Symposium, pp. 9S–16S; Smith *et al.* 1977; Moir and Smith 1985, 1990) and the biochemistry of germination (Foster and Johnstone 1989b, 1990) for this purpose.

Correspondence to: Dr K. Johnstone, Department of Plant Sciences, University of Cambridge, Downing Street, Cambridge CB2 3EA, UK.

2. GERMINATION AND THE GERMINANT RECEPTOR

In this article the definition of germination as 'a series of degradative events triggered by specific germinants which leads to the loss of typical spore properties' proposed by Foster and Johnstone (1990) will be adopted. Germination can be induced by a variety of processes including exposure to nutrient germinants such as amino acids and sugars, to non-nutrient germinants including dodecylamine, to enzymes and to hydrostatic pressure (Gould 1969). This article will focus on the nutrient germinants, the biochemistry of which has been most intensively studied and which represent the physiological germination pathway. The nutrient germinants range from the simple amino acid L-alanine commonly required for triggering of germination of *Bacillus* species (Harrell and Halvorson 1955) to complex mixtures including amino acids, sugars and ions typical of *Clostridium* spp. (Bright and Johnstone 1987).

The stereospecific properties of the germinant receptors demonstrate that this component of the trigger mechanism must be a protein, which may be activated allosterically (Wolgamott and Durham 1971). Typically the germinant receptors are 50% saturated by 50–100 μmol l^{-1} concentrations of germinants (Bright and Johnstone 1987; Venkatasubramanian and Johnstone 1989). Other properties of the germinant receptor have been relatively poorly studied.

Evidence for co-operativity of binding of three molecules of L-alanine bind per receptor of *B. subtilis* has been presented (Irie *et al.* 1984). Analysis of *B. subtilis* PCI219 spore germination has revealed that L-alanine, a potent germinant, and D-alanine, its competitive inhibitor, bind to a common receptor; their amino and carboxyl groups may bind to respective common regions and their methyl group may bind to separate regions which differ in size and electrostatic nature (Yasuda and Tochikubo 1985). The stimulatory effect of glucose on L-alanine binding, and the absence of an effect on D-alanine binding, confirm the presence of separate regions with respect to their methyl groups (Yasuda and Tochikubo 1984). As L-alanine-initiated germination is inhibited quantitatively by various hydrophobic compounds, a hydrophobic environment near the L-alanine receptor has been suggested (Yasuda *et al.* 1978a, b, 1982).

In spores which are able to respond to a range of germinants, interaction between the germinant receptors may occur. Spores of *B. subtilis* can be induced to germinate by several combinations of germinants including L-alanine or by a combination of glucose, fructose and asparagine (Wax and Freese 1968). Biochemical analysis of the germination response of mutants defective in germinant receptors has revealed that functional interaction of germinant receptors which induce triggering in these two pathways may occur (Fig. 1; Venkatasubramanian and Johnstone 1993). Binding of glucose and/or fructose to their respective receptors is essential for activation of the asparagine/alanine receptor, but can also stimulate L-alanine receptor in spores of this organism. Thus these receptors must be located in close proximity to each other within the spore.

The location of the germinant receptor within the spore is uncertain. There is biochemical evidence that the receptor is located in the inner membrane (summarized by Foster and Johnstone 1989b). This includes changes in anisotropy (Skomurski *et al.* 1983) and fluidity (Janoff *et al.* 1979) of isolated inner membranes of *B. megaterium* QMB1551 by the germinant L-proline, specific labelling of a 10·2 kDa inner membrane protein in spores on *B. megaterium* QMB1551 by the germinant analogue L-proline chloromethyl ketone (Rossignol and Vary 1979) and the ability of the germination inhibitor $^{203}Hg^{2+}$ to label a 10·5-kDa inner membrane protein of spores of *B. megaterium* KM (Foster 1986). This view is also supported by sequence analysis of the cloned *gerA* genes of *B. subtilis* which suggests the presence of a membrane-bound multi subunit complex (Feavers *et al.* 1985; Zuberi *et al.* 1985, 1987). Definitive evidence for the location of the germinant receptor, however, remains to be established.

3. THE COMMITMENT REACTION

Activation of the germination receptor by germinants causes the spore to undergo an irreversible reaction which commits the spore to undergo germination even if the germinants are subsequently removed. This commitment reaction has been demonstrated in a number of species including *B. megaterium* KM (Stewart *et al.* 1981) and *B. subtilis* (Venkatasubramanian and Johnstone 1989). The rates of both commitment and germination are stimulated by prior activation of the spore population, e.g. by sublethal heat treatment (Keynan and Evenchik 1969). Little is known of the mechanism by which activation stimulates the germinant receptor (Foster and Johnstone 1989b). The kinetics of the commitment and germination reactions of spores of *B. megaterium* KM are shown in Fig. 2. They are derived from analysis of commitment and germination of spore populations based on the assumption that the time taken for individual spores to undergo these reactions is small in relation to the corresponding time for the whole population. This is consistent with the microlag and microgermination times observed for individual spores of *B. cereus* (Vary and Halvorson 1965). Such analysis demon-

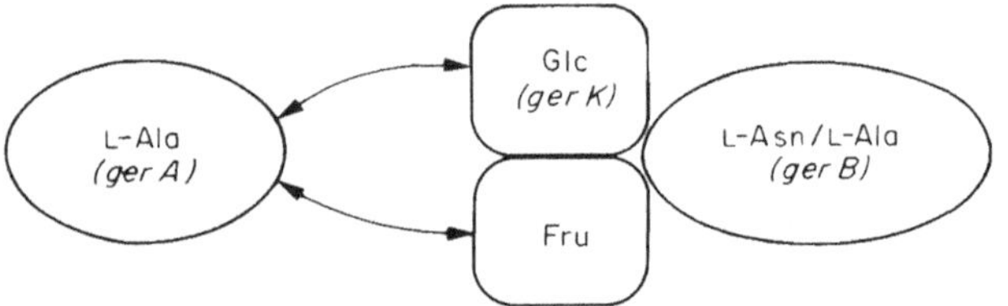

Fig. 1 Model for functional interaction between germinant receptors of spores of *Bacillus subtilis*. For details see text. After Venkatasubramanian and Johnstone (1993)

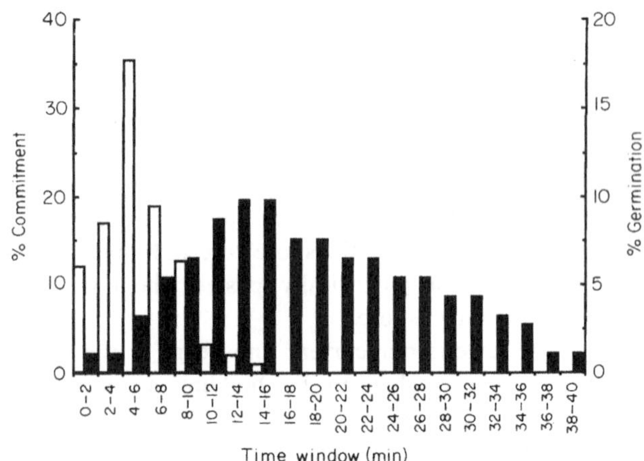

Fig. 2 Kinetics of commitment (open bars) and germination (closed bars) of spores of *Bacillus megaterium* KM at 30°C in 1 mmol l^{-1} L-alanine. Percentage commitment was determined as described by Stewart *et al.* (1981) and percentage germination was measured by loss of $O.D._{600}$

strates that the spore population rapidly undergoes the commitment reaction but that the time window between commitment and germination is much more variable. Identification of the mechanism of germination triggering is thus concerned with analysis of the events which take place within the first minutes of exposure to specific germinants.

4. BIOCHEMICAL CHARACTERIZATION OF THE COMMITMENT REACTION

Several approaches have been employed to identify the biochemical events which occur during germination triggering. The knowledge gained from three successful approaches is summarized below.

4.1 Germinant metabolism

The question as to whether metabolism of germinants is required for germination triggering has generated much controversy. Three strategies have been employed to address this question. First, a range of non-metabolizable germinant analogues has been shown to induce germination in spores of several organisms. These include L-proline chloromethyl ketone in place of L-proline in *B. megaterium* QMB1551 (Rossignol and Vary 1978, 1979), L-alanine chloromethyl ketone as a substitute for L-alanine in *B. megaterium* KM (Foster 1986), 2-deoxy-D-glucose and D-allose as a substitute for glucose in *B. subtilis* (Prasad *et al.* 1972) and allylglycine and cycloleucine as a substitute for L-alanine in *B. subtilis* (Woese *et al.* 1958; Irie *et al.* 1980, 1982; Sammons *et al.* 1981; Yasuda and Tochikubo 1985; Kanda *et al.* 1988). Second, the ability of mutants deficient in enzymes in germinant metabolic pathways to germinate has been studied. Mutants of *B. subtilis* lacking L-alanine dehydrogenase (Freese and Cashel 1965), pyruvate dehydrogenase (Freese and Fortnagel 1969) or glucose dehydrogenase (Rather and Moran 1988) all germinate normally. The third approach has been to analyse metabolism of the germinants themselves. However no incorporation of radioactivity into other compounds from labelled L-alanine and from D-glucose in *B. megaterium* KM (Scott and Ellar 1978b) and *B. megaterium* QMB1551 (Shay and Vary 1978) spore germination respectively was detected. It may therefore be concluded that metabolism of the germinants is not required for germination triggering. There is however one documented exception to this general rule: in spores of *B. fastidiosus*, uricase has been shown to be required for uric-acid induced germination (Salas *et al.* 1985).

4.2 Identification of metabolic events during germination triggering

An alternative approach which has been used to examine events associated with the germination triggering and commitment reactions in spores of *B. megaterium* KM, is to look for biochemical and physiological changes during the first minutes of germination. These changes can be conveniently divided into two categories (Foster and Johnstone 1989b): (1) those events including loss of heat resistance, commitment and dipicolinic acid (DPA) release which are detected within the first minute of germination triggering and which may therefore be associated with the trigger reaction; and (2) those events including $O.D._{600}$ loss, selective cortex hydrolysis and the onset of spore metabolism which are initiated at a later time and are presumably not therefore directly associated with triggering.

No significant changes in spore metabolite pools including tricarboxylic acid intermediates, ATP, NADH and NADPH, were identified during triggering of *B. megaterium* KM spore germination (Scott and Ellar 1978a). Furthermore no detectable irreversible incorporation of protons from tritiated water into spore metabolites was observed during the first minutes of germination (Scott and Ellar 1978b; Johnstone *et al.* 1982b). Thus activity of the major metabolic pathways is not required for germination triggering. This view is supported by the lack of effect of a wide range of metabolic inhibitors on germination (see below).

4.3 Inhibitor studies

Metabolic inhibitors are potentially powerful tools to identify essential metabolic events during germination triggering. For this purpose it is necessary to establish that a metabolic inhibitor blocks the commitment reaction. This has been achieved in studies of *B. megaterium* KM spore germination (see below). In several species, germination has been shown to occur in the presence of a wide range of metabolic inhibitors including inhibitors of DNA, RNA and protein synthesis, of glycolysis and of the respiratory chain (Dills and Vary 1978; Scott *et al.* 1978; Rossignol and Vary 1979). The results of such metabolic inhibitor experiments must however be interpreted with caution since the lack of effect of an inhibitor may be due to its inability to access the target site. In contrast, $HgCl_2$ is a potent reversible inhibitor of germination in several species including *B. megaterium* (Levinson and Hyatt 1966; Foster and Johnstone 1986). Evidence has also been presented that protease inhibitors block early steps of germination in spores of *B. cereus* T (Boschwitz *et al.* 1983, 1985) and germination triggering in spores of *B. megaterium* KM (Foster and Johnstone 1986). These findings therefore suggest that a component of the germination trigger contains essential sulphydryl groups and may have a proteolytic activity.

The $HgCl_2$ sensitivity of *B. megaterium* KM spore germination has been extensively characterized. Two $HgCl_2$ sensitive sites are present in the germination pathway

(Foster and Johnstone 1986). The first (site I) is a pre-commitment event and can be protected from $HgCl_2$ by 50 mmol l^{-1} D-alanine, which suggests that it is part of the trigger reaction. The second (site II) is a post-commitment event and cannot be protected by D-alanine. Due to the differential sensitivity of these two sites it was demonstrated that in the presence of 1 mmol l^{-1} $HgCl_2$, 25% of the spore population becomes committed to germinate whereas DPA, Ca^{2+}, Zn^{2+} and peptidoglycan release as well as loss of refractility and selective cortex hydrolysis are >95% inhibited. This commitment reaction, which occurs in the presence of 1 mmol l^{-1} $HgCl_2$ was, however, inhibited by the presence of protease inhibitors. Thus in *B. megaterium* KM, the commitment reaction can be identified as one which occurs in the presence of 1 mmol l^{-1} $HgCl_2$, but which is protease inhibitor sensitive.

5. PEPTIDOGLYCAN HYDROLYSIS DURING GERMINATION TRIGGERING

Given the central role of the spore cortex in maintaining spore dormancy by maintaining the dehydrated state of the spore core (Ellar 1978; Warth 1978), hydrolysis of the spore cortex peptidoglycan might be expected to occur early during germination. Activation of a cortex lytic enzyme as a primary event in spore germination was originally suggested by Powell and Strange (1956). This view is supported by the observation that germination can be initiated by peptidoglycan lytic enzymes if the spore coats are naturally permeable to the enzyme (Suzuki and Rode 1969) or are rendered permeable by chemical treatment (Gould and Dring 1972). The spore cortex peptidoglycan is structurally distinct from that of the vegetative cell and in particular it is less extensively cross-linked (Rogers 1977). Its structural integrity is therefore likely to be significantly altered by selective hydrolysis of either the glycan chains or the peptide cross-links as a result of glycosylase or peptidase activities respectively. Although $HgCl_2$ prevents release of soluble peptidoglycan fragments during germination (Rossignol and Vary 1978), cortex hydrolysis has been considered to be a late event during germination (Dring and Gould 1971; Hsieh and Vary 1975). Such analysis was based on the release of *soluble* peptidoglycan fragments and would be unlikely to detect selected limited hydrolysis of the spore cortex which could occur early in germination. Measurement of reducing termini during germination of spores of *B. megaterium* KM showed that selective cortex hydrolysis could be detected within 2 min of addition of germinants (Johnstone and Ellar 1982), providing evidence that cortex hydrolysis might constitute an early germination event.

Two classes of cortex lytic enzymes have been purified from dormant and germinating spores. First, surface bound enzymes which are able to hydrolyse isolated spore cortex but are unable to induce germination-like changes in permeablized spores (Srivastava and Fitz-James 1981; Brown *et al.* 1982; Foster and Johnstone 1987). Since spores germinate normally after extraction of such enzymes, these surface bound enzymes appear to play no direct role in the germination process. Secondly, enzymes which are able to induce germination in permeablized spores have been extracted from both germinating and broken spores (Gombas and Labbe 1981; Brown *et al.* 1982; Ando and Tsuzuki 1984). The best characterized of this second class of enzymes is the germination-specific cortex-lytic enzyme (GSLE) of *Bacillus megaterium* KM (Foster and Johnstone 1987). There is substantial biochemical evidence that GSLE plays a key role in germination and that proteolytic activation of GSLE from a 68-kDa cortex-associated pro-form to yield a 29-kDa active enzyme constitutes part of the germination trigger mechanism (Foster and Johnstone 1989a, 1989b). This includes:

(1) Substrate specificity: GSLE is only active on the cortex of intact spores. No activity is observed on isolated spore peptidoglycan or on vegetative cell walls. GSLE thus has a specific requirement for intact stressed cortex peptidoglycan as a substrate.
(2) Mechanism of action: incubation of GSLE with permeablized spores in the absence of germinants results in an increase in spore cortex muramic acid δ-lactam reducing termini as is observed during initiation of germination.
(3) Inhibitor profile: purified GSLE shows the same Hg^{2+} inhibition characteristics as the post-commitment site II $HgCl_2$ sensitive site.
(4) Activation during germination triggering: activation of GSLE occurs in the presence of 1 mmol l^{-1} $HgCl_2$ and its activation is inhibited by 1 mmol l^{-1} PMSF. This inhibition profile parallels that of the commitment reaction as described above.

Activation of GSLE therefore conforms to the biochemical criteria for it to be considered part of the germination triggering mechanism.

6. A MODEL FOR GERMINATION TRIGGERING

A model for the germination triggering reaction in spores of *B. megaterium* KM which is based on models described previously (Stewart *et al.* 1981; Foster and Johnstone 1989b, 1990) and which combines the key findings described above, is shown in Fig. 3. In this model the germination receptor (R) is altered conformationally by heat shock such that it is more responsive to the presence of the

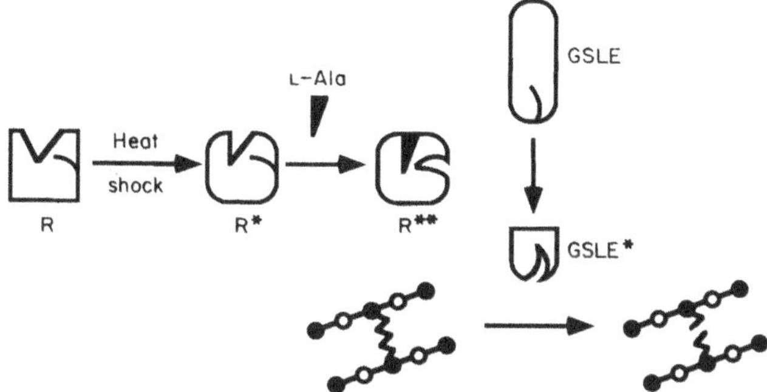

Fig. 3 Model for germination triggering in spores of *Bacillus megaterium* KM. For details see text. Based on previous models described by Stewart *et al.* (1981) and Foster and Johnstone (1989a, b)

germinant L-alanine. The rates of both commitment and germination have been shown to be stimulated by heat shock in this organism (Stewart *et al.* 1981). Binding of L-alanine induces a further conformational change in the germinant receptor which activates its PMSF-sensitive proteolytic site (R**). It is deduced that the activity of the protease is Hg^{2+}-sensitive and constitutes site I of Hg^{2+} inhibition described above. The active protease cleaves the 68-kDa pro-form of peptidoglycan-immobilized GSLE to release the active heat sensitive 29-kDa enzyme. Depolymerization of the spore cortex is catalysed by GSLE, which is also Hg^{2+} sensitive and represents site II of Hg^{2+} inhibition described above. Cortex depolymerization results in uptake of water into the spore core, the release of spore ions including DPA and Ca^{2+}, and the onset of spore metabolism. In this model, germination triggering requires no metabolism of the germinant L-alanine and takes place in the absence of general spore metabolism; it is driven by a series of exergonic hydrolytic reactions. The first irreversible event in the pathway is the activation of GSLE; commitment thus represents the proteolysis of sufficient GSLE such that a spore will germinate within the time window defined for commitment. Once GSLE has been activated, the subsequent germination pathway is an inevitable process. Since GSLE is Hg^{2+}-sensitive, it probably corresponds to the post-commitment Hg^{2+}-sensitive site described above. It is not known whether there are additional steps between R** and GSLE*; for example a proteolytic cascade might be involved.

Although this germination pathway has been extensively characterized in spores of *B. megaterium* KM, the question arises as to whether it is common to spores of other organisms. The general principle of germinant-mediated proteolytic activation of a peptidoglycan hydrolytic enzyme to initiate the germination response is supported by a number of observations in spores of other organisms. These include the conserved structure of spore cortex peptidoglycan, the identification of pre- and post-Hg^{2+}-sensitive sites in other germination pathways (Venkatasubramanian and Johnstone 1989), the observed release of peptidoglycan fragments during germination and identification of a commitment reaction. There is, however, a growing body of evidence that the mechanism of cortex hydrolysis during germination of spores of other organisms differs from that observed in *B. megaterium* KM. An increase in cortex muramic acid δ-lactam content was not found during germination of spores of *B. subtilis* (Venkatasubramanian and Johnstone 1989), *B. fastidiosus* (Salas *et al.* 1985), *B. megaterium* ATCC 12872 (Nakatani *et al.* 1985) and *Clostridium bifermentans* (Bright and Johnstone, unpublished). In the case of *B. megaterium* ATCC 12872, an increase in cortical glucosaminol content was observed during germination, which suggests activation of a glucosamidase during triggering; in the other spores no change in the reducing termini of residual spore peptidoglycan was detected. It is therefore likely that cortex depolymerization occurs by endopeptidase or transpeptidase activity in these organisms.

7. CONCLUSIONS

During the past decade the signal transduction pathways whereby micro-organisms detect and respond to environmental stimuli have been extensively studied and a number of response mechanisms have been identified (Parkinson 1993). These include protein phosphorylation via conserved two-component sensors and regulators which induce changes in protein function at the level of transcription (e.g. osmoregulation) or by directly influencing protein function (e.g. the chemotactic response). Alternatively, substitution of sigma factors may alter patterns of gene expression (e.g. during sporulation; Kaiser and Losick 1993). Although allosterically activated proteolytic cleavage is well established as a mechanism of intracellular signalling in bacteria (e.g. the *lonA* activity in the SOS response; Little and Mount 1982), a germination triggering mechanism based on allosterically-induced proteolytic cleavage represents a novel prokaryotic environmental sensing mechanism.

There are several key questions which remain to be answered concerning the spore germination trigger reaction. These include identification of the location of the receptor in the spore and demonstration *in vitro* of germinant-dependent catalytic activity of the germinant receptor. In addition, in order to establish the validity of the model proposed above it will also be necessary to examine the effects on germination of mutations in the

GSLE structural gene. Experiments are currently in progress to these ends.

8. REFERENCES

Ando, Y. and Tsuzuki, T. (1984) The role of surface charge in ionic germination of *Clostridium perfringens* spores. *Journal of General Microbiology* **130**, 267–273.

Boschwitz, H., Milner, Y., Keynan, A., Halvorson, H.O. and Milner, Y. (1983) Effect of inhibitors of trypsin-like enzymes on *Bacillus cereus* T spores. *Journal of Bacteriology* **153**, 700–708.

Boschwitz, H., Halvorson, H.O., Keynan, A. and Milner, Y. (1985) Trypsin-like enzymes from dormant and germinated spores of *Bacillus cereus* T and their possible involvement in germination. *Journal of Bacteriology* **164**, 302–309.

Bright, J.J. and Johnstone, K. (1987) Germination kinetics of spores of *Clostridium bifermentans* M 86b. *Microbios* **52**, 17–28.

Brown, W.C., Vellom, D., Ho, I., Mitchell, N. and McVay, P. (1982) Interaction between *Bacillus* spore hexosaminidase and specific germinants. *Journal of General Microbiology* **149**, 969–976.

Dills, S.S. and Vary, J.C. (1978) An evaluation of respiratory chain associated functions during initiation of germination of *Bacillus megaterium* spores. *Biochimica et Biophysica Acta* **541**, 301–311.

Dring, G.J. and Gould, G.W. (1971) Sequence of events during rapid germination of spores of *Bacillus cereus. Journal of General Microbiology* **65**, 101–104.

Ellar, D.J. (1978) Spore specific structures and their function. *Symposia of the Society for General Microbiology* **28**, 295–325.

Feavers, I.M., Miles, J.S. and Moir, A. (1985) The nucleotide-sequence of a spore germination gene (*gerA*) of *Bacillus subtilis* 168. *Gene* **38**, 95–102.

Foster, S.J. (1986) Biochemistry of *Bacillus megaterium* spore germination. Ph.D. Thesis, University of Cambridge.

Foster, S.J. and Johnstone, K. (1986) The use of inhibitors to identify early events during *Bacillus megaterium* KM spore germination. *Biochemical Journal* **237**, 865–870.

Foster, S.J. and Johnstone, K. (1987) Purification and properties of a germination-specific cortex-lytic enzyme from spores of *Bacilius megaterium* KM. *Biochemical Journal* **242**, 573–579.

Foster, S.J. and Johnstone, K. (1989a) Germination-specific cortex-lytic enzyme is activated during triggering of *Bacillus megaterium* KM spore germination. *Molecular Microbiology* **2**, 727–733.

Foster, S.J. and Johnstone, K. (1989b) The trigger mechanism of bacterial spore germination. In *Regulation of Procaryotic Development* ed. Smith, I., Slepecky, R.A. and Setlow, P. pp. 89–108. Washington, DC: American Society for Microbiology.

Foster, S.J. and Johnstone, K. (1990) Pulling the trigger, the mechanism of bacterial spore germination. *Molecular Microbiology* **4**, 137–141.

Freese, E. and Cashel, M. (1965) Initial stages of germination. In *Spores III* ed. Campbell, L.L. and Halvorson, H.O. pp. 144–151. Ann Arbor, Michigan: American Society for Microbiology.

Freese, E. and Fortnagel, U. (1969) Growth and sporulation of *Bacillus subtilis* mutants blocked in the pyruvate dehydrogenase complex. *Journal of Bacteriology* **99**, 745–756.

Gerhardt, P. and Marquis, P. (1989) Spore thermoresistance mechanisms. In *Regulation of Procaryotic Development* ed. Smith, I., Slepecky, R.A. and Setlow, P. pp. 43–63. Washington, DC: American Society for Microbiology.

Gombas, D.E. and Labbe, R.G. (1981) Extraction of spore-lytic enzyme from *Clostridium perfringens* spores. *Journal of General Microbiology* **126**, 37–44.

Gould, G.W. (1969) Germination. In *The Bacterial Spore* ed. Gould, G.W. and Hurst, A. pp. 397–444. London: Academic Press.

Gould, G.W. (1983) Germination and the problem of dormancy. *Journal of Applied Bacteriology* **33**, 34–49.

Gould, G.W. and Dring, G.J. (1972) Biochemical mechanisms of spore germination. In *Spores V* ed. Halvorson, H.O., Hanson, R. and Campbell, L.L. pp. 401–408. Washington, DC: American Society for Microbiology.

Harrell, W.R. and Halvorson, H.O. (1955) Studies on the role of L-alanine in the germination of spores of *Bacillus terminalis. Journal of Bacteriology* **69**, 275–279.

Hsieh, L.K. and Vary, J.C. (1975) Germination and peptidoglycan solubilisation in *Bacillus megaterium* spores. *Journal of Bacteriology* **123**, 463–470.

Irie, R., Okamoto, T. and Fujita, Y. (1980) Initiation of germination of *Bacillus subtilis* spores by allylglycine. *Journal of General and Applied Microbiology* **26**, 425–428.

Irie, R., Okamoto, T. and Fujita, Y. (1982) A germination mutant of *Bacillus subtilis* deficient in response to glucose. *Journal of General and Applied Microbiology* **28**, 345–354.

Irie, R., Okamoto, T. and Fujita, Y. (1984) Kinetics of spore germination of *Bacillus subtilis* in low concentrations of L-alanine. *Journal of General and Applied Microbiology* **30**, 109–113.

Janoff, A.S., Coughlin, R.T., Racine, F.M., McGroarty, E.J. and Vary, J.C. (1979) Use of electron spin resonance to study *Bacillus megaterium* spore membranes. *Biochemical and Biophysical Research Communications* **89**, 565–570.

Johnstone, K. and Ellar, D.J. (1982) The role of cortex hydrolysis in the triggering of germination of *Bacillus megaterium* KM endospores. *Biochimica et Biophysica Acta* **714**, 185–191.

Johnstone, K., Ellar, D.J. and Appleton, T.C. (1980) Location of metal ions in *Bacillus megaterium* spores by high-resolution electron probe X-ray microanalysis. *FEMS Microbiology Letters* **7**, 97–101.

Johnstone, K., Stewart, G.S.A.B., Barratt, M.D. and Ellar, D.J. (1982a) An electron paramagnetic study of the manganese environment within dormant spores of *Bacillus megaterium* KM. *Biochimica et Biophysica Acta* **714**, 379–381.

Johnstone, K., Stewart, G.S.A.B., Scott, I.R. and Ellar, D.J. (1982b) Zinc release and the sequence of biochemical events during triggering of *Bacillus megaterium* KM spore germination. *Biochemical Journal* **208**, 407–411.

Kaiser, D. and Losick, R. (1993) How and why bacteria communicate with each other. *Cell* **73**, 873–885.

Kanda, K., Yasuda, Y. and Tochikubo, K. (1988) Germination-initiating activities for *Bacillus* spores of analogues of L-alanine derived by modification at the amino or carboxyl group. *Journal of General Microbiology* **134**, 2747–2755.

Keynan, A. (1978) Spore structure and its relation to resistance, dormancy and germination. In *Spores VII* ed. Chambliss, G. and Vary, J.C. pp. 43–53. Washington, DC: American Society for Microbiology.

Keynan, A. and Evenchik, Z. (1969) Activation. In *The Bacterial Spore* ed. Gould, G.W. and Hurst, A. pp. 358–396. London: Academic Press.

Levinson, H.L. and Hyatt, M.T. (1966) Heat activation kinetics of *Bacillus megaterium* spores. *Biochemical and Biophysical Research Communications* **37**, 909–916.

Little, J.W. and Mount, D.W. (1982) The SOS regulatory system of *Escherichia coli*. *Cell* **29**, 11–22.

Moir, A. and Smith, D.A. (1985) The genetics of spore germination in *Bacillus subtilis*. In *Fundamental and Applied Aspects of Bacterial Spores* ed. Dring, G.J., Ellar, D.J. and Gould, G.W. pp. 89–100. London: Academic Press.

Moir, A. and Smith, D.A. (1990) The genetics of bacterial spore germination. *Annual Review of Microbiology* **44**, 531–553.

Nakatani, Y., Tanida, I., Koshikawa, T., Imagawa, M., Nishihara, T. and Kondo, M. (1985) Collapse of cortex expansion during germination of *Bacillus megaterium* spores. *Microbiology and Immunology* **29**, 689–699.

Parkinson, J.S. (1993) Signal transduction schemes of bacteria. *Cell* **73**, 857–871.

Powell, J.F. and Strange, R.E. (1956) Biochemical changes occurring during sporulation in *Bacillus* species. *Biochemical Journal* **63**, 661–668.

Prasad, C., Diesterhaft, M. and Freese, E. (1972) Initiation of spore germination in glycolytic mutants of *Bacillus subtilis*. *Journal of Bacteriology* **110**, 321–328.

Rogers, H.J. (1977) Peptidoglycans (mucopeptides), structure, form and function. In *Spore Research 1976* ed. Barker, A.N., Wolf, J., Ellar, D.J. and Gould, G.W. pp. 33–54. London: Academic Press.

Rather, P.N. and Moran, C.P. (1988) Compartment-specific transcription in *Bacillus subtilis*, identification of the promotor for *gdh*. *Journal of Bacteriology* **170**, 5086–5092.

Rossignol, D.P. and Vary, J.C. (1978) L-Proline initiated germination in *Bacillus megaterium* spores. In *Spores VII* ed. Chambliss, G. and Vary, J.C. pp. 90–94. Washington, DC: American Society for Microbiology.

Rossignol, D.P. and Vary, J.C. (1979) L-Proline site for triggering *Bacillus megaterium* spore germination. *Biochemical and Biophysical Research Communications* **89**, 547–551.

Russell, A.D. (1982) The bacterial spore. In *The Destruction of Bacterial Spores*, pp. 1–29. London: Academic Press.

Salas, J.A., Johnstone, K. and Ellar, D.J. (1985) The role of uricase in the triggering of germination of *Bacillus fastidiosus* spores. *Biochemical Journal* **229**, 241–249.

Sammons, R.L., Moir, A. and Smith, D.A. (1981) Isolation and properties of spore germination mutants of *Bacillus subtilis* 168 deficient in the initiation of germination. *Journal of General Microbiology* **124**, 229–241.

Scott, I.R. and Ellar, D.J. (1978a) Metabolism and the triggering of germination of *Bacillus megaterium*, concentrations of amino acids, organic acids, adenine nucleotides and nicotinamide nucleotides during germination. *Biochemical Journal* **174**, 627–634.

Scott, I.R. and Ellar, D.J. (1978b) Metabolism and the triggering of germination of *Bacillus megaterium*, use of L-[^{3}H] alanine and tritiated water to detect metabolism. *Biochemical Journal* **174**, 635–640.

Scott, I.R., Stewart, G.S.A.B., Koncewicz, M.A., Ellar, D.J. and Crafts-Lighty, A. (1978) Sequence of biochemical events during germination of *Bacillus megaterium* spores. In *Spores VII* ed. Chambliss, G. and Vary, J.C. pp. 95–103. Washington, DC: American Society for Microbiology.

Shay, L.K. and Vary, J.C. (1978) Biochemical studies on glucose initiated germination in *Bacillus megaterium*. *Biochimica et Biophysica Acta* **538**, 284–292.

Skomurski, J.F., Racine, F.M. and Vary, J.C. (1983) Steady state anisotropy changes of 1-6,-diphenyl 1,3,5-hexatriene in membranes from *Bacillus megaterium*. *Biochimica et Biophysica Acta* **731**, 428–436.

Smith, D.A., Moir, A. and Lafferty, E. (1977) Spore germination genetics in *Bacillus subtilis*. In *Spore Research 1976* ed. Barker, A.N., Dring, G.J., Ellar, D.J., Gould, G.W. and Wolf, J. London: Academic Press.

Srivastava, O.P. and Fitz-James, P.C. (1981) Alteration by heat activation of enzymes localised in spore coats of *Bacillus cereus*. *Canadian Journal of Microbiology* **27**, 408–416.

Stewart, G.S.A.B., Eaton, M.W., Johnstone, K., Barratt, M.D. and Ellar, D.J. (1979) An investigation of membrane fluidity changes during sporulation and germination of *Bacillus megaterium* KM measured by electron spin and nuclear magnetic resonance spectroscopy. *Biochimica et Biophysica Acta* **600**, 270–290.

Stewart, G.S.A.B., Johnstone, K., Hagelberg, E. and Ellar, D.J. (1981) Commitment of bacterial spores to germinate, a measure of the trigger reaction. *Biochemical Journal* **198**, 101–106.

Stewart, M.A., Somylo, A.P., Somylo, A.V., Shurman, H., Lindsay, J.A. and Murrell, W.G. (1980) Distribution of calcium and other elements in cryosectioned *Bacillus cereus* T spores determined by high-resolution scanning electron probe X-ray microanalysis. *Journal of Bacteriology* **143**, 481–491.

Suzuki, Y. and Rhode, L.J. (1969) Effect of lysozyme on resting spores of *Bacillus megaterium*. *Journal of Bacteriology* **98**, 238–245.

Vary, J.C. and Halvorson, H.O. (1965) Kinetics of germination of *Bacillus* spores. *Journal of Bacteriology* **89**, 1340–1347.

Venkatasubramanian, P. and Johnstone, K. (1989) Biochemical analysis of the *Bacillus subtilis* 1604 germination response. *Journal of General Microbiology* **135**, 2723–2733.

Venkatasubramanian, P. and Johnstone, K. (1993) Biochemical analysis of germination mutants to characterise germinant receptors of *Bacillus subtilis* spores. *Journal of General Microbiology* **139**, 1921–1926.

Warth, A.D. (1978) Molecular structure of the bacterial spore. *Advances in Microbial Physiology* **17**, 1–45.

Wax, R. and Freese, E. (1968) Initiation of the germination of *Bacillus subtilis* spores by combination of compounds in place of L-alanine. *Journal of Bacteriology* **95**, 433–438.

Woese, C.R., Morowitz, H.J. and Hutchison, C.A. (1958) Analysis of action of L-alanine analogues in spore germination. *Journal of Bacteriology* **76**, 578–588.

Wolgamott, G.D. and Durham, N.N. (1971) Initiation of spore germination in *Bacillus cereus*, a proposed allosteric mechanism. *Canadian Journal of Microbiology* **17**, 1043–1048.

Yasuda, Y. and Tochikubo, K. (1984) Relation between D-glucose and L-alanine and D-alanine in the initiation of germination of *Bacillus subtilis* spores. *Microbiology and Immunology* **28**, 197–207.

Yasuda, Y. and Tochikubo, K. (1985) Germination-initiation and inhibitory activities of L-alanine and D-alanine analogs for *Bacillus subtilis* spores-modification of the methyl-group of L-alanine and D-alanine. *Microbiology and Immunology* **29**, 229–241.

Yasuda, Y., Namiki-Kanie, S. and Hachisuka, Y. (1978a) Inhibition of germination of *Bacillus subtilis* spores by alcohols. In *Spores VII* ed. Chambliss, G. and Vary, J.C. pp. 104–108. Washington, DC: American Society for Microbiology.

Yasuda, Y., Namiki-Kanie, S. and Hachisuka, Y. (1978b) Inhibition of *Bacillus subtilis* spore germination by various hydrophobic compounds, demonstration of hydrophobic character of the L-alanine receptor site. *Journal of Bacteriology* **136**, 484–490.

Yasuda, Y., Tochikubo, K., Hachisuka, Y., Tomida, H. and Ikeda, K. (1982) Quantitative structure–inhibitory activity relationships of phenols and fatty acids for *Bacillus subtilis* spore germination. *Journal of Medicinal Chemistry* **25**, 315–320.

Zuberi, A.R., Feavers, I.M. and Moir, A. (1985) Identification of three complementation units in the *gerA* spore germination locus of *Bacillus subtilis*. *Journal of Bacteriology* **162**, 756–762.

Zuberi, A.R., Moir, A. and Feavers, I.M. (1987) The nucleotide-sequence and gene organization of the *gerA* spore germination operon of *Bacillus subtilis* 168. *Gene* **51**, 1–11.

Journal of Applied Bacteriology Symposium Supplement 1994, **76**, 25S–39S

The role and regulation of cell wall structural dynamics during differentiation of endospore-forming bacteria

S.J. Foster
Department of Molecular Biology and Biotechnology, University of Sheffield, Sheffield, UK

1. Introduction, 25S
2. Life cycle of the bacilli, 25S
3. Cell wall structure
   3.1 Vegetative cell wall, 26S
   3.2 Endospore cell wall, 27S
4. Cell wall synthesis
   4.1 Vegetative growth, 28S
   4.2 Sporulation, 28S
   4.3 Germination and outgrowth, 30S
5. Cell wall autolysis
   5.1 Vegetative growth, 30S
   5.2 Sporulation, 31S
   5.3 Germination and outgrowth, 32S
6. Role of cell wall structure and structural dynamics
   6.1 Vegetative growth, 33S
   6.2 Sporulation, 33S
   6.3 Dormancy, 34S
   6.4 Germination and outgrowth, 34S
7. Conclusions and future prospects, 35S
8. Acknowledgement, 36S
9. References, 36S

1. INTRODUCTION

Bacterial endospores are formed mainly by the genera *Bacillus* and *Clostridium* as the result of a relatively simple differentiation system (Slepecky and Leadbetter 1983). Much academic interest has focused on the control of the life cycle and the function of its various components. Several endospore forming bacteria are also major food-poisoning organisms and their highly resistant spores pose a huge problem to the food preservation industry (Crowther and Baird-Parker 1983).

The bacterial cell wall is responsible for the maintenance of cellular integrity and shape determination. During vegetative growth it has a dynamic structure that is continually being synthesized, modified and hydrolysed to allow cell growth and division. Complete control of wall compositional flux is essential for the survival of the organism. Sporulation results in the production of the dormant, highly resistant endospore which, under favourable conditions, can germinate and outgrow into a new vegetative cell. This differentiation pathway is characterized by distinct morphological changes which define the place of any given cell in the life cycle. Most of these morphological markers appear as a direct result of changes in cell wall structure. Differential gene expression is coupled to morphogenesis and the cell wall is therefore essential not only for survival, but also for the ability of the organism to complete the life cycle. It is surprising, therefore, that relatively little is known about the structure of the cell wall during differentiation and the structural modifications which occur apart from basic chemical composition. Cell wall compositional flux requires specific enzymes for biosynthesis, remodelling and autolysis. Several of these enzymes, some unique to the differentiation process, have been identified and their possible roles proposed.

In this review, differentiation is approached from the perspective of the cell wall and its importance highlighted in all parts of the life cycle, concentrating on the most studied organisms (mainly *Bacillus subtilis*) to illustrate general principles.

Correspondence to: Dr S. J. Foster, Department of Molecular Biology and Biotechnology, University of Sheffield, PO Box 594, Firth Court, Western Bank, Sheffield S10 2UH, UK.

2. LIFE CYCLE OF THE BACILLI

Under favourable conditions, cells of the genera *Bacillus* and *Clostridium* will grow and divide by binary fission with the new transverse division septum forming at the middle of the cell. This cycle of growth and division will continue until the organism becomes nutritionally deprived and then the cells may go on to sporulate. The basic differentiation process and the structures involved are common to all endospore formers and the life cycle has been often reviewed from various aspects (Losick and Youngman 1984; Doi 1989; Errington 1993). Features which are salient to cell walls are the main considerations below.

The process of sporulation is divided into several stages based on morphological studies and is shown diagrammatically in Fig. 1. The first visible marker that a cell has

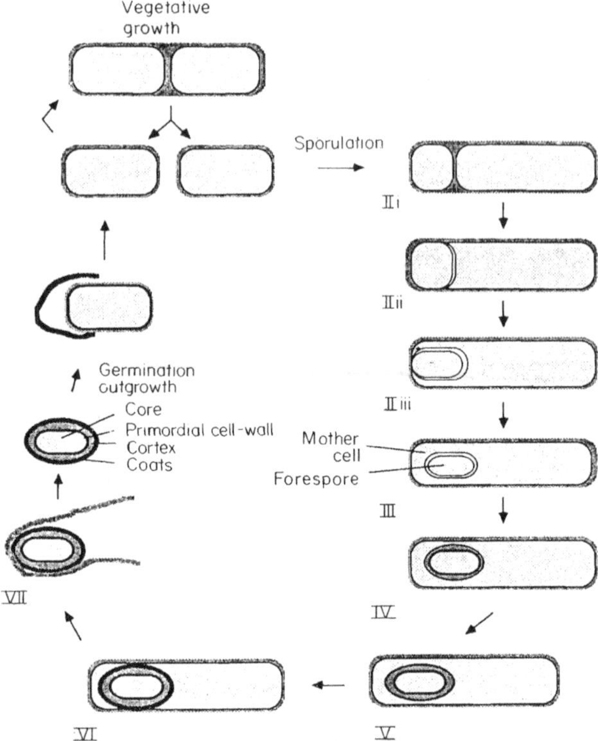

Fig. 1 Morphological changes during differentiation of *Bacillus* spp. During vegetative growth, cells divide by binary fission. Under conditions of nutrient deprivation asymmetric septation takes place (stage IIi). The forespore is then engulfed by the mother cell membrane (stages IIii and IIiii) and becomes completely surrounded by mother cell cytoplasm (stage III). During stage IV the primordial cell wall is laid down surrounding the forespore inner membrane (black line) and then the cortex (shaded area) on the outside of the primordial cell wall. The coats (black line) begin to be deposited outside the cortex during stage V. The spore matures (stage VI) and is released on lysis of the mother cell (stage VII). Germination of the endospore occurs in the presence of germinants and the cortex is hydrolysed, the geminated spore can then outgrow to form a new vegetative cell

started to sporulate is the appearance of an asymmetric septum near one pole of the cell (stage II). This septum presumably contains peptidoglycan which is hydrolysed prior to engulfment of the smaller compartment by the mother cell. This seems to occur by the continued formation of septal membrane and the movement of the septal initiation ring towards the cell pole. Engulfment is important as it results in the nascent forespore being surrounded by two membranes of opposing polarity (stage III), designated the inner (closest to the forespore cytoplasm) and outer forespore membranes.

The next major morphological event to occur is the synthesis of a layer of peptidoglycan between the two forespore membranes (stage IV). Proximal to the forespore inner membrane a layer of peptidoglycan of a structure similar to that of vegetative cell wall is deposited. This is called the primordial cell wall. After this a much thicker electron-transparent layer of peptidoglycan known as the spore cortex is laid down. The cortex has a composition and structure unique to bacterial endospores. After cortex formation the forespore becomes surrounded by proteinaceous coats and matures into a phase bright endospore with all its typical resistance properties (stages V and VI). The final landmark event during sporulation is the lysis of the mother cell, resulting in the release of the mature dormant endospore (stage VII).

Endospores are characterized by their extreme dormancy as shown by their dehydration, lack of metabolism and resistance to many treatments including heat, u.v. and deleterious chemicals. The spore cortex is responsible for maintenance of dormancy and thus is of paramount importance to the survival of the organism in this quiescent state.

Despite this extreme dormancy, spores retain an alert sensory mechanism which is able to respond within minutes to specific germinants. Germination is characterized by the change from a phase-bright dormant endospore to a phase-dark germinated spore due to the release of spore components and the uptake of water. During germination the spore cortex is hydrolysed and released as peptidoglycan fragments. The primordial cell wall remains and prevents the loss of cellular integrity of the germinated spore. This layer of peptidoglycan also probably forms the template for the formation of the new vegetative cell wall during outgrowth. The germinated spore elongates as new wall material is synthesized prior to the first post-germination division, the production of two new vegetative daughter cells and the completion of the differentiation process (Fig. 1).

3. CELL WALL STRUCTURE

3.1 Vegetative cell wall

All endospore-forming bacteria are Gram-positive. The cytoplasmic membrane of the vegetative cell is surrounded by a reasonably amorphous layer of cell wall material 20–50 nm thick (Shockman and Barrett 1983). Cell walls can be isolated as a morphologically identifiable insoluble residue from mechanically disrupted cells. The major shape-determining polymer in the wall is peptidoglycan which can account for 40% or more of the wall mass. Peptidoglycan consists of linear chains of glycan held together by cross-linked peptides (Fig. 2a). The glycan is composed of repeating disaccharide units of *N*-acetylglucosamine and *N*-acetylmuramic acid. The carboxyl group of each muramic acid residue forms an amide linkage with the amino-

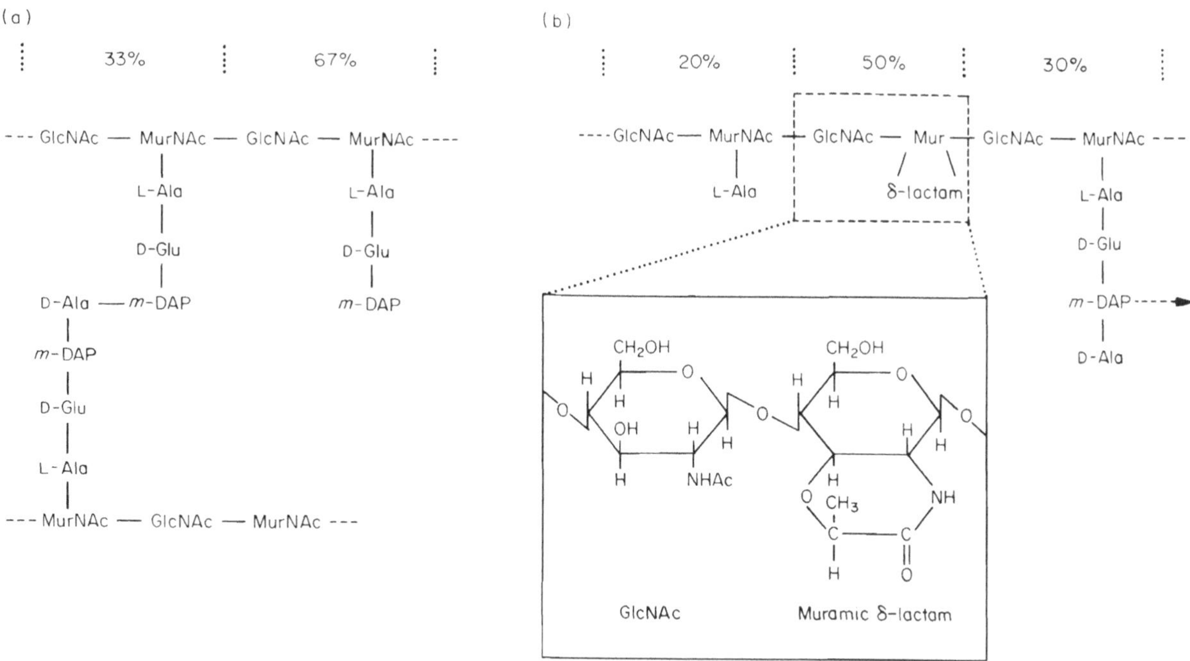

Fig. 2 Structure of vegetative cell and spore cortex peptidoglycan of *Bacillus subtilis*. The basic chemical structure of the wall peptidoglycans is shown, as is the approximate percentage of each type of substituted repeating disaccharide unit (based on Tipper and Gauthier 1972). (a) Vegetative cell wall peptidoglycan; (b) spore cortex. GlcNAc, *N*-acetylglucosamine; MurNAc, *N*-acetylmuramic acid; L-Ala, L-alanine; D-Glu, iso-D-glutamic acid; m-DAP, *meso*-diaminopimelic acid; D-Ala, D-alanine. The dotted arrow from the tetrapeptide substituted disaccharide of spore cortex (b) shows where possible cross-linking may occur to another peptide side chain

terminal residue of a short peptide, which may be cross-linked to the corresponding peptide on another glycan strand (Fig. 2a). Glycan chains can become linked to more than one other and so a large macromolecular structure can be formed. Secondary anionic wall polymers such as teichoic or teichuronic acids are also present, covalently bound to the peptidoglycan. During vegetative growth these accessory polymers are essential for cell viability (Mauel *et al.* 1989).

3.2 Endospore cell wall

Sporulation results in the formation of a mature dormant endospore of which one of the major morphological features is a thick cell wall comprised of peptidoglycan. This cell wall consists of two layers. The inner layer has been termed the primordial cell wall because after germination this develops into the cell wall of the outgrowing vegetative cell (Cleveland and Gilvarg 1975; Tipper and Linnett 1976; Warth 1978). The primordial cell wall differs from cortex in that it retains the structure of the vegetative cell wall peptidoglycan (Warth 1978). The outer layer of peptidoglycan, the spore cortex, has a unique spore-specific structure distinct from that of the vegetative cell wall (Fig. 2b; Warth and Strominger 1972). In vegetative cell peptidoglycan, most muramic acid residues are substituted with a peptide side chain. In spore cortex however, approximately 50% of the muramic acid residues are present as the spore-specific muramic acid δ-lactam (Fig. 2b; Warth and Strominger 1969). The δ-lactam is not randomly located in the glycan chain, but occurs predominantly at every alternate disaccharide (Tipper and Gauthier 1972). The remainder of the muramic acid residues are *N*-acetylated and their side chain carboxyls either substituted with a tetrapeptide (30–35%) or a single L-alanine substituent with a free carboxyl group (15–20%) (Fig. 2b; Warth and Strominger 1969, 1972; Tipper and Gauthier 1972). Of the disaccharides bearing a tetrapeptide side chain only 1 in 4 to 1 in 6 are cross-linked (Tipper and Gauthier 1972; Warth and Strominger 1972). This gives a final cross-linkage value of between 6 and 9% which contrasts dramatically with an average of 33% in the vegetative cell walls of *B. subtilis* (Tipper and Gauthier 1972; Forrest *et al.* 1991). It has been suggested that the low cross-linkage value may be due to shear forces disrupting the peptide bonds during mechanical isolation of cortex, as the non-disrupted cortex contains few amino termini (Marquis and Bender 1990), however chemically extracted cortex also shows low cross-linkage (Popham and Setlow 1993a). The uncross-linked peptide sidechains in spore cortex of *B. subtilis* and *B. sphaericus* are tetrapeptides ending in a single D-alanine, whereas the uncross-linked vegetative cell wall peptide side chains are tripeptides (Tipper and Gauthier 1972). Also, in vegetative cell wall peptidoglycan of *B. subtilis* but not in spore cortex, the uncross-linked diaminopimelic acid residues are amidated (Warth and Strominger 1972). The peptidoglycan chain length of spore cortex has been variously described as being between 40 and 320 disaccharide residues (Warth 1978; Johnstone and Ellar 1982).

The accessory polymers such as teichoic acid are not present in the spore cortex or the primordial cell wall (Warth and Strominger 1972; Johnstone and Ellar 1982). A sporulation-specific teichoicase has been identified and purified from *B. subtilis* and may be responsible for removal of teichoic acid from existing cell wall material during differentiation (Kusser and Fiedler 1983).

4. CELL WALL SYNTHESIS

4.1 Vegetative growth

The strength and integrity of the vegetative cell wall are determined by a complex of peptidoglycan and anionic polymer (teichoic or teichuronic acid). The polymers are assembled by membrane-bound enzyme systems, covalently attached to each other and incorporated into the wall (Shockman and Barrett 1983). The biosynthesis of peptidoglycan occurs in three phases: in the cytoplasm, in the membrane and external to the membrane. *N*-acetylmuramic acid pentapeptide is assembled and transferred to an undecaprenylpyrophosphate lipid carrier at the cytoplasmic membrane. In most *Bacillus* spp. the pentapeptide has the sequence L-alanine, D-glutamic acid, meso-diaminopimelic acid, D-alanine, D-alanine (Schleifer and Kandler 1972). *N*-acetyl glucosamine is then added and the growing glycan chain polymerized. Insertion of this new wall material occurs by cross-linking of the nascent glycan chain to existing peptidoglycan by transpeptidation reactions in which the terminal D-alanine is removed from the pentapeptide and the D-alanine at position 4 is linked to the free amino group of the diaminopimelic acid residue of another peptide. D-Alanine carboxypeptidases can remove the terminal D-alanine residues from the pentapeptides and are involved in wall structure determination (Shockman and Barrett 1983).

The transpeptidases and carboxypeptidases involved in the final stages of peptidoglycan biosynthesis are the sites of action of the β-lactam antibiotics including penicillin, and can be identified as penicillin-binding-proteins (PBPs; Blumberg and Strominger 1972). If cells of *B. subtilis* are treated with penicillin then linear uncross-linked cell wall glycan strands are secreted substituted predominantly with pentapeptide side chains (Waxman *et al.* 1980). The penicillin sensitivity of these enzymes has been exploited in their identification and purification (Blumberg and Strominger 1972; Todd *et al.* 1985). PBPs can be identified after labelling with radioactive penicillin, separation by denaturing SDS-polyacrylamide gel electrophoresis and fluorography (Todd and Ellar 1982). PBPs are located in the cytoplasmic membrane of vegetative cells. *Bacillus subtilis* 168 and *B. megaterium* KM have at least six and seven respectively ranging from 122 to 50 kDa (Fig. 3, lane a; Blumberg and Strominger 1972; Todd and Ellar 1982; Neyman and Buchanan 1985). Some of these apparent PBPs may be artefacts due to proteolytic digestion of higher molecular weight forms. The low molecular weight PBPs seem to be non-essential for normal vegetative growth and PBP 5 (50 kDa) of *B. subtilis* has been shown to have D,D-carboxypeptidase activity (Todd *et al.* 1986; Buchanan 1987).

New wall material is added all over the cell cylinder to allow elongation to occur and septum formation at cell division proceeds from an annular zone of synthesis (Archibald 1989).

4.2 Sporulation

During sporulation, asymmetric separation takes place to create the new mother cell and forespore and presumably

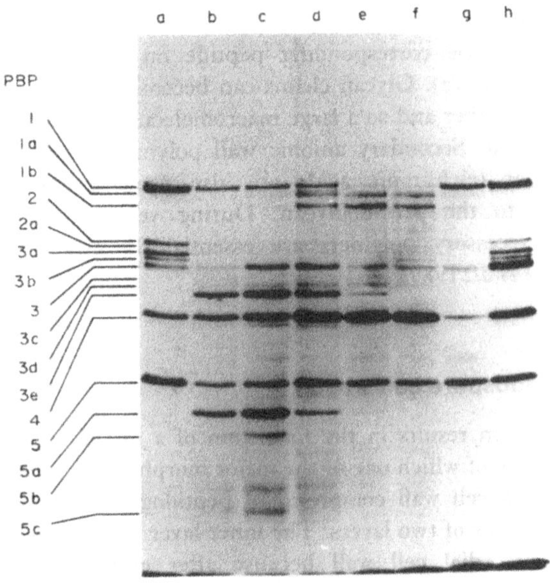

Fig. 3 Fluorogram of PBPs of membranes during the life cycle of *Bacillus megaterium* KM (from Todd and Ellar 1982 with permission). The designated number of each PBP is shown on the left. Lane a, vegetative cell membrane; b, stage III inner and outer forespore membranes (ifm and ofm); c, stage IV ifm and ofm; d, stage V ifm; e, stage VI ifm; f, dormant spore inner membrane; g and h, cell membrane of germinated spores at 85 and 200 min after the addition of germinant, respectively

involves the modified use of the existing vegetative cell division apparatus. The product of the *ftsZ* gene which plays a pivotal role in vegetative cell division is also essential for asymmetric septation (Beall and Lutkenhaus 1991). The asymmetric septum is thinner than the vegetative septum but seems to contain peptidoglycan based on electron microscopic data (Illing and Errington 1991) and the fact that continued peptidoglycan synthesis is necessary for septation (Jonas *et al.* 1990). The structure of the asymmetric septum peptidoglycan is unknown.

Two forespore-specific peptidoglycan structures are synthesized during sporulation. These are the primordial cell wall and the spore cortex, which have a different composition and which are synthesized by separate biosynthetic machinery. The primordial cell wall retains the structure of the vegetative cell wall but does not seem to contain accessory polymers. The spore cortex has the same basic composition as vegetative cell wall but with some important modifications.

Peptidoglycan biosynthesis occurs with two maxima in *B. cereus*, the first coincident with primordial cell wall synthesis and the second associated with cortex formation (Warth 1978). Muramic acid δ-lactam synthesis occurs only during the latter phase (Wickus *et al.* 1972). The primordial cell wall lies closest to the forespore membrane and within the thicker spore cortex layer (Fig. 1; Tipper and Gauthier 1972). Peptidoglycan biosynthesis is in part membrane associated and the reversed polarity of the two forespore membranes allows a possible mechanism for synthesis of the two peptidoglycan types (Ellar 1978). The hypothesis that primordial cell wall and cortex are synthesized at the inner and outer forespore membranes, respectively, was tested by Tipper and coworkers, using *B. sphaericus*. This organism is unusual in that instead of diaminopimelic acid residues in vegetative cell wall peptidoglycan it possesses L-lysine and the cross-links contain an additional spacer D-isoasparaginyl residue (Linnett and Tipper 1976). In contrast the spore cortex contains the standard diaminopimelic acid residues with direct cross-links to D-alanine (Linnett and Tipper 1976). Two peaks of synthesis during sporulation of the enzymes that produce peptidoglycan precursors were found, which correspond to those maxima shown above. In the first phase (stages II–III) only lysyl ligase (vegetative cell wall-specific synthesis enzyme) was produced and conversely in the second phase (stage IV) only diaminopimelyl ligase (cortex-specific synthetic enzyme) was found (Linnett and Tipper 1976). Further work showed that the diaminopimelyl ligase was restricted to the mother cell and although L-lysine ligase activity was found in both compartments it was present at a much higher specific activity in the forespore (Tipper and Linnett 1976). From these results it is apparent, certainly in *B. sphaericus*, that the primordial cell wall is synthesized at the inner forespore membrane and the spore cortex at the outer forespore membrane.

No direct evidence exists for the pathway of muramic acid δ-lactam residue formation, but it could occur by either the action of an *N*-acetylmuramyl-L-alanine amidase followed by transacylation or by de-*N*-acetylation followed by transpeptidation (Tipper and Gauthier 1972). It is interesting to note that the germination-specific cortex-lytic enzyme of *B. megaterium* KM causes an increase in δ-lactam residues in the spore cortex and may have a role in sporulation (Foster and Johnstone 1987, 1988). An L-alanine-D-glutamic acid endopeptidase has been isolated from lysed mother cells of *B. thuringiensis* and may be responsible for the creation of the muramic acid residues substituted with a single L-alanine moiety by the cleavage of the pentapeptide side chain (Kingan and Ensign 1968). Alternatively, sequential cleavage of the pentapeptide side chain by the action of a D,D-carboxypeptidase, D-glutamyl-*meso*-diaminopimelate endopeptidase and an L,D-carboxypeptidase capable of hydrolysing L-Ala-D-Glu could result in the L-alanine C-terminus (Guinand *et al.* 1979). A sporulation associated D-Glu-*meso*-DAP endopeptidase has been identified in sporulating cultures of *B. sphaericus* and *B. subtilis* (Guinand *et al.* 1976, 1979). The above categories of peptidoglycan hydrolases may lead to cortex maturation resulting in the final dormant endospore cortex structure.

Given the structure of the spore cortex and the fact that penicillin inhibits its synthesis, transpeptidase and carboxypeptidase activities seem essential for cortex formation (Tipper and Gauthier 1972). Todd and Ellar (1982) showed that during sporulation of *B. megaterium* KM the PBP profile changes (Fig. 3). Two spore-specific PBPs of 73·5 and 44·5 kDa, called PBP 3e and 5a respectively, are produced (Fig. 3, lanes b, c and d), with unique peptide digest maps, during stages II–III of sporulation (Todd and Ellar 1982). Two sporulation-specific PBPs corresponding to PBP 3e and 5a designated PBP 4a and 5a (or PBP 4* and 5*) were found in *B. subtilis* (Sowell and Buchanan 1983; Todd *et al.* 1983). PBP 5a (5*) from *B. subtilis* was purified by affinity chromatography and found to have D,D-carboxypeptidase activity *in vitro* (Todd *et al.* 1985). This enzyme is located primarily in the outer forespore membrane and was thus proposed to have a role in cortex synthesis (Buchanan and Neyman 1986). Recently the structural gene for PBP 5* (5a) designated *dacB* has been isolated from *B. subtilis*, cloned, sequenced and insertionally inactivated (Buchanan and Gustafson 1992; Buchanan and Ling 1992). The effect of inactivating PBP 5* on cortex structure was not determined but the spores were very heat-sensitive (Buchanan and Gustafson 1992). The *dacB* mutant also revealed the presence of another unrelated spore-specific PBP of a similar size to PBP 5* (Buchanan and Gustafson

1992). The transcription of *dacB* seems to be under the control of the sporulation-specific sigma factor, σ^E (Buchanan and Ling 1992). This sigma factor is responsible for the expression of early mother cell-specific genes during sporulation (Losick and Stragier 1992). The structural gene for PBP 4*, termed *pbpE*, has also been characterized, however a mutation in the gene had no effect on growth, differentiation or endospore heat resistance (Popham and Setlow 1993b).

The vegetative cell PBPs are also likely to contribute to the synthesis of the cortex. PBP 3 of *B. subtilis* is present in both inner and outer forespore membranes and may have a role in primordial cell wall and cortex synthesis (Buchanan and Neyman 1986). The gene encoding PBP 5 of *B. subtilis*, designated *dacA*, encodes the major D,D-carboxypeptidase activity in vegatative cells (Todd *et al.* 1986). Inactivation of this gene led to the formation of dormant endospores with an altered cortex structure as seen by more free amino groups, which is indicative of a less cross-linked peptidoglycan network (Todd *et al.* 1986).

Two other putative spore-specific PBPs have been identified based on sequence homology (Wu *et al.* 1992; Errington 1993). The *dacF* gene is cotranscribed with the sporulation-specific sigma factor, σ^F, in the developing forespore and thus may have a role in primordial cell wall formation, although a mutant in *dacF* has no distinguishable phenotype (Wu *et al.* 1992). The *spoVD* gene encodes a PBP homologue and a mutation in this gene leads to a block in sporulation at the time of cortex biosynthesis (Errington 1993).

The sequence of the *spoVE* gene is related to the *Escherichia coli ftsW* gene which is involved in septum formation although the function of *spoVE* is unknown (Errington 1993).

4.3 Germination and outgrowth

Germination is an irreversible series of degradative reactions that results in the breaking of spore dormancy and loss of typical spore properties (Foster and Johnstone 1989). No wall synthesis occurs during this phase and penicillin has no effect on germination (Foster 1986).

Outgrowth follows spore germination and during this phase considerable macromolecular synthesis occurs, resulting in swelling of the germinated spore. This is followed by cell elongation, emergence from the spore coats and the first separation event. Outgrowth in several species occurs synchronously and study of the changes in PBP profile during this phase has given a suggestion as to the likely role of individual enzymes in cell wall growth and division (Neyman and Buchanan 1985; Todd and Ellar 1985). PBP 2a of *B. subtilis* may be involved in cell elongation and PBP 2b seems to have a role in septation (Neyman and Buchanan 1985).

5. CELL WALL AUTOLYSIS

5.1 Vegetative growth

Rapid autolysis can often by observed when cultures of *Bacillus* spp. are incubated under conditions that are unfavourable to growth (Archibald 1989). This is due to hydrolysis of the peptidoglycan by autolysins present in the wall. Bacterial autolysins are potentially lethal enzymes and are characterized *in vivo* by their hydrolytic bond specificity as *N*-acetylmuramidases (muramidases), *N*-acetylglucosaminidases (glucosaminidases), *N*-acetylmuramyl-L-alanine amidases (amidases) and endopeptidases (Fig. 4; Ghuysen *et al.* 1966). Autolysins have been implicated in several important cellular functions including cell wall growth and turnover, cell separation, flagellation, competence, differentation, phage lysis and the lytic action of

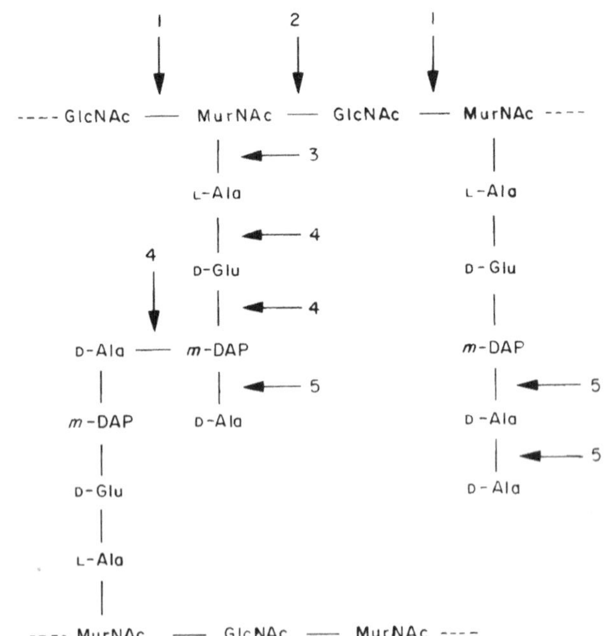

Fig. 4 Autolysin hydrolytic bond specificity. The basic structure of newly synthesized vegetative cell wall peptidoglycan of *Bacillus subtilis* is shown. The types of peptidoglycan hydrolases are as indicated: 1, *β*-*N*-acetylglucosaminidase (glucosaminidase); 2, *N*-acetylmuramidase (muramidase); 3, *N*-acetylmuramyl-L-alanine amidase (amidase); 4, endopeptidase; 5, carboxypeptidases. Only 1–4 can be classed as autolysins as carboxypeptidases only modify, but do not disrupt, the peptidoglycan structure

antibiotics (Pooley and Karamata 1984; Ward and Williamson 1984). However, despite abundant speculation, the exact role of these enzymes has remained elusive.

Bacillus subtilis has two major vegetative cell autolysins, a 90-kDa glucosaminidase and a 50-kDa amidase, which have been studied both biochemically and genetically although their function is unknown (Herbold and Glaser 1975; Rogers *et al.* 1984; Margot and Karamata 1992 and unpublished). The amidase is associated with a modifier protein which partially regulates its activity (Herbold and Glaser 1975; Kuroda *et al.* 1992). With renaturing SDS-polyacrylamide gel electrophoresis and substrate-containing gels, five autolysins present in vegetative cells of *B. subtilis* 168 (Fig. 5, lane a) have been identified (Foster 1992). Of these, the two higher molecular weights are the previously studied 90- and 50-kDa glucosaminidase and amidase, respectively. The expression of both of these enzymes has been shown to be under the control of the alternative sigma factor, σ^D (which controls the flagellar, chemotaxis and motility regulon), although the amidase has another promoter which is recognized by the housekeeping sigma factor, σ^A (Lazarevic *et al.* 1992; Margot 1992). This is confirmed by the reduced levels of these enzymes in the *sigD* mutant (Fig. 5, lane b). Three novel vegetative cell autolysins of 34, 32 and 30 kDa were also identified (Fig. 5, lane a; Foster 1992). Both the 34- and 30-kDa enzymes appear to be lytic enzymes encoded for by the defective prophage PBSX (Foster 1993; Longchamp and Karamata, unpublished). Another distinct 30-kDa amidase, the product of the *cwlA* gene, has been cloned, sequenced and the protein characterized (Kuroda and Sekiguchi 1990; Foster 1991), this enzyme may be encoded for by a cryptic prophage (Foster 1993).

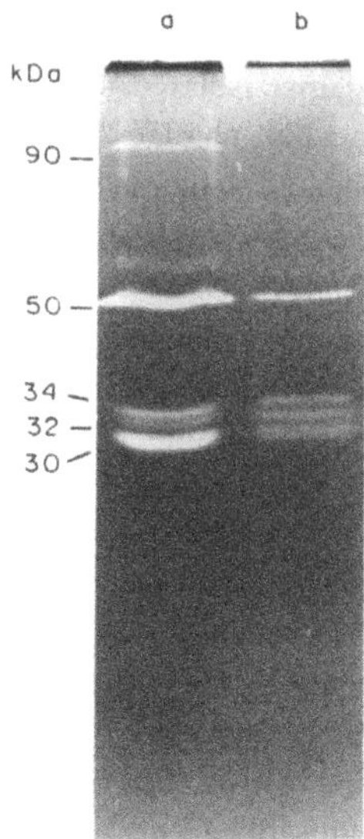

Fig. 5 *Bacillus subtilis* 168 vegetative cell lytic enzyme profile. SDS vegetative cell extracts were analysed by renaturing SDS-PAGE using *B. subtilis* 168 vegetative cell walls as the enzyme substrate (from Foster 1992). Lane a, BG-2 (wild type); b, DP-1 (*sigD*). Molecular masses of the lytic enzyme bands are indicated (kDa)

5.2 Sporulation

Hydrolysis of the asymmetric septum is necessary to allow engulfment of the forespore at stages II–III and the product of the *spoIID* gene is required for this process (Illing and Errington 1991). The *spoIID* gene has homology with the structural gene for the modifier protein of the vegetative cell 50-kDa amidase (Kuroda *et al.* 1992; Lazarevic *et al.* 1992). Thus the *spoIID* product may be involved in the regulation of a spore-specific autolysin that is responsible for the hydrolysis of peptidoglycan in the asymmetric septum allowing engulfment of the developing forespore.

Various autolysins have been found associated with developing forespores, mother cells and isolated mature endospores. The exact function of all of these enzymes during differentiation is unknown. The developmentally associated autolysins of *Bacillus cereus* have been classified as 'sporangial', 'surface bound' and 'spore core' enzymes based on their substrate specificities (Brown 1977), this categorization is also valid for other organisms (Table 1).

Mother cell lysis is the final landmark event to occur during sporulation; the mother cell wall is hydrolysed and the mature endospore released. The 'sporangial' category of autolysins are likely to be involved in this process and they have activity against both isolated vegetative cell walls and spore cortex (Brown 1977). Strange and Dark (1957b) identified an activity from *B. cereus* sporulating cells, called enzyme V, which they proposed was responsible for mother cell lysis. More recently with renaturing SDS-PAGE and substrate-containing gels, Foster (1992) has measured sporulation-specific changes in the autolysin profile for mother cells of *B. subtilis* during sporulation (Fig. 6a, b). At stages V–VI the activity of a 30-kDa enzyme greatly increases and this autolysin may be involved in mother cell

Table 1 Substrate specificities of developmentally associated autolysins

	Activity on various substrates*		
Type of enzyme	Vegetative cell walls	Spore cortex	Permeabilized spores
Sporangial	+	+ +	−
Surface bound	−	+ +	−
Spore core	−	−	+ +

* −, No activity; + or + +, level of activity.

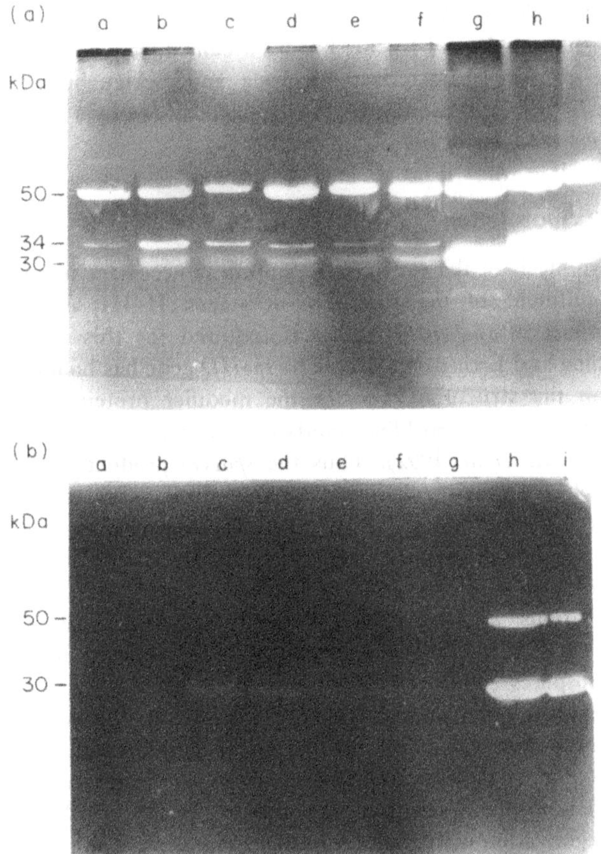

Fig. 6 Analysis of sporulation-associated autolysins of *Bacillus subtilis* 168. Samples were taken during sporulation and analysed by renaturing SDS-PAGE using *B. subtilis* 168 vegetative cell walls as the enzyme substrate (from Foster 1992). (a) SDS cell extracts; (b) TCA precipitate of culture supernatants. Molecular masses of autolytic bands (kDa) are shown on the left. Samples were taken at various times after the induction of sporulation. Lanes: a, $T = 0$ h; b, $T = 1$ h; c, $T = 2$ h; d, $T = 3$ h; e, $T = 4$ h; f, $T = 5$ h; g, $T = 6$ h; h, $T = 8$ h; i, $T = 22$ h

lysis (Fig. 6a, lanes f and g). The 30-kDa sporulation-specific enzyme accumulates cell associated prior to mother cell lysis (Fig 6a, lane g; $T = 6$, mother cells contain phase bright endospores but with no release). The enzyme only appears in the culture supernatant fluid after lysis (Fig. 6b, lane h; $T = 8$, >95% mature endospore release), and so the cell associated enzyme must be highly post-translationally regulated *in situ* and acts at a specific time to lyse the mother cell. Also present at this time during sporulation is the major vegetative 50-kDa amidase (Fig 6a), but this enzyme cannot be solely responsible for mother cell lysis, as a mutant inactivated in the structural gene for this enzyme lyses normally at the end of sporulation (Foster, unpublished). The 30-kDa sporulation enzyme is not CWLA or encoded for by the prophage PBSX (Foster, unpublished). Initial biochemical results suggest that the sporulation enzyme is under the control of the mother cell-specific sigma factor (σ^K) because a mutant in the sigma factor gene (*spoIIIC*) does not exhibit the 30-kDa enzyme burst.

The second type of sporulation-associated autolysins defined by Brown (1977) are the 'surface bound' enzymes. These autolysins are specific in that they are able to hydrolyse isolated spore cortex but have no activity against vegetative cell walls or permeabilized spores. Permeabilized spores are those which have had their coats disrupted so as to allow the passage of lysozyme to the cortex, and so the spores become lysozyme sensitive (Gould and Hitchins 1963; Foster and Johnstone 1987). 'Surface bound' lytic enzymes have been extracted from mature endospores of a number of species including *B. cereus* and *B. megaterium* (Strange and Dark 1957a; Brown and Cuhel 1975; Foster and Johnstone 1987). Most of these enzymes were isolated by procedures which extracted spore coat protein. Renaturing gel electrophoresis has revealed the presence of an enzyme with equivalent substrate specificity of 41 kDa associated with spores of *B. subtilis* (Foster 1992). This cortex-specific activity could only be visualized in broken spore extracts as high levels of the 30-kDa 'sporangial' enzyme otherwise masked its presence (Foster 1992). A 43-kDa 'surface bound' cortex lytic enzyme was purified from spores of *B. cereus* T and found to be either a glucosaminidase or a muramidase (Brown *et al.* 1978). The potential role(s) of 'surface bound' enzymes is at present unclear.

The final category of spore-associated lytic enzymes are the 'core' enzymes (Brown 1977). These enzymes have only very low activity against isolated spore cortex but are able to cause germination-like changes in permeabilized spores. They have been isolated from germinating and disrupted spores (Gould *et al.* 1966; Ando and Tsuzuki 1984; Foster and Johnstone 1987). A 30-kDa muramic acid δ-lactam forming enzyme has been purified from germinating spores of *B. megaterium* KM and has been postulated to be involved in the germination response and called a germination-specific lytic enzyme (GSLE; Foster and Johnstone 1987).

The above categories (Table 1) are not exclusive to other enzyme specificities. Lysozyme is able to hydrolyse all three substrates and the spore lytic enzymes of *Clostridium perfringens* can hydrolyse both isolated and *in situ* spore cortex (Gombas and Labbe 1981, 1985). Thus, from enzyme substrate specificity alone one cannot ascertain the function of any given enzyme.

5.3 Germination and outgrowth

As part of the germination response the spore cortex is hydrolysed to allow swelling of the germinated spore and eventual outgrowth. It is hard to define whether an autolysin present in an active form in dormant spores has a role in sporulation and/or germination. One criterion for a germination enzyme would be its substrate specificity in the fact that one would expect such an enzyme to 'germinate' permeabilized spores. Also if an enzyme is truly germination-specific then its activity should only be able to be measured during germination and not in dormant spores. Two enzymes which share these properties and would fall into the 'core' enzyme category shown above are the GSLE of *B. megaterium* KM (Foster and Johnstone 1987, 1988) and the germination activated spore lytic enzyme of *Cl. perfringens* (Ando 1979; Ando and Tsuzuki 1984). GSLE has been shown to be activated by proteolytic cleavage from a 63-kDa latent precursor to the 30-kDa active form during germination (Foster and Johnstone 1988). Activation of the *Cl. perfringens* germination enzyme occurs as an energy dependent process from an alkali extractable precursor during germination (Ando and Tsuzuki 1984).

6. ROLE OF CELL WALL STRUCTURE AND STRUCTURAL DYNAMICS

6.1 Vegetative growth

The cell wall peptidoglycan and anionic polymers are essential for cell growth and shape determination (Shockman and Barrett 1983). Protoplasts lacking the cell wall are round and can only be maintained under osmotically stabilizing conditions. The importance of cell wall growth is shown by the action of cell wall synthesis inhibitory antibiotics such as penicillin and vancomycin, which have proved to be such effective antibacterial chemotherapeutic reagents. The importance of individual components of the wall biosynthetic machinery is not always easy to define, strains of *B. subtilis* missing either PBP 4 or 5 have been isolated or constructed and found to have no obvious vegetative abnormalities (Todd *et al.* 1985; Buchanan 1987).

Wall expansion allows growth of the cell and in the bacilli, new wall material is added all over the inside of the cylindrical wall. The surface stress theory for bacterial growth proposes that specific autolysins are active at localized sites in the cell wall and that their activity may be directly regulated by peptidoglycan stress (Koch 1985). New wall material is incorporated into existing peptidoglycan and becomes part of the stress-bearing fabric by the activity of these 'space maker hydrolases' (Koch 1985), although none of these specific enzymes has been identified to date. The role of autolysins during growth and division of *B. subtilis* has been open to much speculation with much of the evidence coming from the study of *lyt* mutants with reduced autolysin levels (Pooley and Karamata 1984; Ward and Williamson 1984). The two most comprehensively analysed *lyt* mutants turn out to carry mutations in either the *sigD* or *sin* regulatory genes (Gaur *et al.* 1986; Marquez *et al.* 1990). Mutations in either of these genes resulted in filamentation. As these strains have a pleiotropic phenotype, however, one cannot ascertain whether a certain defect is due to lack of autolysins or other cellular components. This is underlined by the fact that strains insertionally inactivated in the genes for either of the two major vegetative cell autolysins of *B. subtilis* are not filamentous (Kuroda and Sekiguchi 1991; Margot 1992; Margot and Karamata 1992). A strain missing the 50-kDa amidase is more resistant to lysis at the end of exponential growth and shows diminished motility on swarm plates (Kuroda and Sekiguchi 1991; Margot and Karamata 1992) and a mutant in the 90-kDa glucosaminidase had no distinguishable phenotype (Margot 1992).

There is possible functional redundancy in both cell wall biosynthetic and autolytic components and therefore in order to assign roles for any given component or set of components it is important to use caution and, ideally, sets of individually or multiply insertionally inactivated mutants in component structural genes.

6.2 Sporulation

The initiation of sporulation in response to nutrient deprivation requires the products of several genes, called *spo0* genes. The exact environmental stimuli which cause sporulation are unknown and may require the combination of several factors. The *spo0K* locus of *B. subtilis* has been cloned, sequenced and found to consist of an operon of five genes which are homologous to the oligopeptide transport system (Opp) of Gram-negative species (Perego *et al.* 1991). Toxic peptide studies showed this operon to be functional (Perego *et al.* 1991). The major role of the Opp system is the recycling of cell wall peptides after release from the peptidoglycan, probably by the action of an amidase (Goodell and Higgins 1987). Thus one of the signals for the initiation of sporulation may be the accumulation of peptidoglycan digestion products and the *spo0K* sporulation defect is due to the inability to transport these peptides (Perego *et al.* 1991). The autolysin(s) responsible for release of these peptides from the cell wall peptidoglycan have not been identified. A strain insertionally inactivated in the gene for the major vegetative 50-kDa amidase, sporulates at the same frequency as the parent (Kuroda and Sekiguchi 1991; Margot and Karamata 1992).

During stages II and III of sporulation, the asymmetric septum is synthesized, subsequently autolysed and the forespore engulfed. Sporulation-specific proteins involved in the wall changes have not as yet been identified but they must be essential components in this process (Illing and Errington 1991; Higgins and Piggot 1992). A possible biochemical regulator of an autolysin which is involved in asymmetric septal lysis has been identified as the product of the *spoIID* gene which has homology with the modifier of the major 50-kDa vegetative amidase (Kuroda *et al.* 1992; Lazarevic *et al.* 1992).

Gene expression during sporulation is dependent on a cascade of temporally and spatially regulated σ factors (Losick and Stragier 1992). Shortly after septum formation, compartment-specific gene expression begins to take place. Although σ^F is produced prior to septation, σ^F controlled gene expression only occurs in the forespore compartment (Losick and Stragier 1992). Expression of the operon containing the σ^F gene (*spoIIA*) is induced at two points during sporulation, initially prior to septum formation (stage I) and again after forespore engulfment (stages III–IV) (Wu *et al.* 1992). The later transcript is longer and encodes for an additional protein, the product of the *dacF* gene. By sequence homology this protein is probably a D,D-carboxypeptidase (Wu *et al.* 1992). Expression of *dacF* takes place only in the developing forespore, so it is likely that this putative PBP plays a role in primordial cell wall synthesis. A mutant in *dacF*, however, has no distinguishable phenotype, forming heat resistant endospores which are able to germinate as efficiently as spores of the parent strain (Wu *et al.* 1992). Cotranscription of *dacF* with the σ^F gene indicates a way in which transcriptional regulation can be coupled to morphogenesis.

Functional redundancy in peptidoglycan synthetic and autolytic enzymes has hampered attempts to identify specific roles for differentially expressed enzymes during development. This has favoured the isolation of regulatory mutants which control the expression of more than one component. A mutant in the *dacB* gene which encodes the differentially expressed PBP 5* has no sporulation phenotype (Buchanan and Gustafson 1992). This enzyme does seem to have a role in cortex biosynthesis however, as *dacB* spores were very much more heat sensitive than the parent (Buchanan and Gustafson 1992). A mutant in PBP 5 which is expressed both during vegetative growth and sporulation, produces spore cortex with reduced cross-linking and a 10-fold reduction in heat resistance (Todd *et al.* 1986). A recent study, however, has questioned the involvement of PBP 5 in cortex biosynthesis (Buchanan and Gustafson 1992).

The role of specific autolysins during sporulation has remained elusive although their existence is inferred during asymmetric septum hydrolysis and mother cell lysis. The 'sporangial' enzymes of Brown (1977) are good candidates for a role in mother cell lysis. The 30-kDa differentially expressed autolysin of *B. subtilis* which has been identified is likely to be involved in the process (Foster 1992).

6.3 Dormancy

The spore cortex has been proposed to be involved in the maintenance but not the creation of the dormant state (Ellar 1978). Most theories of spore resistance to heat and the maintenance of the ametabolic state assume a common mechanism that is based on the dehydration of the spore core. This dehydration is maintained by the spore cortex (Gerhardt and Murrell 1978). Mutants of *B. sphaericus* lacking cortex have been found to be both heat and octanol sensitive (Imae and Strominger 1976; Imae *et al.* 1978). About 25 and 90% of wild-type cortex levels were found to be necessary for octanol and heat resistance respectively (Imae and Strominger 1976). The anisotropic swelling theory of cortex expansion during sporulation which results in the maintenance of spore dormancy and heat resistance was first proposed by Alderton and Snell (1963) and modified by Warth (1978). According to this model the cortex layers are first synthesized in a sequential manner and subsequent selective enzymatic cleavage of the peptide cross-links results in radial expansion causing inward mechanical pressure on the spore core. Selective hydrolysis of the cortex during germination would release the pressure on the core and allow uptake of water. A more recent proposal suggests that the cortex is highly cross-linked and forms a restraining structure around the spore protoplast (Marquis and Bender 1990). The observation that the *dacB* mutant missing PBP 5*, which has been proposed to be involved in cortex formation, has many more heat sensitive spores than the wild type gives good evidence that the cortex is involved in the maintenance of spore heat resistance (Buchanan and Gustafson 1992). Also, the appearance of PBP 5* during sporulation of a strain bearing the mutation *gerJ* is delayed, which results in spores that are abnormally heat sensitive (Warburg *et al.* 1986). The spore cortex structural defect in the *dacB* mutant has not been identified, this objective is important as it will begin to define the cortex structural features that are responsible for the upkeep of the characteristic heat resistance properties of the dormant spore.

6.4 Germination and outgrowth

Despite their extreme dormancy, spores retain an alert sensory mechanism which is able to respond to specific germinants within minutes and trigger a series of sequentially interrelated degradative events known as germination. Activation of a cortex lytic enzyme as a primary event in spore germination was originally suggested by Powell and Strange

(1956). Although release of peptidoglycan fragments takes place relatively late during germination (Hsieh and Vary 1975a,b), selective cortex hydrolysis occurs as an early germination event in *B. megaterium* KM (Johnstone and Ellar 1982). New cortical-reducing groups were detectable within 2 min after the addition of germinants to spores of *B. megaterium* KM and these were due to an increase in the number of muramic acid δ-lactam residues (Johnstone and Ellar 1982). Foster and Johnstone (1986) proposed a model for spore germination in which the proteolytic activation of a germination-specific lytic enzyme (GSLE) occurred at an essential point during germination. GSLE was subsequently purified from germinating spores of *B. megaterium* KM and found to be a 30-kDa enzyme capable of causing an increase in the number of muramic acid δ-lactam residues on the spore cortex (Foster and Johnstone 1987). Studies with antisera raised against the purified GSLE showed that it is activated during germination by the proteolytic cleavage of a 63-kDa latent precursor which is covalently bound to the spore cortex peptidoglycan (Foster and Johnstone 1988). Activation of GSLE occurs as part of the commitment response (Foster and Johnstone 1988). Irreversible commitment to germinate due to the action of germinants is the earliest measurable germination event to occur during germination triggering of *B. megaterium* KM (Stewart *et al.* 1981). GSLE is able to cause germination-like changes only to permeabilized spores and has the same inhibitor profile as the reactions which control loss of absorbance by the spore population (Foster and Johnstone 1987). For all of the above reasons it was postulated that GSLE is a key enzyme in germination and its activity may allow selective hydrolysis of the spore cortex, collapse of cortex structure, uptake of water, release of spore components and the onset of all later germination events. Insertional inactivation of the structural gene for this enzyme will identify the probable role for GSLE during differentiation. GSLE is synthesized early during sporulation of *B. megaterium* KM and may function as a δ-lactam-forming enzyme during sporulation as well as germination (Foster and Johnstone 1988).

Cortex hydrolysis is a common feature during germination of different speices, but whether a GSLE equivalent is at work is unknown. During germination of *B. megaterium* ATCC 12872, a glucosaminidase is involved in cortex hydrolysis (Nakatani *et al.* 1985). The 'surface bound' lytic enzyme of *B. cereus* has been shown to interact with germinants but is unable to cause germination alone (Brown *et al.* 1982). It is unlikely that this type of enzyme plays a primary role in germination as spores are still able to germinate after its extraction (Foster and Johnstone 1987). The 'surface bound' enzymes may be involved in more generalized cortex lysis or cortex maturation during sporulation.

The primordial cell wall remains after germination and is resistant to germination associated autolysins (Cleveland and Gilvarg 1975). Primordial cell wall constitutes about 20% of the total spore peptidoglycan and is not hydrolysed during outgrowth but seems to form the basis of the new vegetative cell wall peptidoglycan (Cleveland and Gilvarg 1975).

7. CONCLUSIONS AND FUTURE PROSPECTS

Differentiation of the *Bacillus* spp. from the very earliest days of spore research was seen to follow a common morphological pathway. The elucidation of the chemical composition and the basic structure of the cell wall and spore cortex also revealed common underlying features. Since that time, work on identifying components involved in wall structural dynamics which led to the synthesis or autolysis of the characteristic developmentally associated cell wall features has been slow. This is due to a number of possible reasons. As during vegetative growth, there may be functional redundancy in these components and so the loss of one enzyme may not cause a recognizable defect; none of the *spo* mutants identified by traditional techniques has been shown to have a specific defect in wall synthesis or autolysis. If the components are differentiation-specific, it is likely that their substrates are also, which may create problems in assaying for their activity. The spore is a three-dimensional structure and enzymes may only be active *in situ* and not on isolated substrate. Enzymes with very specific roles may only be present in low amounts, for a short period during differentiation and associated with specific intracellular fractions. Despite all of these problems we are now beginning to identify, isolate and characterize components involved in wall structural dynamics and define their roles during growth and differentiation.

Many of the techniques developed for the study of vegetative cell components have proved very useful in the identification of differentially expressed morphogenes, such as PBP gels (Todd and Ellar 1982) and renaturing SDS-PAGE with substrate containing gels (Foster 1992). These studies and others have revealed the complexity of the machinery involved in the changes in structure of such a relatively simple macromolecule as peptidoglycan. Reverse genetics is the method of choice to elucidate the role of these various components. It is likely that only by the creation of mutants singly and multiply inactivated in cell wall machinery components will their combined roles in wall flux be determined.

The 30 kDa autolysin of *B. subtilis* which has been proposed to be responsible for mother cell lysis (Foster 1992) is an ideal candidate for such an approach. It may function alone or require the action of the 50 kDa vegetative cell amidase to cause lysis (Foster 1992). As a mutant in the 50 kDa amidase gene has already been constructed (Margot and Karamata 1992), a possible double mutant would be feasible. The GSLE of *B. megaterium* has been proposed to be solely responsible for initial cortex hydrolysis during germination. Disruption of the gene for the enzyme will elucidate its role.

One of the most surprising things about peptidoglycan structure during both vegetative growth and differentiation is that apart from its basic chemical composition we know very little about the features which lead to the final wall architecture. This is especially true of dormant spores because it is the spore cortex which maintains spore dormancy and the heat resistant state. New technology that uses HPLC has revealed a much higher level of peptidoglycan complexity in several species and has opened up the possibility of in-depth structural studies (Glauner and Schwarz 1983). Subtle changes in peptidoglycan architecture in response to stimuli or in different mutants can be resolved using this method (Kohlrausch and Holtje 1991; De Jonge *et al.* 1992). By the application of this approach to differentiation of *B. subtilis*, particularly using the range of mutants which are available, an exciting opportunity exists to further elucidate important peptidoglycan features in a system where cell wall structure and structural dynamics are known to be so crucial.

8. ACKNOWLEDGEMENT

Research in this laboratory is supported by the Royal Society and the SERC.

9. REFERENCES

Alderton, G. and Snell, N.S. (1963) Base exchange and heat resistance in bacterial spores. *Biochemical and Biophysical Research Communications* **10**, 139–143.

Ando, Y. (1979) Spore lytic enzyme release from *Clostridium perfringens* spores during germination. *Journal of Bacteriology* **140**, 59–64.

Ando, Y. and Tsuzuki, T. (1984) Energy-dependent activation of spore-lytic enzyme precursor by germinated spores of *Clostridium perfringens. Biochemical and Biophysical Research Communications* **123**, 463–467.

Archibald, A.R. (1989) The *Bacillus* cell envelope. In *Bacillus* ed. Harwood, C.R. pp. 217–254. London: Plenum Press.

Beall, B. and Lutkenhaus, J. (1991) FtsZ in *Bacillus subtilis* is required for vegetative septation and for asymmetric septation during sporulation. *Genes and Development* **5**, 447–455.

Blumberg, P.M. and Strominger, J.L. (1972) Five penicillin-binding components occur in *Bacillus subtilis* membranes. *Journal of Biological Chemistry* **247**, 8107–8113.

Brown, W.C. (1977) Autolysins in *Bacillus subtilis*. In *Microbiology 1977* ed. Schlessinger, D. pp. 75–84. Washington, DC: American Society for Microbiology.

Brown, W.C. and Cuhel, R.L. (1975) Surface-localized cortex-lytic enzyme in spores of *Bacillus cereus* T. *Journal of General Microbiology* **91**, 429–432.

Brown, W.C., Vellom, D., Schnepf, E. and Greer, C. (1978) Purification of a surface-bound hexosaminidase from spores of *Bacillus cereus* T. *FEMS Microbiology Letters* **3**, 247–251.

Brown, W.C., Vellom, D., Ho, I., Mitchell, N. and McVay, P. (1982) Interaction between *Bacillus cereus* spore hexosaminidase and specific germinants. *Journal of General Microbiology* **149**, 969–976.

Buchanan, C.E. (1987) Absence of penicillin-binding protein 4 from an apparently normal strain of *Bacillus subtilis*. *Journal of Bacteriology* **169**, 5301–5303.

Buchanan, C.E. and Gustafson, A. (1992) Mutagenesis and mapping of the gene for a sporulation-specific penicillin-binding protein in *Bacillus subtilis*. *Journal of Bacteriology* **174**, 5430–5435.

Buchanan, C.E. and Ling, M.-L. (1992) Isolation and sequence analysis of *dacB*, which encodes a sporulation-specific penicillin-binding protein in *Bacillus subtilis*. *Journal of Bacteriology* **174**, 1717–1725.

Buchanan, C.E. and Neyman, S.L. (1986) Correlation of penicillin-binding protein composition with different functions of two membranes in *Bacillus subtilis* forespores. *Journal of Bacteriology* **165**, 498–503.

Cleveland, E.F. and Gilvarg, C. (1975) Selective degradation of peptidoglycan from *Bacillus megaterium* spores during germination. In *Spores VI* ed. Gerhardt, P., Costilow, R.N. and Sadoff, H.L. pp. 458–464. Washington, DC: American Society for Microbiology.

Crowther, J.S. and Baird-Parker, A.C. (1983) The pathogenic and toxigenic spore-forming bacteria. In *The Bacterial Spore*, Vol. 2. ed. Hurst, A. and Gould, G. W. pp. 275–311. London: Academic Press.

De Jonge, B.L.M., Chang, Y.-S., Gage, D. and Tomasz, A. (1992) Peptidoglycan composition in heterogeneous Tn551 mutants of a methicillin-resistant *Staphylococcus aureus* strain. *Journal of Biological Chemistry* **267**, 11255–11259.

Doi, R.H. (1989) Sporulation and germination. In *Bacillus* ed. Harwood, C.R. pp. 169–215. London: Plenum Press.

Ellar, D.J. (1978) Spore specific structures and their function. *Symposia of the Society for General Microbiology* **28**, 295–324.

Errington, J. (1993) *Bacillus subtilis* sporulation: regulation of gene expression and control of morphogenesis. *Microbiological Reviews* **57**, 1–33.

Forrest, T.M., Wilson, G.E., Pan, Y. and Schaefer, J. (1991) Characterization of cross-linking of cell walls of *Bacillus subtilis* by a combination of magic-angle spinning NMR and gas chromatography–mass spectrometry of both intact and hydrolyzed ^{13}C- and ^{15}N-labeled cell-wall peptidoglycan. *Journal of Biological Chemistry* **266**, 24485–24491.

Foster, S.J. (1986) Biochemistry of *Bacillus megaterium* spore germination. Ph.D. Thesis, University of Cambridge, UK.

Foster, S.J. (1991) Cloning, expression, sequence analysis and biochemical characterization of an autolytic amidase of *Bacillus subtilis* 168 *trpC2*. *Journal of General Microbiology* **137**, 1987–1998.

Foster, S.J. (1992) Analysis of the autolysins of *Bacillus subtilis* 168 during vegetative growth and differentiation by using renaturing gel electrophoresis. *Journal of Bacteriology* **174**, 464–470.

Foster, S.J. (1993) Analysis of *Bacillus subtilis* 168 prophage-associated lytic enzymes; identification and characterization of CWLA-related prophage proteins. *Journal of General Microbiology* **139**, 3177–3184.

Foster, S.J. and Johnstone, K. (1986) The use of inhibitors to identify early events during *Bacillus megaterium* KM spore germination. *Biochemical Journal* **237**, 865–870.

Foster, S.J. and Johstone, K. (1987) Purification and properties of a germination-specific cortex-lytic enzyme from spores of *Bacillus megaterium* KM. *Biochemical Journal* **242**, 573–579.

Foster, S.J. and Johnstone, K. (1988) Germination-specific cortex-lytic enzyme is activated during triggering of *Bacillus megaterium* KM spore germination. *Molecular Microbiology* **2**, 727–733.

Foster, S.J. and Johnstone, K. (1989) The trigger mechanism of bacterial spore germination. In *Regulation of Procaryotic Development* ed. Smith, I., Slepecky, R.A. and Setlow, P. pp. 89–108. Washington, DC: American Society for Microbiology.

Gaur, N.K., Dubnau, E. and Smith, I. (1986) Characterization of a cloned *Bacillus subtilis* gene that inhibits sporulation in multiple copies. *Journal of Bacteriology* **168**, 860–869.

Gerhardt, P. and Murrell, W.G. (1978) Basis and mechanisms of spore resistance: a brief preview. In *Spores VII* ed. Chambliss, G.H. and Vary, J.C. pp. 18–20. Washington, DC: American Society for Microbiology.

Ghuysen, J.-M, Tipper, D.J. and Strominger, J.L. (1966) Enzymes that degrade bacterial cell walls. *Methods in Enzymology* **8**, 685–699.

Glauner, B. and Schwarz, U. (1983) The analysis of murein composition with high-pressure-liquid chromatography. In *The Target of Penicillin* ed. Hackenbeck, R., Holtje, J.-V. and Labischinski, H. pp. 29–34. Berlin: de Gruyter.

Gombas, D. E. and Labbe, R.G. (1981) Extraction of spore-lytic enzyme from *Clostridium perfringens* spores. *Journal of General Microbiology* **126**, 37–44.

Gombas, D.E. and Labbe, R.G. (1985) Purification and properties of spore-lytic enzymes from *Clostridium perfringens* type A spores. *Journal of General Microbiology* **131**, 1487–1496.

Goodell, E.M. and Higgins, C.F. (1987) Uptake of cell wall peptides by *Salmonella typhimurium* and *Escherichia coli*. *Journal of Bacteriology* **169**, 3861–3865.

Gould, G.W. and Hitchins, A.D. (1963) Sensitization of bacterial spores to lysozyme and to hydrogen peroxide with agents which rupture disulphide bonds. *Journal of General Microbiology* **33**, 413–423.

Gould, G.W., Hitchins, A.D. and King, K.L. (1966) Function and location of a 'germination enzyme' in spores of *Bacillus cereus*. *Journal of General Microbiology* **44**, 293–302.

Guinand, M., Michel, G. and Balassa, G. (1976) Lytic enzymes in sporulating *Bacillus subtilis*. *Biochemical and Biophysical Research Communications* **68**, 1287–1293.

Guinand, M., Vacheron, M.J., Michel, G. and Tipper, D.J. (1979) Location of peptidoglycan lytic enzymes in *Bacillus sphaericus*. *Journal of Bacteriology* **138**, 126–132.

Herbold, D.R. and Glaser, L. (1975) *Bacillus subtilis* *N*-acetylmuramic acid L-alanine amidase. *Journal of Biological Chemistry* **250**, 1676–1682.

Higgins, M.L. and Piggot, P.J. (1992) Septal membrane fusion—a pivotal event in bacterial spore formation? *Molecular Microbiology* **6**, 2565–2571.

Hsieh, L.K. and Vary, J.C. (1975a) Germination and peptidoglycan solubilization in *Bacillus megaterium* spores. *Journal of Bacteriology* **123**, 463–470.

Hsieh, L.K. and Vary, J.C. (1975b) Peptidoglycan autolysis during initiation of spore germination in *Bacillus megaterium*. In *Spores VI* ed. Gerhardt, P., Costilow, R.N. and Sadoff, H.L. pp. 465–471. Washington, DC: American Society for Microbiology.

Illing, N. and Errington, J. (1991) Genetic regulation of morphogenesis in *Bacillus subtilis*: roles of σ^E and σ^F in prespore engulfment. *Journal of Bacteriology* **173**, 3159–3169.

Imae, Y. and Strominger, J.L. (1976) Relationship between cortex content and properties of *Bacillus sphaericus* spores. *Journal of Bacteriology* **126**, 907–913.

Imae, Y., Strominger, M.B. and Strominger, J.L. (1978) Conditional spore cortexless mutants of *Bacillus sphaericus*. In *Spores VII* ed. Chambliss, G. and Vary, J.C. pp. 62–66. Washington, DC: American Society for Microbiology.

Johnstone, K. and Ellar, D.J. (1982) The role of cortex hydrolysis in the triggering of germination of *Bacillus megaterium* KM endospores. *Biochimica et Biophysica Acta* **714**, 185–191.

Jonas, R.M., Holt, S.C. and Haldenwang, W.G. (1990) Effect of antibiotics on synthesis and persistence of σ^E in sporulating *Bacillus subtilis*. *Journal of Bacteriology* **172**, 4616–4623.

Kingan, S.L. and Ensign, J.C. (1968) Isolation and characterization of three autolytic enzymes associated with sporulation of *Bacillus thuringiensis* var. *thuringiensis*. *Journal of Bacteriology* **96**, 629–638.

Koch, A.L. (1985) Bacterial wall growth and division or life without actin. *Trends in Biochemical Science* **10**, 11–14.

Kohlrausch, U. and Holtje, J.-V. (1991) Analysis of murein and murein precursors during antibiotic-induced lysis of *Escherichia coli*. *Journal of Bacteriology* **173**, 3425–3431.

Kuroda, A. and Sekiguchi, J. (1990) Cloning, sequencing and genetic mapping of a *Bacillus subtilis* cell wall hydrolase gene. *Journal of General Microbiology* **136**, 2209–2216.

Kuroda, A. and Sekiguchi, J. (1991) Molecular cloning and sequencing of a major *Bacillus subtilis* autolysin gene. *Journal of Bacteriology* **173**, 7304–7312.

Kuroda, A., Rashid, M.H. and Sekiguchi, J. (1992) Molecular cloning and sequencing of the upstream region of the major *Bacillus subtilis* autolysin gene: a modifier protein exhibiting sequence homology to the major autolysin and the *spoIID* product. *Journal of General Microbiology* **138**, 1067–1076.

Kusser, W. and Fiedler, F. (1983) Teichoicase from *Bacillus subtilis* Marburg. *Journal of Bacteriology* **155**, 302–310.

Lazarevic, V., Margot, P., Soldo, B. and Karamata, D. (1992) Sequencing and analysis of the *Bacillus subtilis lytRABC* divergon: a regulatory unit encompassing the structural genes of the

N-acetyl muramoyl-L-alanine amidase and its modifier. *Journal of General Microbiology* **138**, 1949–1961.

Linnett, P.E. and Tipper, D.J. (1976) Transcriptional control of peptidoglycan precursor synthesis during sporulation in *Bacillus sphaericus*. *Journal of Bacteriology* **125**, 565–574.

Losick, R. and Stragier, P. (1992) Crisscross regulation of cell-type specific gene expression during development in *B. subtilis*. *Nature* **355**, 601–604.

Losick, R. and Youngman, P. (1984) Endospore formation in *Bacillus*. In *Microbial Development* ed. Losick, R. and Shapiro, L. pp. 63–88. Cold Spring Harbor, NY: Cold Spring Harbor Laboratory.

Margot, P. (1992) Genetique des autolysines de *Bacillus subtilis*. Ph.D. Thesis, University of Lausanne, Switzerland.

Margot, P. and Karamata, D. (1992) Identification of the structural genes for *N*-acetyl muramoyl-L-alanine amidase and its modifier in *Bacillus subtilis* 168: inactivation of these genes by insertional mutagenesis has no effect on growth or cell separation. *Molecular and General Genetics* **232**, 359–366.

Marquez, L.M., Helmann, J.D., Ferrari, E., Parker, H.M., Ordal, G.W. and Chamberlin, M.J. (1990) Studies of σ^D-dependent functions in *Bacillus subtilis*. *Journal of Bacteriology* **172**, 3435–3443.

Marquis, R.E. and Bender, G.R. (1990) Compact structure of cortical peptidoglycans from bacterial spores. *Canadian Journal of Microbiology* **36**, 426–429.

Mauel, C., Young, M., Margot, M. and Karamata, D. (1989) The essential nature of teichoic acids in *Bacillus subtilis* as revealed by insertional mutagenesis. *Molecular and General Genetics* **215**, 388–394.

Nakatani, Y., Tanida, I., Koshikawa, T., Imagawa, M., Nishihara, T. and Kondo, M. (1985) Collapse of cortex expansion during germination of *Bacillus megaterium* spores. *Microbiology and Immunology* **29**, 689–699.

Neyman, S.L. and Buchanan, C.E. (1985) Restoration of vegetative penicillin-binding proteins during germination and outgrowth of *Bacillus subtilis* spores: relationship of individual proteins to specific cell cycle events. *Journal of Bacteriology* **161**, 164–168.

Perego, M., Higgins, C.F., Pearce, S.R., Gallagher, M.P. and Hoch, J.A. (1991) The oligopeptide transport system of *Bacillus subtilis* plays a role in the initiation of sporulation. *Molecular Microbiology* **5**, 173–185.

Pooley, H.J. and Karamata, D. (1984) Flagellation and the control of autolysin activity in *Bacillus subtilis*. In *Microbial Cell Wall Synthesis and Autolysis* ed. Nombela, C. pp. 13–19. Amsterdam: Elsevier.

Popham, D.L. and Setlow, P. (1993a) The cortical peptidoglycan from spores of *Bacillus megaterium* and *Bacillus subtilis* is not highly cross-linked. *Journal of Bacteriology* **175**, 2767–2769.

Popham, D.L. and Setlow, P. (1993b) Cloning, nucleotide sequence, and regulation of the *Bacillus subtilis pbpE* operon, which codes for penicillin-binding protein 4* and an apparent amino acid racemase. *Journal of Bacteriology* **175**, 2917–2925.

Powell, J.F. and Strange, R.E. (1956) Biochemical changes occurring during sporulation in *Bacillus* species. *Biochemical Journal* **63**, 661–668.

Rogers, H.J., Taylor, C., Rayter, S. and Ward, J.B. (1984) Purification and properties of an autolytic endo-β-glucosaminidase and the *N*-acetylmuramyl-L-alanine amidase from *Bacillus subtilis* strain 168. *Journal of General Microbiology* **130**, 2395–2402.

Schleifer, K.H. and Kandler, O. (1972) Peptidoglycan types of bacterial cell walls and their taxonomic implications. *Bacteriological Reviews* **36**, 407–477.

Shockman, G.D. and Barrett, J.F. (1983) Structure, function, and assembly of cell walls of Gram-positive bacteria. *Annual Reviews of Microbiology* **37**, 501–527.

Slepecky, R.A. and Leadbetter, E.R. (1983) On the prevalence and roles of spore-forming bacteria and their spores in nature. In *The Bacterial Spore*, Vol. 2. ed. Hurst, A. and Gould, G.W. pp. 79–99. London: Academic Press.

Sowell, M.O. and Buchanan, C.E. (1983) Changes in penicillin-binding proteins during sporulation of *Bacillus subtilis*. *Journal of Bacteriology* **153**, 1331–1337.

Stewart, G.S.A.B., Johnstone, K., Hagelberg, E. and Ellar, D.J. (1981) Commitment of bacterial spores to germinate: a measure of the trigger reaction. *Biochemical Journal* **198**, 101–106.

Strange, R.E. and Dark, F.A. (1957a) A cell-wall lytic enzyme associated with spores of *Bacillus* species. *Journal of General Microbiology* **16**, 236–249.

Strange, R.E. and Dark, F.A. (1957b) Cell-wall lytic enzymes at sporulation and spore germination in *Bacillus* species. *Journal of General Microbiology* **17**, 525–537.

Tipper, D.J. and Gauthier, J.J. (1972) Structure of the bacterial endospore. In *Spores V* ed. Halvorson, H.O., Hanson, R. and Campbell, L.L. pp. 3–12. Washington, DC: American Society for Microbiology.

Tipper, D.J. and Linnett, P.E. (1976) Distribution of peptidoglycan synthetase activities between sporangia and forespores in sporulating cells of *Bacillus sphaericus*. *Journal of Bacteriology* **126**, 213–221.

Todd, J.A. and Ellar, D.J. (1982) Alteration in the penicillin-binding profile of *Bacillus megaterium* during sporulation. *Nature* **300**, 640–643.

Todd, J.A. and Ellar, D.J. (1985) The association of penicillin-binding proteins with cell elongation and septum formation in *Bacillus megaterium*. *Biochemical Journal* **230**, 829–832.

Todd, J.A., Bone, E.J., Piggot, P.J. and Ellar, D.J. (1983) Differential expression of penicillin-binding protein structural genes during *Bacillus subtilis* sporulation. *FEMS Microbiology Letters* **18**, 197–202.

Todd, J.A., Bone, E.J. and Ellar, D.E. (1985) The sporulation-specific penicillin-binding protein 5a from *Bacillus subtilis* is a DD-carboxypeptidase *in vitro*. *Biochemical Journal* **230**, 825–828.

Todd, J.A., Roberts, A.N., Johnstone, K., Piggot, P.J., Winter, G. and Ellar, D.J. (1986) Reduced heat resistance of mutant spores after cloning and mutagenesis of the *Bacillus subtilis* gene encoding penicillin-binding protein 5. *Journal of Bacteriology* **167**, 257–264.

Warburg, R.J., Buchanan, C.E., Parent, K. and Halvorson, H.O. (1986) A detailed study of *gerJ* mutants of *Bacillus subtilis*. *Journal of General Microbiology* **132**, 2309–2319.

Ward, J.B. and Williamson, R. (1984) Bacterial autolysins: specificity and function. In *Microbial Cell Wall Synthesis and Autolysis* ed. Nombela, C. pp. 159–166. Amsterdam: Elsevier.

Warth, A.D. (1978) Molecular structure of the bacterial spore. *Advances in Microbial Physiology* **17**, 1–47.

Warth, A.D. and Strominger, J.L. (1969) Structure of the peptidoglycan of bacterial spores: occurrence of the lactam of muramic acid. *Proceedings of the National Academy of Sciences USA* **64**, 528–535.

Warth, A.D. and Strominger, J.L. (1972) Structure of the peptidoglycan from spores of *Bacillus subtilis*. *Biochemistry* **11**, 1389–1396.

Waxman, D.J., Yu, W. and Strominger, J.L. (1980) Linear, uncross-linked peptidoglycan secreted by penicillin-treated *Bacillus subtilis*. *Journal of Biological Chemistry* **255**, 11577–11587.

Wickus, G.G., Warth, A.D. and Strominger, J.L. (1972) Appearance of muramic lactam during cortex synthesis in sporulating culture of *Bacillus cereus* and *Bacillus megaterium*. *Journal of Bacteriology* **111**, 625–627.

Wu, J.-J., Schuch, R. and Piggot, P.J. (1992) Characterization of a *Bacillus subtilis* sporulation operon that includes genes for an RNA polymerase σ factor and for a putative DD-carboxypeptidase. *Journal of Bacteriology* **174**, 4885–4892.

Journal of Applied Bacteriology Symposium Supplement 1994, **76**, 40S–48S

Molecular mechanisms of resistance to heat and oxidative damage

R.E. Marquis, J. Sim and S.Y. Shin
Department of Microbiology and Immunology, University of Rochester Medical Center, Rochester, NY, USA

1. Introduction, 40S
2. Spore properties important for heat resistance
   2.1 Intrinsic properties, 41S
   2.2 Dehydration, 41S
   2.3 Mineralization and dipicolinate, 42S
   2.4 Roles of the cortex, 42S
   2.5 Roles of small acid-soluble spore proteins, 43S
3. Targets of heat damage
   3.1 Nucleic acids, 43S
   3.2 Proteins and enzymes, 43S
   3.3 Membranes, 43S
4. Oxidative damage and resistance
   4.1 Nature of oxidative killing, 44S
   4.2 Spore properties important for resistance to oxidative killing, 45S
5. Mechanisms of heat and oxidative damage, 45S
6. Summary, 46S
7. Acknowledgements, 46S
8. References, 46S

1. INTRODUCTION

Our knowledge of the relationships between spore composition and heat resistance is sufficiently advanced that it is possible at least roughly to predict resistance from knowledge of key parameters such as: optimal growth temperature of the sporeformer, spore protoplast water content, total and specific mineral content, temperature of sporulation and cortex size (Gerhardt and Marquis 1989). In general, thermophiles produce more heat resistant spores than do mesophiles, which produce more resistant spores than do psychrophiles (Warth 1978a). Spores with less water in the protoplast are more heat resistant than those with higher water levels (Beaman and Gerhardt 1986). Sporulation at higher temperatures within the normal range of the organism generally results in spores with more dehydrated protoplasts and greater heat resistance (Beaman and Gerhardt 1986). Mineralization enhances heat resistance, partly by increasing protoplast dehydration, and partly by other mechanisms protective even against dry heat damage. Some spores are highly mineralized, for example those of *Bacillus subtilis* var. *niger*, and demineralization and remineralization have major effects on resistance (Bender and Marquis 1985; Marquis and Bender 1985). For spores of other species, e.g. *B. megaterium* ATCC 19213, mineralization is less important. Demineralized, so-called 'hydrogen-form' (H-form) spores of *B. megaterium* ATCC 19213 and *B. subtilis* var. *niger* have about the same heat resistance. However, mineralized native spores of the latter organism have markedly higher heat resistance than the less mineralized, native spores of the former. For spores of *B. stearothermophilus* ATCC 7953, the high level of intrinsic resistance which is characteristic of thermophiles contributes to resistance associated with mineralization and dehydration to yield very highly resistant spores.

The cortex is considered to play a major role in heat resistance. This role includes maintaining the state of osmotic dehydration achieved early in sporulation, after engulfment, when the contra-workings of the two oppositely oriented spore membranes result in loss of potassium and water from the forespore (Marquis *et al.* 1983). At this time the forespore acquires a darkened image in positive phase contrast microscopy, indicative of low water content. The cortex is then formed around this partially dehydrated core and functions to resist movement of water into the forespore. More active roles have been proposed for the cortex in protoplast dehydration as either a contracting (Lewis *et al.* 1960) or an expanding (Gould and Dring 1975; Warth 1985) envelope acting to squeeze water from the protoplast.

The question of the relationship of dipicolinate (DPA) to heat resistance is controversial, and there is growing evidence, reviewed by Gerhardt and Marquis (1989), that DPA has little or no role in thermoresistance.

Correspondence to: R.E. Marquis, Box 672, University of Rochester Medical Center, Rochester, NY 14642-8672, USA.

In contrast to the extensive knowledge of how spore composition is related to heat resistance, our knowledge of the specific molecular mechanisms by which dehydration and mineralization protect against heat damage is woefully inadequate. At present, we really do not know the major targets for the type of heat damage that leads to spore death. For some spores, such as those of *Clostridium perfringens* type A, a major site of heat damage appears to be the germination apparatus so that heat renders the spores superdormant (Ando and Tsuzuki 1983). Subsequent treatment of the heat damaged spores with alkali and plating on medium with added lysozyme restores the cells to life. For most spore types, however, heat damage does not result in superdormancy and treatment with alkali and lysozyme does not subvert the damage.

Other types of conditional death caused by heat have been described for spores, and have been reviewed for example by Gombas (1983) and Gould (1984). Typically, plating of conditionally dead spores on complex medium after heating yields higher counts than plating on minimal medium. In other words, the cells appear to have enzymatic defects that increase nutritional requirements but no non-repairable genetic defects. New gene expression then allows for replacement of damaged enzymes and restoration of full metabolic capability. More prolonged or intense heating results in irreversible damage with unconditional death, and our knowledge of the molecular details of the irreversible damage is minimal.

In general, what are the major damaging effects of high temperatures for biopolymers such as proteins? It is well known that increasing heat causes conformational changes in proteins that can lead to reversible denaturation, then irreversible denaturation, and in the extreme, aggregation. Denaturation of proteins is commonly a co-operative process with a high enthalpy of activation. The conformational changes leading to denaturation result largely from alterations of electrostatic interactions, hydrogen bonds or hydrophobic interactions. However, as shown clearly by Klibanov and Ahern (1987), irreversible denaturation of proteins can occur because of breakage of covalent bonds, usually disulphide bonds, amide bonds and peptide bonds adjacent to aspartyl residues. Also, high temperatures can result in damage to DNA, including depurination (Lindahl and Nyberg 1972). The exact mechanisms of covalent bond breakage in biopolymers at high temperatures are not well defined. They probably involve formation of radicals, including oxygen radicals in aerobic systems, and oxidation–reduction reactions of these radicals with target molecules. Spore killing by ionizing radiation and oxidative agents, such as H_2O_2 or organic peroxides, also involves radical formation. Thus, there may be a commonality among these sporicidal agents. In this paper, mechanisms for heat killing and for oxidative killing will be considered.

2. SPORE PROPERTIES IMPORTANT FOR HEAT RESISTANCE

2.1 Intrinsic properties

Research on the molecular properties of biopolymers, mainly proteins, important for stability and function at high temperatures has been stimulated by the discovery of extreme thermophiles, initially in the region of hydrothermal vents in the ocean floor, and by the needs of industrial enzymology. Thermotolerance of proteins appears to depend on rather small changes in primary structure that result in changes in disulphide bonding, electrostatic interactions and hydrophobic interactions important for maintaining tertiary and quaternary structures of the molecules at high temperatures (Jaenicke 1991). There is a great deal of individuality for specific proteins and some difficulty in formulating generally applicable rules for predicting thermotolerance. Membranes of thermophiles are adapted to function at high temperatures, partly because of changes in lipid composition but also because of increased thermostability of membrane proteins. The properties of thermophilic proteins important for function at high temperatures are presumably also important for the intrinsic or molecular aspect of spore heat resistance.

2.2 Dehydration

The first stage of dehydration of the protoplast of the forespore occurs after engulfment. The phase microscopic image of a darkening forespore within a less dark sporangium gives a straightforward indication that after engulfment the forespore solids content and refractive index increase and the water content decreases. This initial water loss appears to be osmotic, associated with loss of solutes from the protoplast. Subsequent increases in solids content leading to refractility are less well understood. Concentration of minerals into the forespore is thought to be by passive movement from the cytoplasm of the sporangium, which actively takes up minerals from the environment (Ellar 1978). This solute movement into a cell surrounded by a thick cortex that resists water uptake should result in further increases in solids content per unit volume. Subsequent uptake of DPA from the sporangial cytoplasm is thought to involve active uptake (Errington 1993) and will still further increase the solids content and refractive index of the protoplast, so much so that it becomes refractile when viewed with a phase-contrast microscope with a 90° positive phase plate.

The minerals and DPA in spores do not form crystalline phases, as indicated by electron microscopic observations of thin sections, although Johnstone *et al.* (1982) did propose on the basis of electron paramagnetic resonance studies that Mn in spores may be in a crystalline lattice. Lundin and

Sacks (1988) interpreted their data on high-resolution, solid-state ^{13}C nuclear magnetic resonance (NMR) of DPA more in terms of packing of DPA around Mn in complexes with cell constituents. Presumably, the spore cytoplasm does not contain the types of proteolipids required to initiate the crystal formation that occurs in bone or calculus. However, because of the high mineral contents of spores and the immobilized state of the minerals, it is reasonable to think that amorphous precipitates form with spore polymers. Thus, when minerals and DPA first move into the spore cytoplasm, they may be in free solution and osmotically active, but subsequent precipitation would put them in another chemical phase and sharply reduce osmotic activity.

The water activity (a_w) within spores is thought to be much reduced. For example, Warth (1985) estimated that the internal a_w would have to be about 0·7 to achieve the degree of thermal stabilization of enzymes found within spores. As spores are freely permeable to water, as are all cells, this low a_w could be maintained only with very high turgor pressure, some tens of megapascals. Such high pressures would be remarkable for a cell, even one with a greatly thickened peptidoglycan envelope and could contribute to thermotolerance and dormancy. No one has been able directly to assess the turgor pressures of spores. In fact, no one has been able to measure directly a_w of the spore interior, but a_w can be manipulated in controlled ways by drying spores in constant humidity chambers (Watt 1981). The water within spores appears to be highly mobile, as indicated by the results of NMR and dielectric studies (Carstensen *et al.* 1971; Bradbury *et al.* 1981). Thus the water does not appear to be in a glassy state, although Sapru and Labuza (1993) have recently presented arguments for a glassy state in spores. The amount of water within the spore protoplast (28–57% on a wet weight basis) is sufficient to hydrate the cytoplasmic polymers at least to the levels of proteins in crystals. Thus, the molecules should have a great deal of flexibility, and the exact mechanism of stabilization is not clear at this time. Certainly, the polymers should not be in a dry state nor in organic solvents. However, they probably are in an aggregated state with spore minerals.

2.3 Mineralization and dipicolinate

Although minerals and dipicolinate (DPA) are commonly thought to be associated within spores, there is clear evidence that they can go their separate ways. DPA uptake tends to occur somewhat later than the major uptake of minerals (Gorman *et al.* 1984), and in demineralization/remineralization procedures, minerals and DPA can be extracted separately. DPA has been found to enhance formation of spore DNA photoproduct in response to u.v. irradiation (Setlow and Setlow 1993). DPA also is a known germinant (Riemann and Ordal 1961) and a chelator (Chung *et al.* 1971). However, it does not appear to be involved in heat resistance. Perhaps the best evidence on this point is that minerals and DPA can be extracted from spores to yield heat sensitive H-form spores. Then the spores can be remineralized with Ca^{2+} in the absence of DPA to regain fully the level of resistance of native spores (Gerhardt and Marquis 1989).

2.4 Roles of the cortex

It is generally accepted that an intact cortex is required for heat resistance (Imae and Strominger 1976), and heat resistance has been related to cortex size (Murrell and Warth 1965). Major functions of the cortex are to maintain the state of osmotic dehydration of the forespore protoplast achieved early in sporulation and to resist water movements into the protoplast later during mineralization and DPA uptake. How the cortex performs these functions is still open to debate. Currently, there are two very different views of cortex electrochemistry. The view of the cortex based on chemical analyses of fragments isolated from mechanically damaged spores is that it is a very open, loosely cross-linked structure with a high number of electrically charged groups (Warth 1978b). Thus, the structure is looser than that of most vegetative peptidoglycans, and the cortical peptidoglycan should swell and shrink greatly in response to changes in environmental ionic strength. This view is central to proposals for a contracting or expanding cortex able to squeeze water from the protoplast. The view of the cortex that derives from dielectric analyses is that it has a very low level of mobile ions and is a tightly linked, restraining structure. Thus, the cortex would be suited more to maintaining protoplast dehydration than to causing dehydration. Results of recent light-scattering analyses of spores (Ulanowski and Ludlow 1993) support this view. Moreover, decoated spores with only the cortex-encased protoplast show very little swelling or shrinking in response to changes in environmental pH or ionic strength. Decoated spores chemically extracted to obtain intact murein sacculi enclosing remnants of the protoplast also show little tendency to swell or shrink (Marquis and Bender 1990). Recently, an intermediate view has been proposed by Popham and Setlow (1993) with a cross-linking level about equal to that of vegetative cell peptidoglycan. These divergent views are difficult to resolve partly because of inability to isolate intact, undamaged cortices from spores for chemical analyses without heavy contamination by noncortical material. Mechanical isolation procedures such as sonication clearly cause breakage of peptide bonds in peptidoglycans and possibly other structural rearrangements. However, advances in knowledge of cortical peptidoglycan synthesis and work on penicillin-binding proteins should yield new information.

Spore coats are considered to have little, if any, role in heat resistance. Chemically decoated spores or spores with defective coats are often slightly less resistant to heat killing, but the change is minor compared to differences in resistance between vegetative cells and spores.

2.5 Roles of small acid-soluble spore proteins

Spores of *B. subtilis* genetically deficient for synthesis of the α and β small acid-soluble spore proteins (SASPs) have been found (Mason and Setlow 1986; Setlow, this Symposium, pp. 49S–60S) to be more sensitive to heat than SASP sufficient spores. However, the deficient spores were also less stable even at low temperatures. Thus, although SASPs are important in regard to u.v. damage and affect the type of radiation damage produced, their roles in heat resistance are less well defined. However, they may affect the type of heat damage incurred by spore DNA, and Setlow (this Symposium, pp. 49S–60S) have shown that they protect against depurination.

3. TARGETS OF HEAT DAMAGE

3.1 Nucleic acids

Multiple lines of evidence suggest that DNA could be a prime target in heat killing. Dry heat has been shown to be mutagenic for spores (Chiasson and Zamenhof 1966; Northrop and Slepecky 1967). Kadota *et al.* (1978) have reported that moist heat also is mutagenic, apparently because of depurination rather than strand breakage. Others have not found moist heat to be mutagenic, except for spores defective in SASP (see Setlow, this Symposium, pp. 49S–60S). Deficiencies in DNA repair appear to cause spores to be more heat sensitive (Hanlin *et al.* 1985). This finding suggests that heat-induced DNA damage can be repaired during germination and outgrowth and repairable DNA damage may not necessarily result in spore death. More work is needed to determine whether DNA damage is critical for spore killing. The availability of many repair-deficient mutants should help in determining which repair enzymes play the major roles in repairing heat-damaged DNA.

3.2 Proteins and enzymes

Proteins and enzymes are considered to be the major targets for heat killing of spores. The extensive studies of Warth (1980) indicated a range of sensitivities for a variety of spore enzymes. In general, enzymes in extracts of spores were inactivated at temperatures some 24–46°C lower than those required to inactivate the same enzymes within intact spores. It is not clear just which enzymes or other proteins are the critical targets for killing, except for organisms with very heat sensitive enzymes for cortex lysis. DNA repair enzymes are possible critical targets, but at this time, there is insufficient information for other than speculation.

The recent findings of Belliveau *et al.* (1992) based on differential scanning calorimetry (DSC) suggest that proteins are the main targets for killing. DSC detects co-operative phenomena leading to large changes in heat capacity. For example, melting of spore DNA was reflected in DSC scans of spores as an endothermic peak at 90–91°C. However, DSC would not detect damage to a single type of protein in the spore unless it were present in large amounts, and damage resulted in co-operative phenomena such as denaturation involving extensive unfolding. Breakage of covalent bonds generally would not be detected by DSC unless the breakage led to a co-operative response. For example, breakage of a single disulphide bond would not greatly affect heat capacity unless it also resulted in protein unfolding or dissociation of protein arrays. However, techniques complementary to DSC can be used to detect covalent bond breakage in proteins, for example, two-dimensional gel separation and identification of specific proteins with covalent damage.

Differential calorimetric scans of spores of *B. megaterium* ATCC 33727 indicated three main endothermic peaks and one main exothermic valley, all of which were not evident when the original spore sample was rescanned. In other words, changes produced by the initial heating from 20 to 135°C were not reversible on cooling. Belliveau *et al.* (1992) then used decoated spores to show that the peak centred around 114°C was related to changes in the coat–outer-membrane complex. They used heat activated spores to show that the peak centred around 56°C was related to the activation process. Loss of the peak centred around 100°C was a good indicator of heat killing and was considered to be caused by protein damage. The exothermic valley centred around 119°C appeared to be related to aggregation phenomena.

3.3 Membranes

Flowers and Adams (1976) suggested that damage to cell membranes is the major cause of heat killing. Sublethal heating is known to sensitize spores to a variety of inhibitors, including antibiotics (Barach *et al.* 1975) and nitrite (Chumney and Adams 1980), probably because of changes in membrane permeability. However, the extent of damage done by lethal heating is not well defined. Damage could be to lipids or to proteins. The membrane of the dormant spore appears to be in a very condensed state (Stewart *et al.* 1980) and probably at least somewhat protected against damage. Moreover, in demineralization procedures with titration to pH values as low as 2 and heating to 60°C, all of the minerals of the spore pass out through the membrane.

The spores remain viable, and any acid damage to the membrane must be reversible. Heat damage also may be reversible. Heat killed spores usually remain refractile when viewed with the phase microscope, although they may subsequently release DPA and minerals, possibly because of membrane damage. However, as Belliveau *et al.* (1992) found, killing occurs prior to such release. Thus, loss of DPA and minerals is not a requisite for killing.

4. OXIDATIVE DAMAGE AND RESISTANCE

4.1 Nature of oxidative killing

Hydrogen peroxide is used widely to kill spores, especially in the food industry for sterilization of packaging materials. Because aseptic packaging and processing are becoming more and more popular for a wide variety of products, use of H_2O_2 for spore killing is increasing. Spores can be killed also by organic peroxides such as tertiary butyl hydroperoxide (*t*-BOOH). The authors have found that dormant spores are much more resistant to *t*-BOOH than are vegetative cells or germinated spores. Sample data are shown in Fig. 1 for dormant spores, germinated spores and vegetative cells of *B. megaterium* ATCC 19213 exposed to 720 mmol l^{-1} *t*-BOOH at 37°C. The resistance of dormant spores depends in part on mineralization (Marquis and Shin 1993), but also on other spore properties. Demineralized H-form spores were found to be more resistant than vegetative cells or germinated spores but less resistant than native spores or H-spores remineralized with calcium. The process of germination in response to inosine and alanine was found to be relatively insensitive to *t*-BOOH and occurred even at concentrations of the agent lethal for vegetative cells or germinated spores. However, killing of dormant spores by *t*-BOOH can occur without any signs of germination.

Killing of spores by oxidative chemicals such as H_2O_2 or organic peroxides is highly temperature dependent. Figure 2 shows this dependence for killing of spores of *B. megaterium* ATCC 19213 by *t*-BOOH over the temperature range from 30 to 70°C, below the temperature threshold for heat killing. The plot is of log dose (mmol l^{-1} concentration multiplied by exposure time in minutes) required for 90% killing *vs* temperature. The relationship is nearly linear indicating that for every 15 K increase in temperature there was a 10-fold decrease in the dose required for 90% killing. Thus, the killing process has a high temperature characteristic (μ) of about 120 kJ mol^{-1} and a high Q_{10} of about 4·6. At a temperature of 30°C, killing of dormant spores by 720 mmol l^{-1} *t*-BOOH was preceded by a long lag phase of some hours and was slow with a D value (time required for killing of 90% of the population) of about 180 min. At 70°C, 7·2 mmol l^{-1} *t*-BOOH was lethal with little lag and a D value of about 45 min. At a temperature of 80°C, heat killing of the spores occurred slowly with a D_{80} value of 90 min. The addition of as little as 0·014 mmol l^{-1} *t*-BOOH decreased the D_{80} value to about 24 min. Thus, at higher temperatures, micromolar levels of the peroxide greatly enhanced killing. The authors have obtained similar results for H_2O_2 killing at 80°C. Overall, it appeared that heat and oxidative chemicals are mutually potentiating for spore killing. Effects of temperature on spore killing by H_2O_2 have been investigated by Toledo *et al.* (1973) and are well known in the sterilization industry. Recently, Szweda and Stadtman (1993)

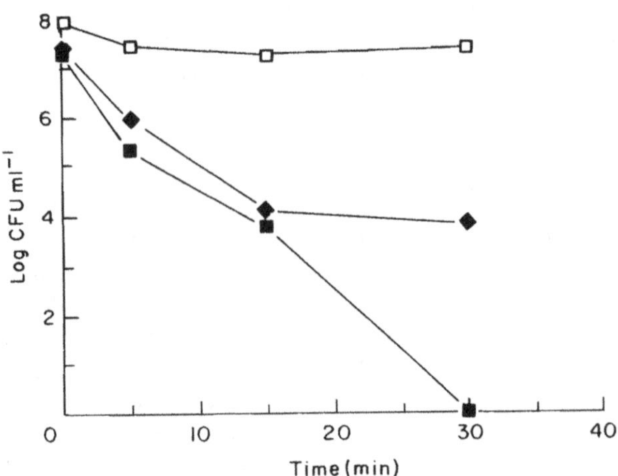

Fig. 1 Killing of dormant spores (□) germinated spores (◆) and vegetative cells (■) of *Bacillus megaterium* ATCC 19213 by 720 mmol l^{-1} *t*-BOOH at 37°C and pH 7. The spores were germinated with inosine and alanine after heat shocking at 60°C for 1 h as described by Bender and Marquis (1982)

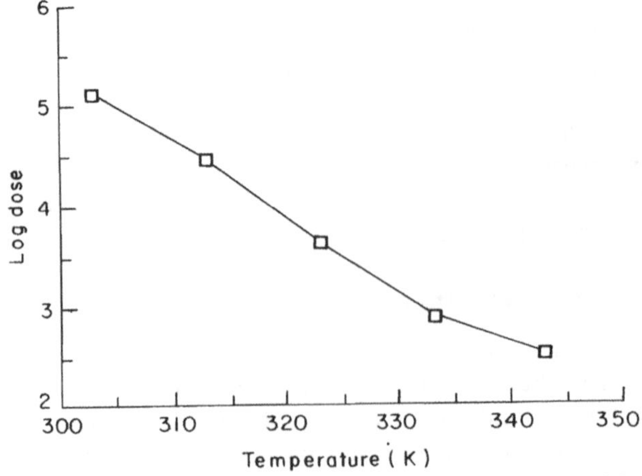

Fig. 2 Effect of temperature (K) in the range from 30 to 70°C on killing of spores of *Bacillus megaterium* ATCC 19213 by *t*-BOOH. The dose of *t*-BOOH is equal to the concentration (mmol l^{-1}) times the exposure time in minutes

found that oxidative modification of glucose-6-phosphate dehydrogenase sensitizes the enzyme to heat inactivation, and their findings suggest a molecular basis for the potentiating actions of heat and oxidative chemicals.

Oxidative killing of metabolically active cells by H_2O_2 or *t*-BOOH generally involves metabolic formation of radicals. In respiration, one-electron reductions of O_2 yield $O_2^{-}\cdot$ and H_2O_2, and in reactions catalysed by transition metal ions, principally Fe^{2+}/Fe^{3+}, the two react to form $OH\cdot$. (Transition metals include Fe, Co, Cu, Mo, Mn, Ni and V.) *t*-BOOH can be converted metabolically or chemically to a variety of radicals, and the alkyl peroxyl radical *t*-BOO · has been found to be most effective for killing vegetative bacteria (Akaike *et al.* 1992). Spore killing probably also involves radical formation because H_2O_2 and *t*-BOOH are not likely to be directly toxic. As spores are completely dormant, however, radical formation must be non-metabolic. In fact, the high Q_{10} for killing may be related to non-metabolic formation of radicals at higher temperatures. Radical formation appears to be catalysed by transition metals, and Cu^{2+} has been found to enhance killing of some types of spores (King and Gould 1969; Bayliss and Waites 1979). However, spores are highly mineralized, in part with transition metals, particularly Mn, and supplementation often does not enhance killing. Chelators such as *o*-phenanthroline do protect against oxidative killing by either *t*-BOOH or H_2O_2 (Marquis and Shin 1993). Chelators were not very protective against heat killing and would not be expected to penetrate the spore protoplast, where the heat-induced formation of lethal radicals is likely to occur. Killing by oxidative chemicals probably involves formation of radicals outside of the protoplast in regions penetrated by *o*-phenanthroline. Results of extensive studies (Gerhardt *et al.* 1972) of spore permeability indicate that molecules with molecular weights of about 200 can penetrate some 40% of the spore volume. Thus *o*-phenanthroline should penetrate through spore coats into the cortex space but not into the protoplast. Chemicals such as H_2O_2 and *t*-BOOH may be able to penetrate even further because of their small molecular sizes. However, protection by *o*-phenanthroline against damage suggests radical formation outside of the protoplast. The radicals would then have to penetrate the protoplast to cause lethal damage. Previous studies with vegetative cells have shown that radicals generated outside the cell enzymatically with xanthine plus xanthine oxidase, or chemically with haeme proteins plus *t*-BOOH (Akaike *et al.* 1992) can cause lethal damage.

Protection by chelators appears not to be due entirely to chelation. For example, the authors found that ethylenediaminetetraacetate (EDTA) and dipicolinate were not protective against killing by *t*-BOOH. Their chelate complexes with iron have lower E_0 values than the *t*-BOO·H^+/*t*-BOOH couple, while the *o*-phenanthroline–iron couple has a higher E_0' value (Buettner 1993). Thus, in addition to chelation, properties such as oxidation–reduction potentials of the chelate complexes may be important for protection.

4.2 Spore properties important for resistance to oxidative killing

Spore coats are generally considered to be barriers to the penetration of damaging enzymes such as lysozyme but would not pose major barriers to penetration of H_2O_2 or *t*-BOOH. Moreover, damage to the coat–outer-membrane complex caused by radicals should not be lethal nor affect heat resistance. Totally decoated spores are fully viable and have nearly full heat resistance. However, they are sensitive to lysozyme germination. The main targets for oxidative killing are probably in the protoplast, and so the agents themselves or radicals formed from them must penetrate the protoplast or react with and damage the protoplast membrane. Hydrogen peroxide at very high concentrations can cause dissolution of dormant spores, but neither lysis nor loss of refractility is an absolute requirement for killing (King and Gould 1969; Ando and Tsuzuki 1986). The authors found that *t*-BOOH at high concentrations does not bring about such extreme damage. Dissolution may occur during industrial sterilization for which solutions of H_2O_2 as concentrated as 35% are routinely used. Decoated spores were found (Ando and Tsuzuki 1986) to undergo lysis after milder oxidative damage induced with enzymes or through chemical oxidation–reduction reactions, but this lysis was preceded by germination-like changes. Overall, it seems that oxidative killing without lysis must involve protoplast damage.

There is a spectrum of resistance to oxidative agents among different spores. For example, the following hierarchy has been found for resistance to killing by *t*-BOOH: *B. stearothermophilus* ATCC 7953 > *B. subtilis* var. *niger* > *B. megaterium* ATCC 33729 > *B. megaterium* ATCC 19213. This order is the same as that for heat resistance or for resistance to H_2O_2. Although the authors have studied too few organisms to date, there is still a suggestion that spore properties important for heat resistance may also be important for resistance to oxidative damage.

5. MECHANISMS OF HEAT AND OXIDATIVE DAMAGE

There is a reasonable basis for thinking that heat killing of spores and oxidative killing are related in that both probably involve free radical damage. The proposal for free radical damage in oxidative killing has a good base of experimental evidence, especially the finding that the transition-metal chelator *o*-phenanthroline had so marked a protective effect. Oxidative chemicals such as H_2O_2 and *t*-BOOH probably are not directly toxic but are converted to toxic radicals such as OH· and *t*-BOO·. Radical formation commonly involves Fenton-type reactions catalysed by transition metal ions, and spores have high levels of Mn^{2+}, which in its complexed, but not precipitated, state could serve as a catalyst for radical formation. Mn^{2+} may be an effective catalyst also when leached from the spore. Killing by ionizing radiation also involves formation of damaging radicals especially OH· and derived radicals. The lethal effects of radiation are much more pronounced under aerobic than anaerobic conditions, and agents such as N_2O greatly enhance lethality by enhancing radical production.

The targets and mechanisms for heat killing are not known. In his review, Molin (1992) indicated that 'Dry heat inactivation has generally been considered to be primarily an oxidation process ...' However, there are seemingly contradictory findings such as increased resistance in the presence of O_2. As indicated above, dry heat has been shown to be mutagenic for spores, and DNA is a likely target for dry heat damage leading to death. Cells are generally able to repair DNA damage, and so such damage would not necessarily lead to killing.

Killing by moist heat may involve different targets or different mechanisms, although increased mutation of spores due to moist heat has been reported (Kadota *et al.* 1978). However, Setlow (this Symposium, pp. 49S–60S) found that spores of *B. subtilis* defective in SASP but not normal spores were mutagenized by exposure to moist heat.

Radiation is known to sensitize spores to heat killing (Gombas and Gomez 1978) and also to killing by H_2O_2 (Bayliss and Waites 1979). Waites *et al.* (1988) suggested that the interaction of u.v. light with H_2O_2 results in the production of OH·. Ionizing radiation produces OH· radicals from water as well as from H_2O_2. The authors have found, as indicated above, that *t*-BOOH or H_2O_2 and heat are mutually potentiating for killing of spores of *B. megaterium* ATCC 19213. Moreover, Scruton (1989) reported that the presence of air actually enhances spore killing under autoclaving conditions, presumably because of the O_2 in the moist air. Kihm and Johnson (1990) found that a hydrogen atmosphere also enhanced heat killing, possibly because the H_2 could act like an inert gas to enhance radical production (Thom 1992). Clearly, we need much more information before coming to very firm conclusions about specific targets and mechanisms for spore killing by heat and oxidative chemicals.

6. SUMMARY

Spore heat resistance can be predicted within reasonable limits from knowledge of optimal growth temperature of the sporeformer, the temperature of sporulation, water content of the spore protoplast, cortex size, total mineralization and specific mineralization. The molecular mechanisms by which dehydration and mineralization act to stabilize spores aginst heat damage are unknown. A major need for further progress is to identify the principal targets for lethal damage.

In this review the hypothesis was explored that heat killing may be related to oxidative killing. The proposed common denominator for the two is the formation of radicals able to react with, and irreversibly damage, spore polymers such as proteins or DNA.

7. ACKNOWLEDGEMENTS

The work of the authors was supported by award DAAL03-90-G-0146 from the US Army Research Office and funds from the Center for Aseptic Processing and Packaging Studies at North Carolina State University.

8. REFERENCES

Akaike, T., Sato, K., Ijiri, S., Miyamoto, Y., Kohno, M., Ando, M. and Maeda, H. (1992) Bactericidal activity of alkyl peroxyl radicals generated by heme–iron-catalyzed decomposition of organic peroxides. *Archives of Biochemistry and Biophysics* **294**, 55–63.

Ando, Y. and Tsuzuki, T. (1983) Mechanism of chemical manipulation of heat resistance of *Clostridium perfringens* spores. *Journal of Applied Bacteriology* **54**, 197–202.

Ando, Y. and Tsuzuki, T. (1986) The effect of hydrogen peroxide on spores of *Clostridium perfringens*. *Letters in Applied Microbiology* **2**, 65–68.

Ando, Y. and Tsuzuki, T. (1986) Changes in decoated spores of *Clostridium perfringens* caused by treatment with some enzymatic and non-enzymatic systems. *Letters in Applied Microbiology* **3**, 61–64.

Barach, J.F., Flowers, R.S. and Adams, D.M. (1975) Repair of heat-injured *Clostridium perfringens* spores during outgrowth. *Applied Microbiology* **30**, 873–875.

Bayliss, C.E. and Waites, W.M. (1976) The effect of hydrogen peroxide on spores of *Clostridium bifermentans*. *Journal of General Microbiology* **96**, 401–407.

Bayliss, C.E. and Waites, W.M. (1979) The synergistic killing of spores of *Bacillus subtilis* by hydrogen peroxide and ultra-violet light irradiation. *FEMS Microbiology Letters* **5**, 331–333.

Beaman, T.C. and Gerhardt, P. (1986) Heat resistance of bacterial spores correlated with protoplast dehydration, mineralization, and thermal adaptation. *Applied and Environmental Microbiology* **52**, 1242–1246.

Belliveau, B.H., Beaman, T.C., Pankratz, S. and Gerhardt, P. (1992) Heat killing of bacterial spores analyzed by differential scanning calorimetry. *Journal of Bacteriology* **174**, 4463–4474.

Bender, G.R. and Marquis, R.E. (1982) Sensitivity of various salt forms of *Bacillus megaterium* spores to the germinating action of hydrostatic pressure. *Canadian Journal of Microbiology* **28**, 643–649.

Bender, G.R. and Marquis, R.E. (1985) Spore heat resistance and specific mineralization. *Applied and Environmental Microbiology* **50**, 1414–1421.

Bradbury, J., Foster, J.R., Hammer, B., Lindsay, J. and Murrell, W.G. (1981) The source of heat resistance of bacterial spores; study of water in spores by NMR. *Biochimica et Biophysica Acta* **678**, 157–164.

Buettner, G.R. (1993) The pecking order of free radicals and antioxidants: lipid peroxidation, alpha-tocopherol, and ascorbate. *Archives of Biochemistry and Biophysics* **300**, 535–543.

Carstensen, E.L., Marquis, R.E. and Gerhardt, P. (1971) Dielectric study of the physical state of electrolytes and water within *Bacillus cereus* spores. *Journal of Bacteriology* **107**, 106–113.

Chiasson, L.P. and Zamenhof, S. (1966) Studies on induction of mutants by heat in spores of *Bacillus subtilis. Canadian Journal of Microbiology* **12**, 43–46.

Chumney, R.K. and Adams, D.M. (1980) Relationship between the increased sensitivity of heat injured *Clostridium perfringens* spores to surface active antibiotics and to sodium chloride and sodium nitrite. *Journal of Applied Bacteriology* **49**, 55–63.

Chung, L., Rajan, K.S., Merdinger, E. and Grecz, N. (1971) Coordinative binding of divalent cations with ligands related to bacterial spore equilibrium studies. *Biophysical Journal* **11**, 469–482.

Ellar, D.J. (1978) Spore specific structures and their function. In *Relations Between Structure and Function in the Prokaryotic Cell* ed. Stanier, R.Y., Rogers, H.J. and Ward, J.B. pp. 295–325. Cambridge: Cambridge University Press.

Errington, J. (1993) *Bacillus subtilis* sporulation: regulation of gene expression and control of morphogenesis. *Microbiological Reviews* **57**, 1–33.

Flowers, R.S. and Adams, D.M. (1976) Spore membrane(s) as the site of damage within heated *Clostridium perfringens* spores. *Journal of Bacteriology* **125**, 429–434.

Gerhardt, P. and Marquis, R.E. (1989) Spore thermoresistance mechanisms. In *Regulation of Procaryotic Development* ed. Smith, I., Slepecky, R. and Setlow, P. pp. 17–63. Washington, DC: American Society for Microbiology.

Gerhardt, P., Scherrer, R. and Black, S.H. (1972) Molecular sieving by dormant spore structures. In *Spores V* ed. Halvorson, H.O., Hanson, R. and Campbell, L.L. pp. 68–74. Washington, DC: American Society for Microbiology.

Gombas, D.E. (1983) Bacterial spore resistance to heat. *Food Technology* **37**, 105–110.

Gombas, D.E. and Gomez, R.F. (1978) Sensitization of *Clostridium perfringens* spores to heat by gamma radiation. *Applied and Environmental Microbiology* **36**, 403–407.

Gorman, S.P., Scott, E.M. and Hutchinson, E.P. (1984) Emergence and development of resistance to antimicrobial chemicals and heat in spores of *Bacillus subtilis. Journal of Applied Bacteriology* **57**, 153–163.

Gould, G.W. (1984) Injury and repair mechanisms in bacterial spores. In *The Revival of Injured Microbes* ed. Andrew, M.H.E. and Russell, A.D. Society for Applied Bacteriology Symposium Series No. 12, pp. 199–220. London: Academic Press.

Gould, G.W. and Dring, G.J. (1975) Heat resistance of bacterial endospores and concept of an expanded osmoregulatory cortex. *Nature London* **258**, 402–405.

Hanlin, J.H., Lombardi, S.J. and Slepecky, R.A. (1985) Heat and UV light resistance of vegetative cells and spores of *Bacillus subtilis* Rec$^-$ mutants. *Journal of Bacteriology* **163**, 774–777.

Imae, Y. and Strominger, J.L. (1976) Cortex content of asporogenous mutants of *Bacillus subtilis. Journal of Bacteriology* **126**, 914–918.

Jaenicke, R. (1991) Protein stability and molecular adaptation to extreme conditions. *European Journal of Biochemistry* **202**, 715–728.

Johnstone, K., Stewart, G.S.A.B., Barratt, M.D. and Ellar, D.J. (1982) An electron paramagnetic resonance study of the manganese environment within dormant spores of *Bacillus megaterium* KM. *Biochimica et Biophysica Acta* **714**, 379–381.

Kadota, H., Uchida, A., Sako, Y. and Harada, K. (1978) Heat-induced DNA injury in spores and vegetative cells of *Bacillus subtilis.* In *Spores VII* ed. Chambliss, G. and Vary, J.C. pp. 27–30. Washington, DC: American Society for Microbiology.

Kihm, D.J. and Johnson, E.A. (1990) Hydrogen gas accelerates thermal inactivation of *Clostridium botulinum* 113B spores. *Applied Microbiology and Biotechnology* **33**, 705–708.

King, W.L. and Gould, G.W. (1969) Lysis of bacterial spores with hydrogen peroxide. *Journal of Applied Bacteriology* **32**, 481–490.

Klibanov, A.M. and Ahern, T.J. (1987) Thermal stability of proteins. In *Protein Engineering* ed. Oxender, D.L. and Fox, C.F. pp. 213–218. New York: Alan R. Liss.

Lewis, J.C., Snell, N.S. and Burr, H.K. (1960) Water permeability of bacterial spores and the concept of a contractile cortex. *Science* **132**, 544–545.

Lindahl, T. and Nyberg, B. (1972) Rate of depurination of native deoxyribonucleic acid. *Biochemistry* **11**, 3610–3618.

Lundin, R.E. and Sacks, L.E. (1988) High-resolution solid-state ^{13}C nuclear magnetic resonance of bacterial spores: identification of the alpha-carbon signal of dipicolinic acid. *Applied and Environmental Microbiology* **54**, 923–928.

Marquis, R.E. and Bender, G.R. (1985) Mineralization and heat resistance of bacterial spores. *Journal of Bacteriology* **161**, 789–791.

Marquis, R.E. and Bender, G.R. (1990) Compact structure of cortical peptidoglycans from bacterial spores. *Canadian Journal of Microbiology* **36**, 426–429.

Marquis, R.E. and Shin, S.Y. (1994) Mineralization and resistance of bacterial spores to heat and oxidative agents. In *Metals and Microorganisms: Relationships and Applications* ed. Bauda, P. and Ferard, J.F. Amsterdam: Elsevier (in press).

Marquis, R.E., Bender, G.R., Carstensen, E.L. and Child, S.Z. (1983) Dielectric characterization of forespores isolated from *Bacillus megaterium. Journal of Bacteriology* **153**, 436–442.

Mason, J.M. and Setlow, P. (1986) Evidence for an essential role for small, acid-soluble, spore proteins in the resistance of *Bacillus subtilis* spores to ultraviolet light. *Journal of Bacteriology* **167**, 174–178.

Molin, G. (1992) Destruction of bacterial spores by thermal methods. In *Principles and Practice of Disinfection, Preservation and Sterilization* 2nd edn, ed. Russell, A.D., Hugo, W.B. and Ayliffe, G.A.J. pp. 499–511. Oxford: Blackwell Scientific Publications.

Murrell, W.G. and Warth, A.D. (1965) Composition and heat

resistance of bacterial spores. In *Spores III* ed. Campbell, L.L. and Halvorson, H.O. pp. 1–24. Ann Arbor, MI: American Society for Microbiology.

Northrop, J. and Slepecky, R.A. (1967) Sporulation mutations induced by heat in *Bacillus subtilis*. *Science* **155**, 838–839.

Popham, D.L. and Setlow, P. (1993) The cortical peptidoglycan from spores of *Bacillus megaterium* and *Bacillus subtilis* is not highly cross-linked. *Journal of Bacteriology* **175**, 2767–2769.

Riemann, H. and Ordal, Z.J. (1961) Germination of bacterial endospores with calcium and dipicolinic acid. *Science* **133**, 1703–1704.

Sapru, V. and Labuza, T.P. (1993) Glassy state in bacterial spores predicted by polymer glass-transition theory. *Journal of Food Science* **58**, 445–448.

Scruton, M.W. (1989) The effect of air on the moist-heat resistance of *Bacillus stearothermophilus* spores. *Journal of Hospital Infection* **14**, 339–350.

Setlow, B. and Setlow, P. (1993) Dipicolinic acid greatly enhances production of spore photoproduct in bacterial spores upon UV irradiation. *Applied and Environmental Microbiology* **59**, 640–643.

Stewart, G.S.A.B., Eaton, M.W., Johnstone, K., Barrett, M.D. and Ellar, D.J. (1980) An investigation of membrane fluidity changes during sporulation and germination of *Bacillus megaterium* K.M. measured by electron spin and nuclear magnetic resonance spectroscopy. *Biochimica et Biophysica Acta* **600**, 270–290.

Szweda, L.I. and Stadtman, E.R. (1993) Oxidative modification of glucose-6-phosphate dehydrogenase from *Leuconostoc mesenteroides* by an iron(II)-citrate complex. *Archives of Biochemistry and Biophysics* **301**, 391–395.

Thom, S.R. (1992) Inert gas enhancement of superoxide radical production. *Archives of Biochemistry and Biophysics* **295**, 391–396.

Toledo, R.T., Escher, F.E. and Ayers, J.C. (1973) Sporicidal properties of hydrogen peroxide against food spoilage organisms. *Applied Microbiology* **26**, 592–597.

Ulanowski, Z. and Ludlow, I.K. (1993) The influence of the cortex on protoplast dehydration in bacterial spores studied with light scattering. *Current Microbiology* **26**, 31–35.

Waites, W.M., Harding, S.E., Fowler, D.R., Jones, S.H., Shaw, D. and Martin, M. (1988) The destruction of spores of *Bacillus subtilis* by the combined effects of hydrogen peroxide and ultraviolet light. *Letters in Applied Microbiology* **7**, 139–140.

Warth, A.D. (1978a) Relationship between the heat resistance of spores and the optimum and maximum growth temperatures of *Bacillus* species. *Journal of Bacteriology* **134**, 699–705.

Warth, A.D. (1978b) Molecular structure of the bacterial spore. *Advances in Microbial Physiology* **17**, 1–45.

Warth, A.D. (1980) Heat stability of *Bacillus cereus* enzymes within spores and in extract. *Journal of Bacteriology* **143**, 27–34.

Warth, A.D. (1985) Mechanisms of heat resistance. In *Fundamental and Applied Aspects of Bacterial Spores* ed. Dring, G.J., Ellar, D.J. and Gould, G.W. pp. 209–225. London: Academic Press.

Watt, I.C. (1981) Water vapour adsorption by *Bacillus stearothermophilus* endospores. In *Sporulation and Germination* ed. Levinson, H.S., Sonenshein, A.L. and Tipper, D.J. pp. 253–255. Washington, DC: American Society for Microbiology.

Journal of Applied Bacteriology Symposium Supplement 1994, **76**, 49S–60S

Mechanisms which contribute to the long-term survival of spores of *Bacillus* species

P. Setlow
Department of Biochemistry, University of Connecticut Health Center, Farmington, CT, USA

1. Introduction, 49S
2. Environment within the spore, 50S
3. Enzymatic dormancy, 51S
   3.1 3-Phosphoglycerate (3PGA) and phosphoglycerate mutase (PGM), 52S
   3.2 Small, acid-soluble spore proteins (SASPs) and germination protease (GPR), 52S
   3.3 Spore cortex and cortex lytic enzyme, 53S
4. Spore chemical resistance, 54S
5. Spore radiation resistance
   5.1 γ-Radiation resistance, 55S
   5.2 u.v.-Radiation resistance, 55S
6. Spore heat resistance, 56S
   6.1 Spore preparation temperature, 57S
   6.2 α/β-Type SASPs, 57S
   6.3 Mineralization, 57S
   6.4 Water content, 58S
7. Conclusions, 58S
8. Acknowledgement, 59S
9. References, 59S

1. INTRODUCTION

Bacteria of the genera *Bacillus* and *Clostridium* can initiate the process of sporulation, generally when one or more nutrients becomes limiting for growth. An early event in sporulation is an unequal division of the cytoplasm, giving rise to small and large progeny, each with a complete genome. The small compartment, termed the 'forespore', is destined to become the mature spore; the large compartment, termed 'the mother cell' engulfs the forespore resulting in a cell (forespore) within a cell (mother cell). Eventually, after a series of further morphological and biochemical changes, the mother cell lyses, releasing the mature spore into the environment. Mature spores have no detectable metabolism, i.e. are dormant, and can survive for extremely long periods in the absence of exogenous nutrients, even when they are in a fully hydrated environment. While there are few carefully controlled long-term studies of the latter phenomenon, survival of $>80\%$ of populations of spores of *Bacillus* species for periods of up to a year is not uncommon (Slepecky and Leadbetter 1983; Setlow 1992). Similarly, there are studies which indicate that spores can survive in nature for even longer periods (100–2000 years) (Gould 1983; Slepecky and Leadbetter 1983). However, there are obviously no detailed population studies over such long time periods.

One reason for spore survival over extended periods is undoubtedly their extreme metabolic dormancy. They neither contain nor make ATP. Spores are much more resistant than their growing cell counterparts to a variety of environmental stresses, including heat, radiation and chemicals. Resistance to these various agents is acquired by the developing forespore in a defined temporal sequence, as is metabolic dormancy (Fig. 1). Strikingly, although the mature spore is metabolically dormant, it contains a number of substrates as well as the enzymes that might be expected to act completely on these substrates within 10–20 min. Examples of such substrate/enzyme pairs are: (1) 3-phosphoglycerate (3PGA) and phosphoglycerate mutase (PGM); (2) small, acid soluble, spore proteins (SASPs) and germination protease (GPR); and (3) spore cortex and cortex lytic enzyme. The lack of interaction in the dormant spore between these substrates and enzymes which could act on them is an example of the enzymatic dormancy of the mature spore, undoubtedly another factor in long term spore survival.

Although the dormant spore has these striking properties of resistance and enzymatic dormancy, these properties are lost upon spore germination. Within minutes of initiating spore germination: (1) spore resistance returns to that of growing cells; (2) there is complete reaction between the substrate/enzyme pairs noted above; (3) large scale production of ATP and other high-energy molecules begins; and (4) macromolecular synthesis is initiated.

The main focus of this review will be the mechanisms which contribute to long-term survival of spores, concentrating on spores of *Bacillus* species for which there is much detailed molecular data. However, it seems likely that conclusions from work on spores of *Bacillus* species will also apply to those of *Clostridium* species. As the nature of the

Correspondence to: P. Setlow, Department of Biochemistry, University of Connecticut Health Center, Farmington, CT 06030-3305, USA.

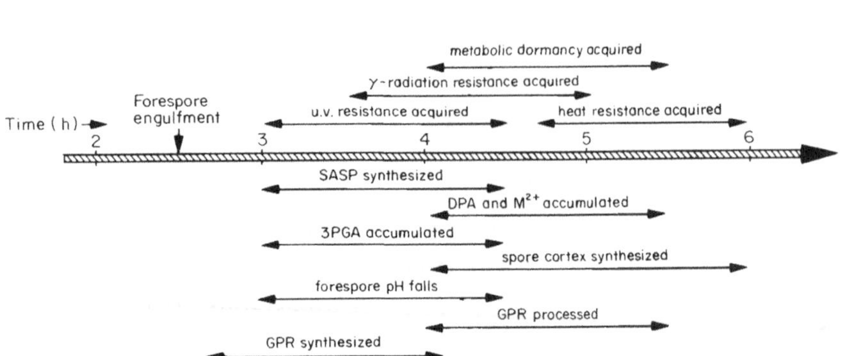

Fig. 1 Approximate time course of major events in sporulation of *Bacillus* species. The time of initiation of sporulation is taken as time zero, and the times shown are the hours after time zero. The ends of the arrows denote the approximate time periods for the events listed. The period when metabolic dormancy is acquired is taken as the time when forespore ATP pools are depleted. Data in support of the kinetic scheme shown here are from both *B. megaterium* and *B. subtilis*, and are from Murrell (1981), Russell (1982), Gerhardt and Marquis (1989), Setlow (1992), and Sanchez-Salas and Setlow (1993)

environment within the spore is a primary cause of spore survival, knowledge about this environment will be reviewed first, followed by discussion of spore enzymatic dormancy, and then spore resistance to chemicals, radiation and heat. In several of the sections attention will be given not only to the situation within the dormant spore, but also to the mechanism by which the dormant spore state is achieved during sporulation, and lost during spore germination.

2. ENVIRONMENT WITHIN THE SPORE

A major difference between the environments within a growing cell and a dormant spore is in their complement of small molecules. The dormant spore lacks most of the common high-energy compounds found in growing cells, such as ATP and NADH, although the low-energy forms of these compounds (AMP, ADP, NAD, etc.) are present (Setlow 1983; Table 1). Spores do contain a significant amount of 3PGA (0·1–0·3% dry wt) which is accumulated within the developing forespore ~1½ h before spores become heat resistant (Table 1; Fig. 1), but no detectable 2PGA is present in spores (see Section 3.1). Spores also contain a very large amount (5–15% dry wt) of pyridine-2, 6-dicarboxylic acid, termed dipicolinic acid (DPA), which probably exists as a 1–1 chelate with a divalent cation, usually Ca^{2+} (Table 1). In large part because of their high level of DPA, spores contain a large amount of divalent cations, predominantly Ca^{2+}, but with significant amounts of Mg^{2+}, Mn^{2+} and other ions as well (Table 1). While many of the divalent ions in the dormant spore are undoubtedly chelated with DPA, a significant amount may be bound to other major spore small molecules such as 3PGA or glutamic acid. DPA and divalent cations are accumulated from the mother cell at or slightly before the time spores become heat resistant (Fig. 1). However, the mechanism of uptake of these molecules is not known. It appears certain that 3PGA, DPA and the majority of the divalent cations are located in the spore core, which is equivalent to a cell's protoplast (Gerhardt and Marquis 1989).

Another important difference between the environments within a growing cell and a spore is in the hydrogen ion concentration (Table 1). The pH within cells of *Bacillus* species is generally between 7·5 and 8·2 (Decker and Lang

Table 1 Levels of small molecules in cells and spores of *Bacillus* species

Molecule	Cells*	Spores†
	(μmol g⁻¹ dry spores)	
ATP	3·6	≤0·005
ADP	1	0·2
AMP	1	1·2–1·3
3PGA	<0·2	5–18
2PGA	<0·1	<0·05
NADH	1·95	<0·002‡
NAD	0·35	0·11‡
NADPH	0·52	<0·001‡
NADP	0·44	0·018‡
Glutamic acid	38	24–30
DPA	<0·1	410–470
Ca^{2+}	—	380–916
Mg^{2+}	—	86–120
Mn^{2+}	—	27–56
H^+	7·5–8·2§	6·3–6·5§

* Values for cells are for *B. megaterium* in mid-log phase and are from Decker and Lang (1978), Shioi *et al.* (1980), Setlow (1983), and Magill, Cowan and Setlow, unpublished.

† Values are the range from spores of *B. cereus*, *B. megaterium* and *B. subtilis*, and are from Setlow (1983) and Loshon and Setlow (1993).

‡ Values are only from *B. megaterium*.

§ These values are expressed as pH.

1978; Setlow and Setlow 1980; Shioi *et al.* 1980; Magill, Cowan and Setlow, unpublished). Late in sporulation, however, approximately during the period of 3PGA accumulation, the forespore pH falls by ~1·2 units. Measurements of the pH in the dormant spore core have also indicated that this value (pH 6·3–6·5) is at least 1–1·2 units more acidic than that in cells (Setlow and Setlow 1980). While it has been possible to elevate spore core pH to ~8·0, this change has no effect on dormant spore properties (Swerdlow *et al.* 1981). However, the fall in forespore pH during sporulation may play a key role in modulating the activity of PGM, thus allowing accumulation of 3PGA in the forespore (see Section 3.1). At present it is not clear how the forespore pH is lowered during sporulation.

The third, and probably most important difference between the environment in a growing cell and a dormant spore is in the amount of water present (Gerhardt and Marquis 1989). Growing cells are 75–80% water, or 3–4 g water per g dry wt. Values for the water content of the spore coat and cortex are similar to those for intact growing cells. However, the spore core has much less water per gram dry weight; values determined by Gerhardt and coworkers using a variety of techniques are between 0·5–1 g water per g dry wt depending on the species examined (Gerhardt and Marquis 1989). The large difference in the amount of water per gram dry weight in cells and the spore core indicates that the spore core has much less free water than cells. Indeed, as much of the water in the spore core may be intimately associated with macromolecules or small molecules, there may actually be little if any free water in the spore core. Unfortunately, there are no good measurements of the amount of free water in spores. However, there are data from a variety of different techniques that indicate that mineral ions in spores are quite immobile, consistent with spores having little if any free water (Gerhardt and Marquis 1989). The great majority of the reduction in water content in the spore core takes place approximately at the time of DPA accumulation (Fig. 1). This reduction in spore water requires the presence of the spore cortex, a structure composed of peptidoglycan which is outside the inner spore membrane. However, the precise mechanism whereby forespore core water is removed and the manner in which the spore cortex participates in this process are not clear, despite a number of theories (Gerhardt and Marquis 1989).

With the exception of pH, which can readily be raised or lowered in the dormant spore (Swerdlow *et al.* 1981), other unique features of the dormant spore environment are remarkably stable. Thus spore DPA, divalent ions, and 3PGA are not lost on prolonged storage in water, although treatment at low pH can remove DPA and divalent ions without killing the spore (Gerhardt and Marquis 1989). Under normal conditions the spore core is impermeable to small (molecular weight $\leq$100) hydrophilic molecules, although hydrophobic molecules up to 280 in molecular weight enter the spore core (Gerhardt *et al.* 1972). While water apparently freely permeates the spore core, no significant increase in *dormant* spore core water content has ever been observed; presumably an intact spore cortex prevents significant uptake of water into the spore core.

Despite the relative stability of the environment within the dormant spore, this environment changes back to that of a growing cell within minutes of initiating spore germination (Gould 1969; Setlow 1983; Foster and Johnstone 1989). One of the early steps in this process is the initiation of hydrolysis of the spore cortext. The breakdown of this structure allows a tremendous influx of water into the spore, increasing the volume of the spore core more than twofold. Accompanying this influx of water, and probably preceding much of it, is the ejection of monovalent ions including H^+, as the spore pH rises to 7·5–8·0. This is followed by: (1) excretion of DPA and divalent cations; (2) utilization of 3PGA to generate ATP; and (3) initiation of macromolecular synthesis. While much is known about the events which take place in the first minutes of spore germination, little is known about the mechanism for a number of these processes, including initiation of spore cortex hydrolysis, and the water and ion movements.

3. ENZYMATIC DORMANCY

In order for spores to survive for long periods of time, a number of enzymes must remain inactive within them. These enzymes include those such as PGM which can catabolize the 3PGA used for ATP production early in spore germination, as well as enzymes that degrade spore components such as the cortex and small, acid-soluble spore proteins which are essential for spore survival. As spores can survive for months or years at ambient temperatures, mechanisms that inhibit enzymes by only a few orders of magnitude seem unlikely to be sufficient for the enzymatic dormancy of spores. For example, *Bacillus megaterium* spores germinating at 30°C degrade their 3PGA depot in ~5 min, but in the dormant spore the 3PGA depot is stable (<10% loss) for at least 3 months at 30°C (Setlow 1983). Thus PGM must be inhibited by at least a factor of $\sim 2 \times 10^5$ to retain 3PGA in the dormant spore. Given the huge magnitude of enzyme inhibition required for enzymatic dormancy, such inhibition seems unlikely to be due to variations in parameters such as pH or ionic composition. Similarly, the presence of specific low molecular weight enzyme inhibitors seems unlikely; indeed there is no evidence for them. The only two mechanisms that do seem likely to give the almost absolute enzyme inhibition required for a spore's enzymatic dormancy are: (1) the presence in the spore of an inactive form of an enzyme, i.e.

a zymogen; and (2) inhibition of enzyme activity by the absence of water. The latter mechanism seems most likely to be the key one, as dehydration could also preclude zymogen activation. However, there is evidence for synthesis of some key spore enzymes as zymogens. One feature of spore enzymatic dormancy which may be quite informative is how this dormancy is achieved during sporulation, and then broken during spore germination. For the three substrate and enzyme pairs discussed below, a variety of strategies are used to achieve and then maintain enzymatic dormancy.

3.1 3-Phosphoglycerate (3PGA) and phosphoglycerate mutase (PGM)

Approximately 90 min before spore core dehydration, the developing forespore accumulates a significant amount of 3PGA—up to 0·3% of the dormant spore dry wt (Fig. 1; Setlow 1983). As the PGA depot is >99·9% 3PGA, with no detectable 2PGA, PGM is the enzyme which is regulated to allow 3PGA accumulation. The 3PGA depot is stable in the dormant spore, but in the first minutes of germination 3PGA is catabolized via PGM, enolase and pyruvate kinase to generate ATP. Clearly, PGM must be inhibited significantly in developing forespores in order to allow 3PGA accumulation, and must be inhibited essentially completely in dormant spores for the 3PGA to be stable. Both of these types of inhibition must be removed in the first minutes of germination.

Phosphoglycerate mutase has been purified from both vegetative cells and spores of *Bacillus* species; the enzyme from both stages of growth appears identical by a wide variety of criteria, and never exists as a zymogen (Singh and Setlow 1979; Watabe and Freese 1979; Kuhn *et al.* 1993). The enzyme does not require 2,3-bis-phosphoglycerate for activity, but has an absolute and specific requirement for Mn^{2+} to maintain the enzyme in an active conformation. While Mn^{2+} is not required for catalytic activity, in its absence PGM is rapidly converted to an inactive form. This observation led to the suggestion that a fall in the free Mn^{2+} concentration in the developing forespore might decrease PGM activity, thus causing 3PGA accumulation. This suggestion was strengthened by the finding that incubation of 3PGA-rich developing forespores with Mn^{2+} and a divalent cation-specific ionophore caused a rapid utilization of forespore 3PGA (Setlow 1983). While the idea that changes in the concentration of forespore Mn^{2+} may regulate PGM remains viable, there is no evidence for changes in free Mn^{2+} concentration during sporulation. Indeed, obtaining data pertaining to this question may be difficult.

More recently, studies of purified PGM *in vitro* have shown that this enzyme is remarkably pH sensitive (Kuhn *et al.* 1993). At a Mn^{2+} concentration of 20 μmol l^{-1}, PGM activity falls ≥30-fold in going from pH 7·5 to 6·5. While the magnitude of the change in PGM activity over this pH range depends on the precise Mn^{2+} concentration, being less at higher [Mn^{2+}] and more at lower [Mn^{2+}], regulation of forespore PGM by changes in pH presents an attractive alternative model to regulation by changes in [Mn^{2+}]. This latter model is made even more attractive by the demonstration of a fall in forespore pH (from ~8·2 to 7·0) at about the time of 3PGA accumulation (Magill, Cowan and Setlow, unpublished). In this model, the fall in forespore pH would result in a substantial (~30–50-fold) decrease in PGM activity, which then allows significant 3PGA accumulation. Approximately 90 min later, forespore core dehydration results in essentially complete inhibition of PGM activity, independent of Mn^{2+} or pH. It is this latter dehydration which maintains PGM in an inactive state in the dormant spore. As noted above (Section 2) estimates of spore water content range between 0·5 and 1 g per g dry wt. There are significant data suggesting that enzymes with ~0·3 g water per g dry wt can exhibit some catalytic activity. However, as much of the spore core's dry weight consists of molecules other than protein, it is not clear how much water is associated with enzymes in spores. Indeed, given the high level of small, hydrophilic molecules in the spore core (i.e. 3PGA, DPA, divalent cations) much spore core water may be associated with these molecules, leaving very little to associate with proteins and thus essentially no enzymatic activity. In the first minutes of spore germination, the rehydration and alkalinization of the spore core results in restoration of PGM activity and rapid 3PGA utilization.

3.2 Small, acid-soluble spore proteins (SASPs) and germination protease (GPR)

Ten to twenty percent of the protein in dormant spores is a group of small, acid-soluble proteins termed SASPs (Setlow 1988a). These proteins are of two types: (1) α/β-type SASP, the products of a multigene family synthesized beginning at about the third hour of sporulation (t_3) and associated with spore DNA (see Sections 4, 5 and 6); and (2) γ-type SASP, the products of a single gene, also synthesized beginning at ~t_3, localized in the spore core, but not thought to be associated with any spore macromolecule. In the first 20–30 min of spore germination all SASPs are degraded to amino acids, which support much protein synthesis during this period. This degradation is initiated by an endoprotease termed GPR (for germination protease), which is specific for a highly conserved sequence present in all SASPs (Fig. 2). GPR cleaves SASP within this sequence, generating 2–3 large oligopeptides which are degraded further by peptidases. GPR is required for rapid SASP degradation during germination, and fully active

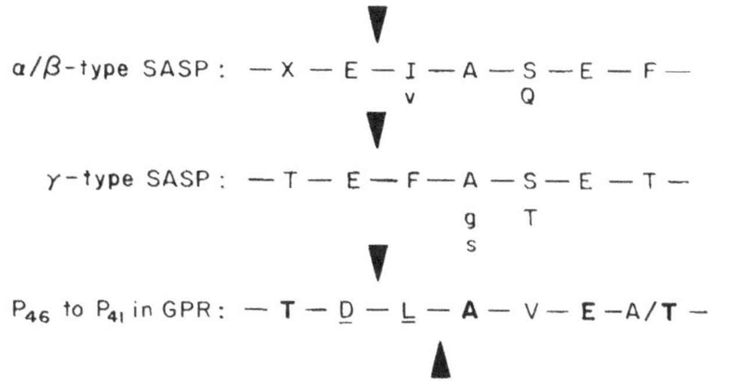

α/β-type SASP : — X — E — I — A — S — E — F —
v Q

γ-type SASP : — T — E — F — A — S — E — T —
g T
s

P_{46} to P_{41} in GPR : — **T** — D — L — **A** — V — **E** — A/**T** —

Fig. 2 Comparison of the amino acid sequences in the P_{46} to P_{41} and P_{41} to P_{40} processing sites of *Bacillus megaterium* and *B. subtilis* GPR, with the GPR cleavage sites in α/β-type and γ-type SASPs of *Bacillus* species. Amino acid sequences are given in the one letter code. SASP-cleavage sequences are from Setlow (1988a) and are from *B. cereus*, *B. megaterium* and *B. subtilis* SASP (total of 19 cleavage sites). P_{46} to P_{41} and P_{41} to P_{40} cleavage sequences in GPR are from Sanchez-Salas and Setlow (1993) and are from *B. megaterium* and *B. subtilis* GPR. The downward pointing vertical arrows denote the bond cleaved in the three types of proteins. The upward pointing arrow denotes the additional bond cleaved when P_{41} is processed to P_{40}. Single residues shown around GPR cleavage sites in SASP are the only residues found at these positions; for positions where there are single variant residues the minor residue(s) is given in small lower case letters below the predominant residue. The small capital letters below the major residue denote the residue found at this position in 4 (Q) or 2 (T) cleavage sites. The letter X in the α/β-type SASP sequence notes that there is no consensus residue at this position. Residues in the P_{46} to P_{41} cleavage site which are identical to those in GPR cleavage sites in γ-type SASP are in bold face; residues which are similar to those in GPR cleavage sites in SASP are underlined

GPR is present in dormant spores (Sanchez-Salas *et al.* 1992). However, GPR does not act in dormant spores, as SASPs are stable until germination is initiated.

GPR is synthesized during sporulation beginning just prior to SASP synthesis. GPR is initially made as a zymogen, termed P_{46}, whose sequence is identical to that encoded by the *gpr* gene; P_{46} is inactive both *in vitro* and *in vivo*. Two hours later in sporulation, and approximately in parallel with accumulation of DPA and forespore core dehydration, P_{46} undergoes proteolytic processing with removal of 15 (*B. megaterium*) or 16 (*B. subtilis*) amino-terminal residues (Sanchez-Salas and Setlow 1993). Strikingly, the sequence cleaved in the P_{46} processing step has much homology with the sequence recognized and cleaved in SASP by GPR (Fig. 2). The product of P_{46} processing (termed P_{41}) has full catalytic activity when assayed *in vitro*. In the first minutes of germination P_{41} is further processed, almost certainly by an aminopeptidase, removing one additional residue (Fig. 2), and the product, termed P_{40} degrades SASP as germination proceeds.

A model has recently been presented based on a variety of types of evidence which attempts to explain regulation of GPR processing and activity (Sanchez-Salas and Setlow 1993). In this model the amino terminal extension in P_{46} sits in the enzyme's active site precluding enzyme action, yet is not itself cleaved. Forespore core dehydration then causes a structural change in the protein such that P_{46} processes itself to P_{41}. Since P_{46}–P_{41} processing is triggered by forespore core dehydration, the conditions when P_{41} is generated preclude both P_{41} attack on SASP and aminopeptidase attack on P_{41}. Spore core dehydration then maintains P_{41} in an inactive state during dormancy. In the first minutes of spore germination, however, spore core rehydration allows aminopeptidase attack on P_{41} and attack of P_{40} (and also probably P_{41}) on SASP.

While the model given above has by no means been proven, it is consistent with a variety of data, and has a number of features that are readily testable. One attractive feature of this model is that the timing of the generation of potentially active GPR (i.e. P_{41}) is coupled to a change in the forespore environment (i.e. dehydration) which precludes P_{41} action. Furthermore, no other factors, i.e. other processing enzymes which themselves would need to be regulated (and which have been looked for but not found) need to be hypothesized to account for regulation of GPR enzymatic dormancy.

3.3 Spore cortex and cortex lytic enzyme

The spore cortex, composed primarily of peptidoglycan, is synthesized between the inner and outer forespore membranes during the fourth–fifth hours of sporulation, and its synthesis coincides reasonably well with forespore core dehydration (Murrell 1981). The cortical peptidoglycan structure is different from that of vegetative cell wall peptidoglycan, and is essential for spore dehydration and thus spore dormancy, survival and heat resistance. The cortex appears stable in the dormant spore, but is degraded to fragments in the first minutes of germination (Gould 1969; Foster and Johnstone 1989).

Maintaining a cortex lytic enzyme in an inactive state presents a problem not encountered in maintaining spore core enzymes in an inactive state, because the spore cortex is outside the inner forespore membrane. Consequently, forespore core dehydration is unlikely to be the direct reason for the inactivity of a cortex lytic enzyme. Our understanding of the regulation of this enzyme is further hampered by the lack of definitive identification of the enzyme responsible for spore cortex breakdown during germination. While a number of candidate enzymes have been identified, none has been proven to play the appropriate role in germination. Indeed, it is possible that there are multiple enzymes which carry out this reaction.

One plausible model has been put forward by Foster and Johnstone (1989) in which a cortex lytic enzyme is made as a zymogen, possibly covalently attached to the cortex itself. Addition of germinants triggers the activation of a protease which cleaves and activates the cortex lytic enzyme, thus initiating spore cortex hydrolysis. There is certainly *in vitro* evidence for this model, but no proof that the proteins identified carry out this role *in vivo*. Furthermore, it is not clear how the proposed protease is maintained in an inactive state. One attractive idea is that the protease is embedded in the inner forespore membrane, possibly with a cytoplasmic domain through which dehydration acts to keep a catalytic domain outside the membrane in an inactive state. Binding of a germinant to a membrane receptor would cause a change in the catalytic domain triggering processing of the cortex lytic enzyme and then spore cortex hydrolysis. One added possibility is that a major function of the huge depot of DPA in the spore core may be to maintain this putative protease in its inactive form. Many studies have now led to the conclusion that the only definitive phenotype of spores which lack DPA is their extreme instability, i.e. they germinate spontaneously, often within the sporangium. Clearly, there is much work left to be done before the regulation of the interaction of this substrate and enzyme pair is well understood.

4. SPORE CHEMICAL RESISTANCE

If a spore is to survive for long periods of time, resistance to chemical agents that might be encountered in its environment could be advantageous. Such chemical agents could include environmental chemicals, as well as free radicals which might be generated by chemical reactions within the spore. Dormant spores are indeed resistant to a variety of chemicals which are normally quite toxic for growing cells (Russell 1982). Unfortunately, for most of these chemicals there is little detailed information at the molecular level concerning either the mechanism of spore resistance or even the chemicals' targets. It seems likely that a major factor in spore resistance to many chemicals is the relative impermeability of the membrane surrounding the spore core, such that penetration of chemicals into the core is very slow. There is also evidence that spore coat layers can contribute to spore chemical resistance (Russell 1982). Spore core dehydration would be expected to further slow reaction of chemical agents with spore core components, further increasing spore chemical resistance. Resistance to various chemicals is acquired at defined times in sporulation. Unfortunately, in most studies the timing of acquisition of chemical resistance has not been rigorously compared with acquisition of known biochemical components of the spore. Given the great increase in knowledge of individual spore components in recent years, reinvestigations in this area could be worthwhile.

While there is little detailed knowledge of the mechanism of spore resistance to most chemicals, recent work has shed some light on spore resistance to hydrogen peroxide. Hydrogen peroxide kills cells by a variety of mechanisms, but a major one is generation of hydroxyl radicals which can cleave the DNA backbone (Imlay and Linn 1988). Dormant spores of *B. subtilis* are much more resistant to hydrogen peroxide than are growing cells (Table 2). While one reason for the increased spore resistance may be decreased permeability to hydrogen peroxide, a second

Table 2 Resistance of spores and cells of *Bacillus subtilis* to various treatments*

Treatment	Cells	Wild-type spores	$\alpha^-\beta^-$ Spores
(1) Hydrogen peroxide resistance			
(% survival after 10% hydrogen peroxide treatment)			
Time of treatment (min)			
1	11	—	—
2·5	0·3	92	70
5	—	88	26
20	—	60	0·1
(2) u.v. Resistance			
(u.v. dose to kill 90% of population—J m^{-2})	40	315	25
(3) Heat resistance (*D* values)			
$D_{95°}$	—	14 min	—
$D_{85°}$	—	320 min	14 min
$D_{65°}$	<15 s	105 h	10 h
$D_{22°}$	—	2·5 year	2·8 month

* Data are taken from Setlow (1988b), Fairhead *et al.* (1993) and Setlow and Setlow (1993b).

appears to be the protection of spore DNA from hydroxyl-radical cleavage by the binding of α/β-type SASP (Setlow and Setlow 1993b). These proteins, described in more detail in Section 5.2, saturate the spore chromosome *in vivo* and protect the DNA backbone against hydroxyl-radical cleavage *in vitro*. Spores deficient in α/β-type SASP (termed $\alpha^-\beta^-$ spores) are much more hydrogen peroxide sensitive than are wild-type spores. Strikingly, the survivors of hydrogen peroxide treatment of $\alpha^-\beta^-$ spores contain a high percentage (10–15%) of mutations, while survivors of hydrogen peroxide treatment of wild-type spores do not (Setlow and Setlow 1993b). These findings suggest that in wild-type spores, DNA is saturated with α/β-type SASP and thus protected against hydroxyl-radical attack. Consequently, in wild-type spores hydrogen peroxide kills by a mechanism(s) other than DNA damage, and this mechanism (which is unknown) requires drastic hydrogen peroxide treatment. In contrast, much of the DNA in $\alpha^-\beta^-$ spores is available for hydroxyl-radical attack, and thus hydrogen peroxide kills $\alpha^-\beta^-$ spores by DNA damage long before it kills wild-type spores by other mechanisms. In addition to hydrogen peroxide, a variety of other compounds or treatments can generate free radicals *in vivo* which can damage molecules such as DNA. Indeed, generation of free radicals with attendant damage to DNA may well contribute significantly to ageing in organisms from yeast to man. Clearly, protection of DNA from free radical damage by α/β-type SASP could be a major component of spore longevity.

5. SPORE RADIATION RESISTANCE

In addition to toxic chemicals, there are a number of other agents with which spores must deal if they wish to survive for long periods. One such agent is radiation, to which spores in water or on the soil surface might be exposed. Unlike growing cells, spores cannot repair radiation damage, as spores lack ATP. Thus for spores to survive radiation treatments they must either minimize radiation damage, or minimize the lethal and mutagenic effects of any radiation damage as much as possible. Not surprisingly, dormant spores exhibit increased resistance to both γ-radiation and u.v. radiation (Russell 1982; Gould 1983; Setlow 1988b, 1992).

5.1 γ-Radiation resistance

Dormant spores of a variety of species are more resistant to γ-radiation of various wavelengths than are their vegetative counterparts (Russell 1982; Gould 1983). When detailed studies were carried out, it was found that γ-radiation resistance is acquired during sporulation 1–2 h prior to acquisition of heat resistance, and may parallel formation of cystine-rich spore coat proteins. While spore core dehydration would also be predicted to be radioprotective, there has been no study attempting to correlate spore γ-radiation resistance with spore core dehydration level, as there has been for spore heat resistance (see Section 6.4). Similarly, the role of most other spore components in γ-radiation resistance has not been systematically evaluated, although neither α/β-type nor γ-type SASP are involved in γ-radiation resistance (Setlow 1988a).

One major deficiency in our knowledge of spore γ-radiation resistance is the precise nature of the γ-radiation damage that kills spores. This is presumably damage to DNA, and much is known about the spectrum of DNA damage induced by γ-radiation in growing cells. However, because of the different environment inside a spore (i.e. reduced hydration) some types of γ-radiation damage (i.e. hydroxyl-radical generation) might be low in spores. Cleary, this is an area which requires further detailed investigation.

5.2 u.v.-Radiation resistance

Spores of *Bacillus* species are 7–50 times more resistant to u.v. radiation at 254 nm than are the corresponding vegetative cells (Setlow 1988b, 1992; Table 2). Spore u.v. resistance is acquired ~90 min before spore core dehydration and in parallel with synthesis of α/β-type SASP. The α/β-type SASPs are essential for spore u.v. resistance, although neither the spore coats nor cortex are required. The major reason for spore u.v. resistance is that u.v. irradiation of spores generates few cyclobutane-type pyrimidine dimers in DNA (Setlow 1988b, 1992). Such dimers include those between adjacent thymine residues (TT; Fig. 3), as well as between adjacent cytosine and thymine residues (CT) and probably dimers between adjacent cytosine residues (CC). Some 6–4 photoproducts may be formed in spore DNA *in vivo* by u.v. irradiation, but their yield as a function of u.v. fluence may be much lower in spores compared to that in vegetative cells (Fairhead and Setlow 1992). The major u.v. photoproduct formed in spore DNA is a thyminyl–thymine adduct termed spore photoproduct (SP; Fig. 3). SP is formed in spore DNA with about the same yield as a function of u.v. fluence as is TT in vegetative cell DNA. However, spores have several repair mechanisms that eliminate SP in the first minutes of spore germination in a process much less error prone than is TT repair. One of the SP repair processes is unique to SP and involves the monomerization of SP back to two thymines. The enzyme which appears responsible for this latter reaction is a spore specific gene product, with some sequence homology to DNA photolyases which use light to monomerize TT (Fajardo-Cavazos *et al.* 1993). However, SP monomerization does not require light.

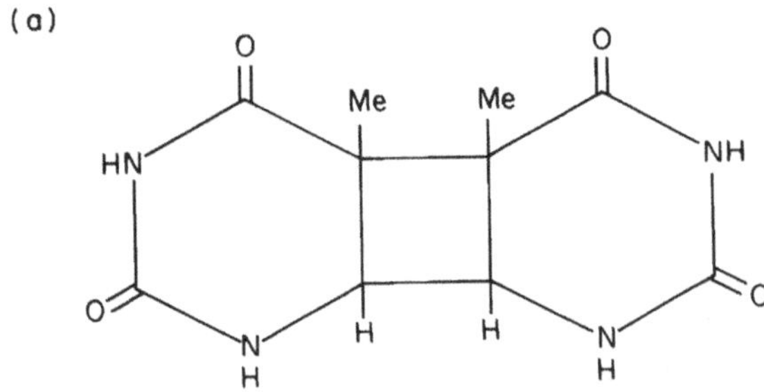

(b)

Fig. 3 Structure of cyclobutane-type thymine dimer (TT) (a) and thyminyl–thymine adduct or spore photoproduct (SP) (b). Me denotes a methyl group

The major factor responsible for the u.v. photochemistry of spore DNA is the presence of α/β-type SASPs which saturate the spore chromosome (Setlow 1988a, b, 1992). These small proteins (60–72 residues) have a highly conserved sequence, both within and across *Bacillus* species (Fig. 4). The sequences of these proteins have also been highly conserved in *Clostridium*, *Sporosarcina* and *Thermoactinomyces* species as well, but show no significant homology to any other protein in available databases (Setlow 1988a; Magill *et al.* 1990; Cabrera-Martinez and Setlow 1991). The α/β-type SASPs are double-stranded DNA binding proteins which bind around the outside of a DNA helix, primarily by interaction with the backbone (Setlow *et al.* 1992). This binding causes a change in the DNA conformation from a B-like helix in protein-free DNA, to an A-like helix in an α/β-type SASP/DNA complex. The properties of DNA in the latter complex are quite different from those of protein-free DNA, as DNA in the complex is completely nuclease resistant, and exhibits altered topology. Strikingly, binding of any wild-type α/β-type SASP to DNA suppresses u.v. induced formation of CC, CT, TT and 6–4 photoproducts, and promotes formation of SP (Fairhead and Setlow 1992; Setlow 1992). The yield of SP as a function of u.v. fluence from an α/β-type SASP/DNA complex *in vitro* is $\sim$10-fold lower than that in intact spores. This difference appears due to the high level of DPA in spores, which sensitizes DNA for SP formation both *in vivo* and *in vitro* (Setlow and Setlow 1993a).

As predicted from the *in vitro* results noted above, *B. subtilis* spores which lack major α/β-type SASP are extremely u.v. sensitive—even more so than are vegetative cells (Setlow 1992; Table 2). Ultraviolet irradiation of $\alpha^-\beta^-$ spores generates reduced amounts of SP and significant levels of TT; the production of TT in $\alpha^-\beta^-$ spores is the reason for their u.v. sensitivity. However, the u.v. resistance of $\alpha^-\beta^-$ spores can be restored to wild-type spore levels by synthesis of a sufficient amount of any α/β-type SASP—even one from a different species. Mutant α/β-type SASPs which no longer bind DNA are ineffective in this regard.

As noted above, spore u.v. resistance is acquired in parallel with α/β-type SASP. At this time the forespores are more u.v. resistant than the dormant spore, probably because the DPA acquired later in sporulation sensitizes DNA to u.v. (Setlow and Setlow 1993a). During spore germination, DPA is lost well before SASP degradation. Consequently, in the first minutes of germination, spores became more u.v. resistant than dormant spores (Setlow 1988b). This elevated u.v. resistance is lost as α/β-type SASPs are degraded. If α/β-type SASP degradation during germination is slowed by a mutation in GPR, the hyper-u.v. resistant state persists much longer during spore germination (Sanchez-Salas *et al.* 1992).

6. SPORE HEAT RESISTANCE

One of the most striking properties of bacterial spores is their extreme resistance to heat: spores of some *Bacillus*

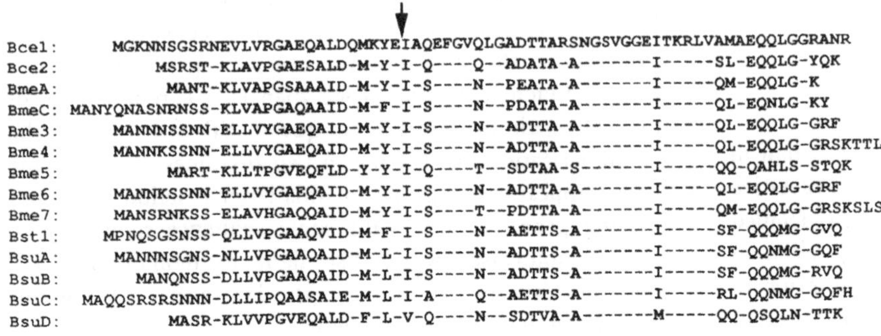

Fig. 4 Comparison of amino acid sequences of α/β-type SASPs from different *Bacillus* species. Amino acids are given in the one letter code and data are from Setlow (1988a). At positions denoted by dashes, the residue present is identical to that in Bce1. The *Bacillus* species represented are: Bce—*B. cereus*; Bme—*B. megaterium*; Bst—*B. stearothermophilus*; and Bsu—*B. subtilis*. The vertical arrow denotes the GPR cleavage site

species can withstand 100°C for many minutes (Russell 1982; Gerhardt and Marquis 1989; Setlow 1992). Heat resistance can be quantitated as a D_x value, defined as the time needed to kill 90% of a cell or spore population at a temperature of x. In general, the D values for spores at a temperature of X°C + 40°C (where X is some other temperature) are approximately equivalent to those for their growing cell counterparts at a temperature of X°C. One feature of spore heat resistance which is often not appreciated, is that their extended survival at elevated temperatures is paralleled by even longer survival times at lower temperatures: spore D values increase about 10-fold for each 10°C fall in temperature (Russell 1982). Thus spores may have a D_{90}° of 30 min and a D_{30}° of many years. Consequently, the mechanisms ensuring spore survival at high temperatures may be the same as those ensuring long term spore survival at more moderate temperatures. Indeed, the latter may have been generally more important in the evolution of spore heat resistance.

At present, the precise mechanism whereby heat kills wild-type spores is not clear. There is evidence suggesting that DNA damage is generally not the mechanism for heat-killing; similarly, there is evidence suggesting that proteins are the target for heat killing (Belliveau *et al.* 1992). However, the nature of this protein, whether a membrane protein or germination enzyme, is not clear; nor is it known whether the protein damage observed upon heat killing is the cause or the effect of heat killing. Because of this lack of knowledge, informed speculation on possible repair of heat damage to spores is difficult. Although our knowledge of the mechanisms by which heat kills spores is limited, much more is known about the factors which increase spore heat resistance. As is probably not surprising, there appear to be a number of factors which contribute to spore heat resistance and thus to spore longevity.

6.1 Spore preparation temperature

Spores of thermophiles are much more heat resistant than are spores of mesophiles. Similarly, the temperature of sporulation influences spore heat resistance, with spores prepared at higher temperatures being generally more heat resistant than those of the same strain prepared at lower temperatures (Gerhardt and Marquis 1989). While all factors causing these results are not known, a major one appears to be that spores prepared at a higher temperature are somewhat more dehydrated (see Section 6.4). However, the mechanism for determining the precise degree of spore core dehydration is not known.

It has been suggested numerous times that the proteins of the heat shock regulon may be involved in spore heat resistance. Indeed, the levels of these proteins might be expected to increase as the sporulation temperature of a strain is increased. While this remains an attractive idea, and one meriting further study, it is still not clear whether this idea is correct.

6.2 α/β-Type SASPs

The α/β-type SASPs are not the major determinant of spore heat resistance as heat resistance is acquired well after α/β-type SASP-synthesis. However, spores lacking α/β-type SASP have D values at any given temperature which are 10–20-fold lower than those of wild-type spores (Fairhead *et al.* 1993; Table 2). Even at cold room temperatures, where wild-type *B. subtilis* spores exhibit no loss in viability over 6 months, $\alpha^-\beta^-$ spores lose almost one log of viability. In general, heat treatment of fully hydrated wild-type spores does not induce mutations (Setlow 1992). The only exception may be in the *gly* region of the chromosome. However, ~20% of the survivors of heat treatment of $\alpha^-\beta^-$ spores carry a variety of mutations, including ones causing auxotrophy, asporogeny and altered colony morphology (Fairhead *et al.* 1993). This suggests that the increased heat sensitivity of $\alpha^-\beta^-$ spores is due to the increased sensitivity of $\alpha^-\beta^-$ spore DNA to heat, and that heat-induced DNA damage is killing these spores. Presumably in wild-type spores α/β-type SASPs protect spore DNA from this damage, such that wild-type spores survive until some other mechanism kills them. The precise nature of the initial DNA damage caused by heat in $\alpha^-\beta^-$ spores is not clear, but DNA from heat killed $\alpha^-\beta^-$ spores has a high frequency of single-strand breaks (Fairhead *et al.* 1993). The damaging event initiating formation of these single-strand breaks has been suggested to be DNA depurination. This chemical reaction takes place in all cells and with all DNAs, and its known rate *in vitro* is not unreasonable for the rate of single-strand break formation during heating of $\alpha^-\beta^-$ spores. As α/β-type SASPs slow DNA depurination *in vitro* at least 20-fold, the mechanism whereby α/β-type SASPs contribute to spore heat resistance and longevity may be to slow DNA depurination *in vivo*.

6.3 Mineralization

Spore heat resistance is acquired approximately in parallel with or slightly after the dehydration of the spore core, as well as the uptake of the spore's large depot of DPA and divalent cations (Gerhardt and Marquis 1989). As noted above, mutants are available which do not make DPA but which sporulate well. While these DPA$^-$ spores are quite unstable, heat resistant spores can be obtained. This suggests that DPA is dispensable for heat resistance. Similarly, DPA can be removed from dormant spores, but they remain viable and heat resistant. It is, however, clear that divalent cation accumulation, presumably over and above that chelated by DPA, is needed for full spore heat resistance. In general the higher the divalent cation level (i.e. the higher the mineralization) the more heat resistant the spore. Both the amount and the nature of the divalent cation accumulated influence spore heat resistance. Analyses of spores of several species from which all monovalent and divalent cations have been removed by acid titration and then restored by back-titration with the appropriate hydroxide, give the following order of increasing spore heat resistance given by different cations: $H^+ < Na^+ < K^+ < Mg^{2+} < Mn^{2+} < Ca^{2+}$ = untreated.

While spore mineralization can have a significant effect on spore heat resistance, the mechanism of this effect is not clear (Gerhardt and Marquis 1989). It might be due to removal of free water which is sequestered in solvating divalent cations. Alternatively, the divalent cations may bind to and stabilize spore macromolecules, possibly by immobilizing them. Similarly, while the great majority of spore cations are in the spore core, the precise mechanism(s) for their uptake late in sporulation are not clear. However, most of this divalent cation load is excreted in the first minutes of spore germination, when heat resistance is lost.

6.4 Water content

The most important factor in determining spore heat resistance is the reduced water content in the spore core (Gerhardt and Marquis 1989). This reduction in spore core water takes place approximately in parallel with the acquisition of heat resistance. Spore heat resistance, and presumbly spore core dehydration, does not require spore coats. While a functional spore cortex is needed for spore core dehydration, neither the mechanism of the dehydration, nor the precise role of the cortex is clear. It seems likely, however, that some mechanical property of the cortex is involved—certainly in restricting a massive influx of water into the spore core. In general, there is a good inverse correlation between spore cortex thickness and core water content.

As noted in previous sections, the reduced spore core water content undoubtedly is the major reason for the enzymatic dormancy of spores, and may be involved in spore chemical and γ-radiation resistance as well. The spore core dehydration may also ensure complete α/β-type SASP/DNA interaction by mass action (Sanchez-Salas *et al.* 1992). There is generally an excellent correlation between spore heat resistance and spore core dehydration, as has been shown by Gerhardt and coworkers using a variety of approaches to measure spore core water content (Gerhardt and Marquis 1989). While these latter studies have given reasonable values for spore core water per gram dry weight, values for free water in the spore core have not been determined. The amount of water per gram dry weight in the spore core does vary somewhat between species, being generally higher in less heat resistant spores. How much of this spore core water, if any, is free water is not clear.

It seems likely that spore core dehydration causes heat resistance and thus long-term survival, by slowing chemical reactions (such as DNA depurination, protein deamidation, etc.) and stabilizing macromolecules against denaturation. The latter type of stabilization has been observed *in vitro* by dehydration, and is due to restriction of macromolecular motion. However, given the lack of quantitative data on free water levels (if any) in spores, as well as the amount of water associated with various macromolecules, a quantitative assessment of the effect of dehydration on spore heat resistance is not presently possible.

7. CONCLUSIONS

One of the most striking features of spores of various *Bacillus* species is their survival for long periods, often up to years, in aqueous environments which lack nutrients. Probably a corollary of this ability to survive long periods without nutrients is that spores lack common high energy compounds such as ATP. As a consequence of the latter, spores cannot use mechanisms available to cells to protect against various damaging agents, i.e. radiation damage to DNA cannot be repaired in dormant spores, but accumulates and could overwhelm repair systems operative in the first minute of spore germination. Therefore, spores utilize a variety of other mechanisms to protect against damage by chemicals and radiation. These include strong permeability barriers which greatly slow access of chemical agents to the spore core, and synthesis of a novel group of DNA-binding proteins which protect spore DNA from damage by free radicals and u.v. radiation. For some of these latter mechanisms we are achieving a rather detailed understanding of how they work.

In order for spores to survive for long periods of time, chemical and enzymatic reactions which normally go on in cells must be slowed almost completely. In the case of DNA depurination, this process appears slowed in large part by the binding of the same group of DNA binding proteins which provide free radical and u.v. resistance. However, the major mechanism to slow reactions in spores, not simply enzyme reactions, but also protein denaturation (very likely a major mechanism for heat killing of organisms), is the relative dehydration of the spore core. The mineralization which accompanies spore core dehydration also appears involved in this process, but dehydration is the major player. While we have a fairly good idea of the relationship between spore water content and spore survival, precise values for water distribution in the spore core, i.e. free *vs* bound water, are lacking. Similarly, the precise mechanism for bringing about and maintaining this dehydration, and then reversing this process during spore germination are still not clear. This latter area remains one of the major mechanistic questions concerning spore survival and resistance.

8. ACKNOWLEDGEMENT

Work in the author's laboratory has received generous support from the Army Research Office and the National Institutes of Health (GM-19698).

9. REFERENCES

Belliveau, B.H., Beaman, T.C., Pankratz, S. and Gerhardt, P. (1992) Heat killing of bacterial spores analyzed by differential scanning calorimetry. *Journal of Bacteriology* 174, 4463–4474.

Cabrera-Martinez, R.M. and Setlow, P. (1991) Cloning and nucleotide sequence of three genes coding for small, acid-soluble proteins of *Clostridium perfringens* spores. *FEMS Microbiology Letters* 77, 127–132.

Decker, S.J. and Lang, D.R. (1978) Membrane bioenergetic parameters in uncoupler-resistant mutants of *Bacillus megaterium*. *Journal of Biological Chemistry* 253, 6738–6743.

Fairhead, H. and Setlow, P. (1992) Binding of DNA to α/β-type small, acid-soluble proteins from spores of *Bacillus* or *Clostridium* species prevents formation of cytosine dimers, cytosine–thymine dimers and bipyrimidine photoadducts upon ultraviolet irradiation. *Journal of Bacteriology* 174, 2874–2880.

Fairhead, H., Setlow, B. and Setlow, P. (1993) Prevention of DNA damage in spores and *in vitro* by small, acid-soluble proteins from *Bacillus* species. *Journal of Bacteriology* 175, 1367–1374.

Fajardo-Cavazos, P., Salazar, C. and Nicholson, W.L. (1993) Molecular cloning and characterization of the *Bacillus subtilis* spore photoproduct lyase (*spl*) gene, which is involved in repair of ultraviolet radiation-induced DNA damage during spore germination. *Journal of Bacteriology* 175, 1735–1744.

Foster, S.J. and Johnstone, K. (1989) The trigger mechanism of bacterial spore germination. In *Regulation of Procaryotic Development* ed. Smith, I., Slepecky, R. and Setlow, P. pp. 89–108. Washington, DC: American Society for Microbiology.

Gerhardt, P. and Marquis, R.E. (1989) Spore thermoresistance mechanisms. In *Regulation of Procaryotic Development* ed. Smith, I., Slepecky, R. and Setlow, P. pp. 43–63. Washington, DC: American Society for Microbiology.

Gerhardt, P., Scherrer, R. and Black, S.H. (1972) Molecular sieving by dormant spore structures. In *Spores V* ed. Halvorson, H.O., Hanson, R. and Campbell, L.L. pp. 68–74. Washington, DC: American Society for Microbiology.

Gould, G.W. (1969) Germination. In *The Bacterial Spore* ed. Gould, G.W. and Hurst, A. pp. 397–444. London: Academic Press.

Gould, G.W. (1983) Mechanisms of resistance and dormancy. In *The Bacterial Spore*, Vol. 2. ed. Hurst, A. and Gould, G.W. pp. 173–210. London: Academic Press.

Imlay, J.A. and Linn, S. (1988) DNA damage and oxygen radical toxicity. *Science* **240**, 1302–1309.

Kuhn, N.J., Setlow, B. and Setlow, P. (1993) Manganese(II) activation of 3-phosphoglycerate mutase of *Bacillus megaterium*: pH-sensitive interconversion of active and inactive forms. *Archives of Biochemistry and Biophysics* **306**, 342–349.

Loshon, C.A. and Setlow, P. (1993) Levels of small molecules in dormant spores of *Sporosarcina* species and comparison with levels in spores of *Bacillus* and *Clostridium* species. *Canadian Journal of Microbiology* 39, 259–262.

Magill, N.G., Loshon, C.A. and Setlow, P. (1990) Small, acid-soluble, spore proteins and their genes from two species of *Sporosarcina*. *FEMS Microbiology Letters* 72, 293–298.

Murrell, W.G. (1981) Biophysical studies on the molecular mechanisms of spore heat resistance and dormancy. In *Sporulation and Germination* ed. Levinson, H.S., Sonenshein, A.L. and Tipper, D.J. pp. 64–77. Washington, DC: American Society for Microbiology.

Russell, A.D. (1982) *The Destruction of Bacterial Spores*. London: Academic Press.

Sanchez-Salas, J.-L. and Setlow, P. (1993) Proteolytic processing of the protease which initiates degradation of small, acid-soluble, proteins during germination of *Bacillus subtilis* spores. *Journal of Bacteriology* 175, 2568–2577.

Sanchez-Salas, J.-L., Santiago-Lara, M.L., Setlow, B., Sussman, M.D. and Setlow, P. (1992) Properties of mutants of *Bacillus megaterium* and *Bacillus subtilis* which lack the protease that degrades small, acid-soluble proteins during spore germination. *Journal of Bacteriology* 174, 807–814.

Setlow, B. and Setlow, P. (1980) Measurements of the pH within dormant and germinated bacterial spores. *Proceedings of the National Academy of Sciences, USA* 77, 2474–2476.

Setlow, B. and Setlow, P. (1993a) Dipicolinic acid greatly enhances the production of spore photoproduct in bacterial spores upon ultraviolet irradiation. *Applied and Environmental Microbiology* 59, 640–643.

Setlow, B. and Setlow, P. (1993b) Binding of small, acid-soluble spore proteins to DNA plays a significant role in the resistance of *Bacillus subtilis* spores to hydrogen peroxide. *Applied and Environmental Microbiology* 59, 3418–3423.

Setlow, B., Sun, D. and Setlow, P. (1992) Studies of the interaction between DNA and α/β-type small, acid-soluble spore proteins: a new class of DNA binding protein. *Journal of Bacteriology* 174, 2312–2322.

Setlow, P. (1983) Germination and outgrowth. In *The Bacterial Spore*, Vol. 2. ed. Hurst, A. and Gould, G.W. pp. 211–254. London: Academic Press.

Setlow, P. (1988a) Small acid-soluble, spore proteins of *Bacillus* species: structure, synthesis, genetics, function and degradation. *Annual Reviews of Microbiology* **42**, 319–338.

Setlow, P. (1988b) Resistance of bacterial spores to ultraviolet light. *Comments on Molecular and Cellular Biophysics* **5**, 253–264.

Setlow, P. (1992) I will survive: protecting and repairing spore DNA. *Journal of Bacteriology* **174**, 2737–2741.

Shioi, J.-I., Matsura, S. and Imae, Y. (1980) Quantitative measurements of proton motive force and motility in *Bacillus subtilis*. *Journal of Bacteriology* **144**, 891–897.

Singh, R.P. and Setlow, P. (1979) Purification and properties of phosphoglycerate phosphomutase from spores and cells of *Bacillus megaterium*. *Journal of Bacteriology* 137, 1024–1027.

Slepecky, R.A. and Leadbetter, E.R. (1983) On the prevalence and roles of spore-forming bacteria and their spores in nature. In *The Bacterial Spore*, Vol. 2. ed. Hurst, A. and Gould, G.W. pp. 79–101. London: Academic Press.

Swerdlow, B.M., Setlow, B. and Setlow, P. (1981) Levels of H^+ and other monovalent cations in dormant and germinated spores of *Bacillus megaterium*. *Journal of Bacteriology* **148**, 20–29.

Watabe, K. and Freese, E. (1979) Purification and properties of the manganese-dependent phosphoglycerate mutase of *Bacillus subtilis*. *Journal of Bacteriology* **137**, 773–778.

Journal of Applied Bacteriology Symposium Supplement 1994, **76**, 61S–66S

Bacillus cereus and its toxins

P.E. Granum
Department of Pharmacology, Microbiology and Food Hygiene, Norwegian College of Veterinary Medicine, Oslo, Norway

1. Introduction, 61S
2. The different *Bacillus cereus* toxins
   2.1 Haemolysins, 61S
   2.2 Phospholipases C, 62S
   2.3 Enterotoxin, 62S
   2.4 Emetic toxin, 64S
3. Significance of *Bacillus cereus* as a food poisoning organism, 65S
4. Conclusions, 65S
5. References, 66S

1. INTRODUCTION

Bacillus cereus is a Gram-positive spore-forming mobile aerobic rod, but grows well anaerobically. It is a common soil saprophyte and has been recognized as an opportunistic pathogen of increasing importance (Turnbull 1986; Kramer and Gilbert 1989). *Bacillus cereus* may be isolated from several types of foods, such as rice, spices, meat, eggs and dairy products (Kramer and Gilbert 1989), and from drugs including oral pharmaceutical products (Arribas *et al.* 1988). *Bacillus cereus* has until recently attracted little attention among clinical microbiologists because it has been presumed to possess a low degree of pathogenicity. The bacterium is known to cause several different types of infection/intoxication. One of the least known is the *B. cereus* ocular infections, where it may cause tissue destruction (O'Day *et al.* 1981). In the ocular infections several enzymes and toxins may be involved, but such enzymes will not be dealt with in this review. *Bacillus cereus* is also the cause of two different types of food poisoning: the emetic type and the diarrhoeal type. The emetic syndrome is caused by a non-protein heat stable component (lipid?), most probably a product made from the growth medium through enzymatic degradation and/or modification. The toxin has a molecular weight of about 5–7 kDa (Turnbull 1986; Kramer and Gilbert 1989; Shinagawa *et al.* 1992). The diarrhoeal type food poisoning is caused by an enterotoxin consisting of one polypeptide chain, with a mature molecular weight of about 40 kDa (Shinagawa *et al.* 1991; Granum and Nissen 1993; Granum *et al.* 1993b).

Bacillus cereus produces a variety of other toxins, including two haemolysins and three phospholipase C proteins (Turnbull 1986). One of the phospholipases, sphingomyelinase is also a haemolysin (Tomita *et al.* 1991). These proteins are, however, not considered to be a major cause of disease, in contrast to what has been found for the phospholipases for other bacteria (for review see Titball 1993). This paper will give a brief survey of the toxins of *B. cereus* with the main emphasis on the diarrhoeal enterotoxin.

Correspondence to: Prof. P.E. Granum, Department of Pharmacology, Microbiology and Food Hygiene, Norwegian College of Veterinary Medicine, PO Box 8146, Dep., 0033 Oslo, Norway. Address to 1 July 1994: *Imperial Cancer Research Fund, University of Oxford, Institute of Molecular Medicine, John Radcliffe Hospital, Oxford OX3 9DU, UK.*

2. THE DIFFERENT *BACILLUS CEREUS* TOXINS

Seven different toxins produced by *B. cereus* are listed in Table 1. Six of these toxins are produced during vegetative growth and secreted by the cells. Thus all of these toxins have lost a signal peptide on their way out of the cells, resulting in mature proteins somewhat smaller than those transcribed. The seventh toxin, the emetic, is most probably produced from components in foods during growth of *B. cereus*. The seven toxins can be divided into four groups as shown in Table 1: enterotoxin, haemolysins, phospholipase C and emetic toxin.

2.1 Haemolysins

Three different haemolysins are known from *B. cereus* and have been characterized quite well (Table 1). Unfortunately only the sphingomyelinase has been cloned and sequenced (Johansen *et al.* 1988). The cereolysin is a thiol activated protein that cross-reacts with streptolysin-O and has a calculated molecular weight of about 55 kDa (518 amino acids) (Cowell *et al.* 1976). It is heat labile but relatively little susceptible to proteolysis (Bernheimer and Grushoff 1967; Cowell *et al.* 1976). It is responsible for the main haemolysis of *B. cereus*, is lethal when injected in mice and is inhibited by cholesterol and serum. Less is known about the secondary haemolysin. It has a molecular weight of about 30 kDa, is heat labile and is easily degraded by proteolytic enzymes (Coolbaugh and Williams 1978). It is not

Table 1 The toxins of *Bacillus cereus*

Toxin	Molecular weight (mature molecule)	Signal peptide size	Characteristics
Enterotoxin	≈40 kDa	nd	Membrane damage, heat labile, susceptible to proteolysis
Haemolysins			
Cereolysin	≈56 kDa	nd	Thiol activated, heat labile, little susceptible to proteolysis
Haemolysin II	≈30 kDa	nd	Heat labile, susceptible to proteolysis
Sphingomyelinase	34 kDa	27 aa	See below
Phospholipase C			
Phosphatidylinositol hydrolase (PIH)	34 kDa	31 aa	Non-metallo enzyme, sequence homology to other pro- and eukaryotic enzymes
Phosphatidylcholine hydrolase (PCH)	27 kDa	38 aa	Stable metallo-enzyme (Zn^{2+}, Ca^{2+})
Sphingomyelinase	34 kDa	27 aa	Stable metallo-enzyme (Mg^{2+}), haemolysin
Emetic toxin	5–7 kDa		Heat stable to 121°C, non-metabolic product, lipid (?)

nd, Not determined.

inhibited by cholesterol. Sphingomyelinase is well characterized, and is most probably responsible for the heat stable haemolysis (for review see Tomita *et al.* 1991). It is a metallo-enzyme (Mg^{2+}) and works best as a haemolysin through a so-called hot–cold incubation (37–4°C). Obviously the haemolytic activity is highest on erythrocytes containing high amounts of sphingomyelin such as sheep, human and rabbit (for molecular detail see Section 2.2).

Several papers have described other multicomponent haemolysins, but at present there is no proof for other haemolysins than those described. It is, however, possible to observe several types of effects by mixing different cyto-active proteins. It has also been questioned whether the enterotoxin possesses haemolytic activity, but it is now quite clear that it does not (Shinagawa *et al.* 1991; Granum and Nissen 1993).

2.2 Phospholipases C

Bacillus cereus produces three different phospholipases C (Table 1). They have all been cloned and are well characterized. Phosphatidylinositol hydrolase (PIH) is a protein that specifically hydrolyses phosphatidylinositol (PI) and PI-glycan-containing membrane anchors, which are important structural components of one class of membrane proteins. The enzyme is synthesized as a 329 amino acids protein with a signal peptide of 31 amino acids, resulting in a mature enzyme of about 34 kDa (Kuppe *et al.* 1989). This enzyme is not genetically linked to the two other phospholipases C, and does not require any ions for biological activity (Kuppe *et al.* 1989). PIH has no haemolytic activity.

Phosphatidylcholine hydrolase (PCH) hydrolyses phosphatidycholine, phosphatidylethanolamine and phosphatidylserine (Little *et al.* 1975; Otnæss *et al.* 1977) while sphingomyelinase (SM) only hydrolyses sphingomyelin (Ikezawa *et al.* 1986). These two enzymes are genetically linked and the two genes are only separated by 79 bases, just leaving enough space for a promoter sequence (RNA polymerase binding sites) and a ribosomal binding site (Gilmore *et al.* 1989). PCH is synthesized as a 283 amino acid protein leaving 245 amino acids in the mature protein. SM is somewhat larger with 333 and 306 amino acids in the synthesized and the mature protein respectively (Gilmore *et al.* 1989). Both enzymes require divalent cations: PCH requires either Zn^{2+} or Ca^{2+}, while SM needs Mg^{2+}, and is inhibited by Zn^{2+} and Ca^{2+} (Tomita *et al.* 1991; Titball 1993). The haemolytic activity of SM has been described in the previous section.

2.3 Enterotoxin

The enterotoxin from *B. cereus* is responsible for the food poisoning symptoms appearing after an incubation period of about 8–16 h: the onset will start with abdominal pain and watery diarrhoea accompanied by rectal spasms and nausea, usually lasting for about 12–24 h. The symptoms do not normally include vomiting and fever. It has recently been shown that the enterotoxin is a single protein without haemolytic activity (Shinagawa *et al.* 1991; Granum and Nissen 1993). The haemolytic activity frequently found in enterotoxin samples is due to small amounts of sphingomyelinase (Granum and Nissen 1993). The enterotoxin is a cytotoxic (membrane disrupting) protein of about 40 kDa (Granum *et al.* 1993b). It has not been clear whether or not the food poisoning is due to preformed enterotoxin or whether *B. cereus* cells are ingested, and the enterotoxin produced in the small intestine. It is however known that the enterotoxin is unstable at pH <4, and that it is degraded by pepsin, trypsin and chymotrypsin (Kramer and Gilbert 1989). It has also recently been shown that *B. cereus* strains can grow quite well anaerobically, and produce enterotoxin at levels comparable to those produced aerobically (Granum *et al.* 1993b). A model experiment of the fate of the enterotoxin passing through the stomach and duodenum has been carried out, and the result was quite clear: no detectable enterotoxin activity was found after exposure to low pH and proteolytic enzymes (Granum *et al.* 1993b). That experiment could, however, be criticized as no food components were added. The experiment was therefore repeated using a 1 : 1 mixture of a crude enterotoxin extract and milk. The results are given in Table 2. It is quite clear that milk does not protect the enterotoxin from degradation. From these experiments and other published data it is concluded that it is likely that the *B. cereus* food poisoning is caused by ingestion of cells or spores rather than by preformed enterotoxin. It must also be pointed out that the total infective dose of *B. cereus* is in the range of about 10^5–10^8 cells (Kramer and Gilbert 1989; Granum *et al.* 1993b). Enterotoxin production is, however, hard to detect before *B. cereus* has reached 10^7 cells ml^{-1} in cultures growing under optimal conditions (Kramer and Gilbert 1989). This would also support the author's view that *B. cereus* food poisoning is caused by ingestion of cells (or spores) and not by preformed enterotoxin. Accepting this assumption, the food industry should be concerned about levels as low as 10^3–10^4 cells ml^{-1} (or cells g^{-1}) of *B. cereus* in any product. The level of enterotoxin produced by *B. cereus* strains varies with a factor of more than 100 (Granum, unpublished results). From the author's experience with *B. cereus* food poisoning in Norway, only high enterotoxin producing strains seem to have caused food poisoning. It is therefore quite possible that only a few of even the enterotoxigenic strains are of public health importance.

The author and coworkers have studied the mode of action of the *B. cereus* enterotoxin using Vero cells as the target [for experimental details see Granum *et al.* (1993b) and figure legends]. These studies gave some surprising results. First of all binding to the target cells is partly

Table 2 A model experiment of the fate of *Bacillus cereus* enterotoxin if preformed enterotoxin is ingested. The activity of the enterotoxin on Vero cells is given [for experimental details see Granum *et al.* (1993b)]

Extract tested	Relative activity (%)	
	Experiment I* without milk	Experiment II† with milk
Supernatant after cell growth in BHIG‡ for 6 h at 32°C, concentrated 10-fold with $(NH_4)_2SO_4$ and dialysed against 5 mmol l^{-1} Tris buffer pH 8·2	100	100
pH adjusted to 3 with HCl, incubated for 20 min at 37°C	20	—
pH adjusted to 2·8 with HCl, incubated for 20 min at 37°C with pepsin (1 mg ml^{-1})	—	12
pH adjusted to 7.6 with bicarbonate	18	10
Incubation with trypsin and chymotrypsin (1 mg ml^{-1}) at 37°C for:		
10 min	5	2
30 min	1	—
60 min	ND§	—

— Not tested.
* From Granum *et al.* (1993b).
† The enterotoxin extract was mixed with milk in a 1 : 1 ratio.
‡ BHIG: brain-heart infusion medium with 1% (v/w) extra glucose.
§ ND: not detectable.

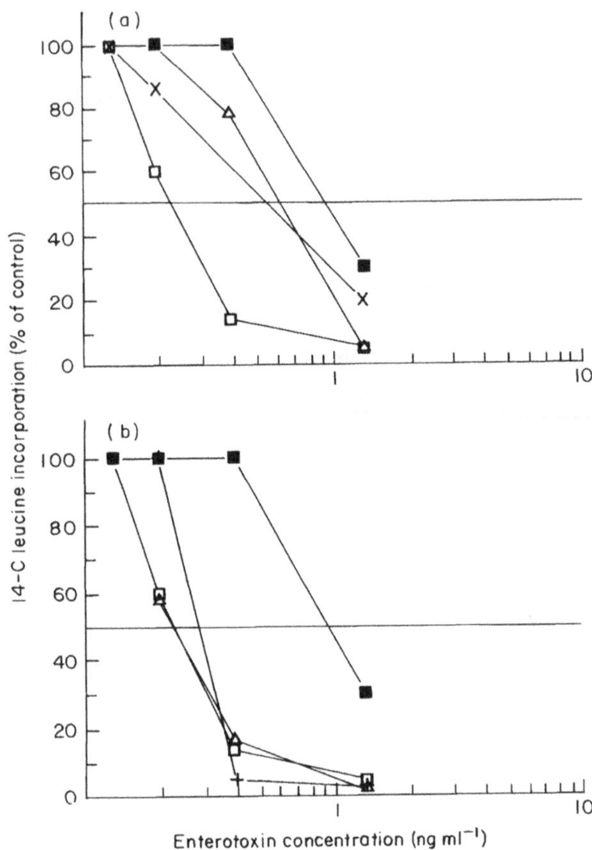

Fig. 1 The influence of Ca^{2+} and Mg^{2+} on the binding of purified enterotoxin from *Bacillus cereus* (Granum *et al.* 1993b) to Vero cells. The cells were incubated with or without ions (EDTA) for 15 min before purified enterotoxin was added at different concentrations. The cells were incubated for 1 h. Unbound enterotoxin was then washed off with HEPES buffer. Leucine-free medium was then added and the cells allowed to grow for 18 h. After addition of ^{14}C-leucine (1 μl) for 15 min the procedure was as described by Granum *et al.* (1993b). (a) □, No ions (10 mmol l^{-1} EDTA); ×, 2 mmol l^{-1} Ca^{2+}; △, 5 mmol l^{-1} Ca^{2+}; ■, 2 mmol l^{-1} Ca^{2+} and 10 mmol l^{-1} Mg^{2+}. (b) □, No ions (10 mmol l^{-1} EDTA); ×, 2 mmol l^{-1} Mg^{2+}; △, 10 mmol l^{-1} Mg^{2+}; ■, 2 mmol l^{-1} Ca^{2+} and 10 mmol l^{-1} Mg^{2+}

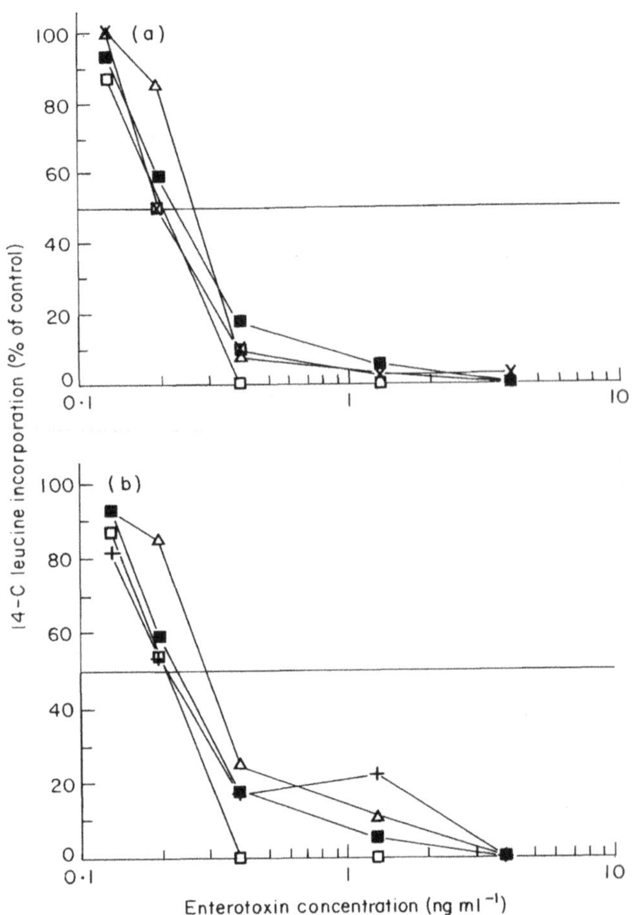

Fig. 2 The influence of Ca^{2+} and Mg^{2+} on inhibition of protein synthesis by purified enterotoxin from *Bacillus cereus* on Vero cells. The procedure was as described by Granum *et al.* (1993b). (a) □, No ions (10 mmol l^{-1} EDTA); ×, 2 mmol l^{-1} Ca^{2+}; △, 5 mmol l^{-1} Ca^{2+}; ■, 2 mmol l^{-1} Ca^{2+} and 10 mmol l^{-1} Mg^{2+}. (b) □, No ions (10 mmol l^{-1} EDTA); ×, 2 mmol l^{-1} Mg^{2+}; △, 10 mmol l^{-1} Mg^{2+}; ■, 2 mmol l^{-1} Ca^{2+} and 10 mmol l^{-1} Mg^{2+}

inhibited by Ca^{2+} ions, while Mg^{2+} ions seem to have little or no effect on binding (Fig. 1a and b). The combination of the two ions is, however, more inhibitory than Ca^{2+} ions by themselves. When the same experiments were carried out just to measure the inhibition of the protein synthesis, no dramatic changes were observed, using different concentrations of Ca^{2+} or Mg^{2+} (Fig. 2 and b). In the third set of experiments the leakage of LDH out of the Vero cells was measured. Cell leakage is the final step in the mode of action of membrane disrupting proteins. These results are shown in Fig. 3a and b. It is obvious that Ca^{2+} and Mg^{2+} ions together inhibit (or delay) the cell leakage to some extent, in hand with the binding studies described above. The cations tested separately did not have any dramatic effect on cell leakage, although complete cell lysis seemed difficult to achieve when no cations were present (EDTA, Fig. 3). The most marked difference between the mode of action of the *B. cereus* and *Clostridium perfringens* enterotoxins, causing similar types of food poisoning, is that Ca^{2+} ions are necessary for the latter enterotoxin to cause cell leakage. The *B. cereus* enterotoxin is also at least 100 times more toxic than the enterotoxin from *Cl. perfringens* (Granum *et al.* 1993b). More detailed studies must be carried out before any speculation on how the cations influence the mode of action of the *B. cereus* enterotoxin.

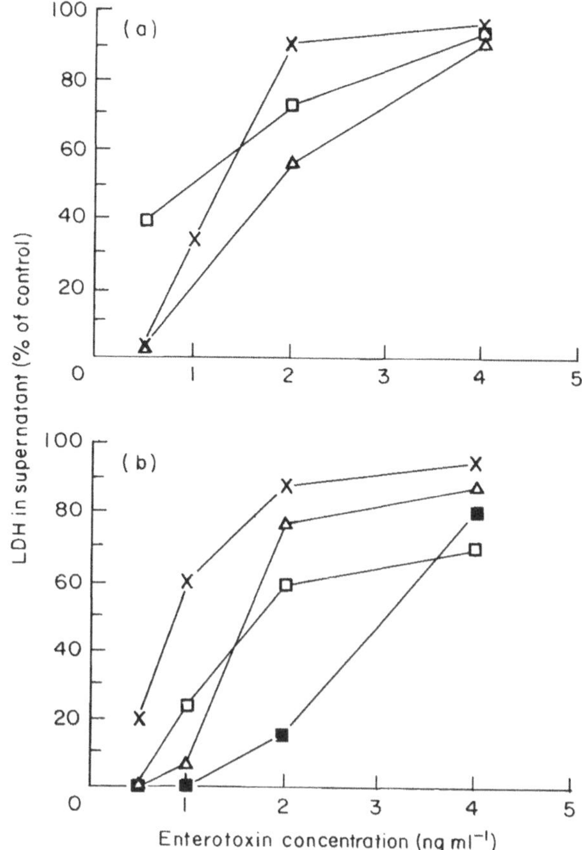

Fig. 3 The influence of Ca^{2+} and Mg^{2+} on cell leakage of LDH from Vero cells, after 60 min of enterotoxin incubation. The procedure was as described by Granum *et al.* (1993b). (a) □, No Ca^{2+} (10 mmol l^{-1} EGTA); ×, 2 mmol l^{-1} Ca^{2+}; △, 5 mmol l^{-1} Ca^{2+}. (b) □, No ions (10 mmol l^{-1} EDTA); ×, 2 mmol l^{-1} Mg^{2+}; △, 10 mmol l^{-1} Mg^{2+}; ■, 2 mmol l^{-1} Ca^{2+} and 10 mmol l^{-1} Mg^{2+}

2.4 Emetic toxin

Very little is known about the *B. cereus* emetic toxin, because until recently the only way to test the biological activity of the toxin was by using monkeys (the toxin is specific to primates). Since cell tests now seem to be in hand (Shinagawa *et al.* 1992), more information on this toxin can be expected in the near future. The emetic toxin was first identified in the UK following several incidents associated with the consumption of cooked rice from Chinese restaurants and take-away outlets, in the early seventies (Kramer and Gilbert 1989). The intoxication was described as: a rapid onset (1–5 h) with nausea and vomiting, occasionally accompanied by diarrhoea (enterotoxin?) and lasting for less than 24 h. The toxic substance may derive from the product that *B. cereus* is growing in, by enzymatic modification, but this is still not known. Outbreaks of emetic *B. cereus* food poisoning include rice, pasta and milk products. The emetic toxin is very stable and is not affected by heat (121°C), proteolysis or pH values from 2–11. The toxin is now believed to be a lipid (?) with a molecular weight in the order of 5–7 kDa (Shinagawa *et al.* 1992). Until further research has been carried out on the emetic toxin, we can only speculate regarding production, structure and mode of action.

3. SIGNIFICANCE OF *BACILLUS CEREUS* AS A FOOD POISONING ORGANISM

Since the diarrhoeal food poisoning caused by *B. cereus* was first described by Hauge (1950) and the emetic type by the Public Health Laboratory Service in 1972 (Kramer and Gilbert 1989), it has become one of the most important causes of food-borne disease in some countries, and it is most probably highly underestimated. In Norway where we are almost free from salmonella, *B. cereus* was the most common cause of food-borne outbreaks in 1990 (Aas *et al.* 1992). We should now be specifically aware of the fact that psychrotrophic strains growing in pasteurized milk products can cause disease (Granum *et al.* 1993a). This kind of food poisoning will most probably not be reported as the cases are spread around in private homes, where products such as milk may not have been stored at temperatures low enough (<4°C) to prevent *B. cereus* growing to a level of about 10^4 ml^{-1} or higher. In recent years some reports on *B. cereus* food poisoning of specific concern have appeared. From Norway comes a report of an outbreak where several people were affected after eating a stew. The infective dose seemed to be about 10^4. Three of the patients were hospitalized, one for three weeks (Granum 1992). For these three patients the onset was late, after more than 24 h. It is quite possible that some strains of *B. cereus* can colonize the small intestine, at least in some patients, and cause a more severe type of disease, by producing the enterotoxin 'on site'. Further research using model experiments, to look closer at this possibility, is in progress.

4. CONCLUSIONS

It has been reported from several countries that at least half of the isolated strains of *B. cereus* are enterotoxin positive (Granum *et al.* 1993a). Only a small number of these strains are probably producing enough enterotoxin to cause food poisoning. Preformed enterotoxin is not important in causing disease, because it is degraded and inactivated on its way to the target. Ingestion of cells or spores (10^4–10^7) are the main course of *B. cereus* food poisoning. A few *B. cereus* strains may have the ability to colonize the small intestine, at least in some people, causing a more severe form of food poisoning. Milk products and products of plant origin are the main source of *B. cereus*. The bacterium may then be transferred to other food products, and may survive, through the spores, in heat-treated product, where competition from other bacteria is usually not present.

5. REFERENCES

Aas, N., Gondrosen, B. and Langeland, G. (1992) *Norwegian Food Control Authority's Report on Food Associated Diseases in 1990*. SNT-report 3, Oslo.

Arribas, M.L.C., Plaza, C.J., de la Rosa, M.C. and Mosso, M.A. (1988) Characterization of *Bacillus cereus* strains isolated from drugs and evaluation of their toxins. *Journal of Applied Bacteriology* **64**, 251–264.

Bernheimer, A.W. and Grushoff, P. (1967) Cereolysin: production, purification and partial characterization. *Journal of General Microbiology* **46**, 143–150.

Coolbaugh, J.C. and Williams, R.P. (1978) Production and characterization of two hemolysins of *Bacillus cereus*. *Canadian Journal of Microbiology* **24**, 1289–1295.

Cowell, J.L., Grushoff-Kosyk, P.S. and Bernheimer, A.W. (1976) Purification of cereolysin and the electrophoretic separation of the active and inactive forms of the purified toxin. *Infection and Immunity* **14**, 144–154.

Gilmore, M.S., Cruz-Rodz, A.L., Leimeister-Wächter, M., Kreft, J. and Goebel, W. (1989) A *Bacillus cereus* cytotoxin determinant, cereolysin AB, which comprises the phospholipase C and sphingomyelinase genes: nucleotide sequence and genetic linkage. *Journal of Bacteriology* **171**, 744–753.

Granum, P.E. (1992) Utbrudd av *Bacillus cereus* matforgiftning i Norge (in Norwegian). *FORUM* **2**, 31.

Granum, P.E. and Nissen, H. (1993) Sphingomyelinase is part of the 'enterotoxin complex' produced by *Bacillus cereus*. *FEMS Microbiology Letters* **110**, 97–100.

Granum, P.E., Brynestad, S. and Kramer, J.M. (1993a) Analysis of enterotoxin production by *Bacillus cereus* from dairy products, food poisoning incidents and non-gastrointestinal infections. *International Journal of Food Microbiology* **17**, 269–279.

Granum, P.E., Brynestad, S., O'Sullivan, K. and Nissen, H. (1993b) The enterotoxin from *Bacillus cereus*: production and biochemical characterization. *Netherlands Milk Dairy Journal* **47**, 63–70.

Hauge, S. (1950) Matforgiftninger framkalt av *Bacillus cereus* (in Norwegian). *Nordisk Hygienisk Tidskrift* **31**, 189–206.

Ikezawa, H., Matsushita, M., Tomita, M. and Taguchi, R. (1986) Effects of metal ions on sphingomyelinase activity of *Bacillus cereus*. *Archives of Biochemistry and Biophysics* **249**, 588–595.

Johansen, T., Holm, T., Guddal, P.H., Sletten, K., Haugli, F.B. and Little, C. (1988) Cloning and sequencing of the gene encoding the phosphatidylcholine-preferring phospholipase C of *Bacillus cereus*. *Gene* **65**, 293–304.

Kramer, J.M. and Gilbert, R.J. (1989) *Bacillus cereus* and other *Bacillus* species. In *Foodborne Bacterial Pathogens* ed. Doyle, M.P. pp. 21–70. New York and Basel: Marcel Dekker.

Kuppe, A., Evans, L.M., McMillen, D.A. and Griffith, O.H. (1989) Phosphatidylinositol-specific phospholipase C of *Bacillus cereus*: cloning sequencing and relationship to other phopholipases. *Journal of Bacteriology* **171**, 6077–6083.

Little, C., Aurebekk, B. and Otnæss, A.-B. (1975) Purification by affinity chromatography of phospholipase C from *Bacillus cereus*. *FEBS Letters* **52**, 175–179.

O'Day, D.M., Ho, P.C., Andrews, J.S., Head, W.S., Ives, J. and Turnbull, P.C.B. (1981) Mechanisms of tissue destruction in ocular *Bacillus cereus* infections. In *The Cornea in Health and Disease* ed. Roper, T. pp. 403–407. New York: Academic Press.

Otnæss, A.-B., Little, C., Sletten, K., Wallin, R., Johansen, S., Flengsrud, R. and Prydz, H. (1977) Some characteristics of phospholipase C from *Bacillus cereus*. *European Journal of Biochemistry* **79**, 459–468.

Shinagawa, K., Ueno, S., Konuma, H., Matsusaka, N. and Sugii, S. (1991) Purification and characterization of the vascular permeability factor produced by *Bacillus cereus*. *Journal of Veterinary Medical Science* **53**, 281–286.

Shinagawa, K., Otake, S., Matsusaka, N. and Sugii, S. (1992) Production of the vacuolation factor of *Bacillus cereus* isolated from vomiting-type food poisoning. *Journal of Veterinary Medical Science* **54**, 443–446.

Titball, R.W. (1993) Bacterial phospholipase C. *Microbiological Reviews* **57**, 347–366.

Tomita, M., Taguchi, R. and Ikezawa, H. (1991) Sphingomyelinase of *Bacillus cereus* as a bacterial hemolysin. *Journal of Toxicology—Toxin Reviews* **10**, 169–207.

Turnbull, P.C.B. (1986) *Bacillus cereus* Toxins. In *Pharmacology of Bacterial Toxins* ed. Dorner, F. and Drews, J. International Encyclopedia of Pharmacology and Therapeutics, Section 119, pp. 397–448. Oxford: Pergamon Press.

Journal of Applied Bacteriology Symposium Supplement 1994, **76**, 67S–80S

Spore resistance and ultra heat treatment processes

K.L. Brown
Food Hygiene Department, Campden Food and Drink Research Association, Chipping Campden, Glos, UK

1. Introduction, 67S
2. High temperature processes
   2.1 Food products, 67S
   2.2 Packaging, 69S
   2.3 Equipment, 69S
3. Methods for determination of spore heat resistance
   3.1 Wet heat resistance methods
      3.1.1 Indirect heating, 70S
      3.1.2 Mixing, 73S
      3.1.3 Particles, 73S
      3.1.4 Direct steam heating, 73S
   3.2 Dry heat resistance methods, 74S
4. Resistance of spores at high temperatures
   4.1 Wet heat resistance, 74S
   4.2 Dry heat resistance, 75S
5. Commercial process requirements, 77S
6. Conclusion, 77S
7. References, 77S

1. INTRODUCTION

In the food industry, ultra-high temperature (UHT) food processes employ heat treatments within the temperature range 130–150°C with holding times of 1 s or more (Burton 1969). Products which have been processed at high temperatures for short periods are generally superior in quality to those processed at lower temperatures for longer times, as in traditional canning (Hersom 1985). The higher the processing temperature the greater the rate of destruction of bacterial spores compared with the rate of destruction of desirable food components.

Coupled with the development of in-line UHT processing, manufacturers have developed a range of aseptic filling systems that use a wide range of containers including laminated paperboard cartons, cans, drums, pots, bottles, bags and even rail cars (Brown and Ayres 1982). Several of the aseptic systems rely on the use of high temperatures to sterilize either the packaging material itself or the filling and packaging machinery. Unlike product processing, where the presence of water results in the exposure of spores to 'wet heating' conditions, many of the methods used to sterilize packaging material or filling and packaging machinery employ dry heat. As described below, bacterial spores are much more resistant to dry heat than to wet heat and for this reason, dry heat sterilization temperatures between 150 and 300°C are commonly used (Holdsworth 1992).

In order to define the processing times and temperatures required for product, packaging and machinery sterilization, a knowledge of the heat resistance characteristics of bacterial spores at high temperatures is necessary.

Correspondence to: K.L. Brown, Food Hygiene Department, Campden Food and Drink Research Association, Chipping Campden, Glos GL55 6LD, UK.

Several methods have been devised for the determination of the heat resistance of bacterial spores but only a limited number are suitable for UHT studies. In this paper examples of the use of high temperatures for food processing and packaging and machinery decontamination will be described, together with examples of the laboratory methods used to determine the heat resistance of bacterial spores under both wet and dry heat conditions at temperatures between 120 and 300°C. In addition, information about spore high temperature resistance will be presented. In this article, the term 'sterilization' is used to describe a heat process which is sufficient to reduce the numbers of bacterial spores to a level which is commercially acceptable (commercial sterilization).

2. HIGH TEMPERATURE PROCESSES

2.1 Food products

Ultra-high temperature sterilization of food products may be divided on the basis of method of heat transfer into two categories, namely indirect and direct heating (Brown and Ayres 1982). In indirect systems the product does not come into direct contact with the heating medium but relies on heat transfer through the surface of a heat exchanger. In direct systems the heating medium is mixed directly with the product, for example, steam-into-product or product-into-steam. Other direct systems use direct heating of the product by electrical, fluidized bed or friction heating.

Temperatures used in indirect systems (Table 1) are generally a few degrees lower than in direct heating systems to avoid product burning onto the surface of the heat exchanger. The simplest form of indirect heating is the

Table 1 Examples of high temperature product sterilization systems

Sterilization system	Equipment	Process temperature (°C)
Indirect		
Tubular	Nuova Frau Sterflux	136–138
	Stork Sterideal	135–150
Plate	Ahlborn	130–140
	Alfa Laval Steritherm	137
	APV Ultramatic	138–140
	Sordi Steriplak	138
Scraped surface	Alfa Laval Contherm	140
Flame sterilization	Steriflamme	130
Direct		
Steam–product	Alfa Laval VTIS	140
	APV Uperiser	144–150
	Cherry Burrell Arovac	145
	Rossi Catelli	142
	Stork Steritwin	138
Product–steam	Laguilharre	140
	Pasilac Palarisator	145
	DASI Falling Film System	140
Electrical	Elecster	140
	Ohmic	120–140
Fluidized bed	Steriglen	133
Friction	Atad	140

tubular heat exchanger which consists of a concentric tube system where the inner tube contains the product and the outer tube contains the heating medium. Plate heat exchangers consist of a series of corrugated plates which are clamped together in such a way as to allow heating fluid to pass down one side of the plate and product down the other. Scraped surface heat exchangers consist of an externally heated tubular heat exchanger inside which is an axially mounted rotor with blades. The blades scrape the surface of the heat exchanger to keep it clean. This type of heat exchanger is primarily used for viscous products. Typically, processes used in tubular, plate and scraped surface systems are between 135 and 140°C for 3–5 s.

Another type of high temperature indirect processing is flame sterilization in which filled, closed cans are transported across gas burners with continuous axial rotation to avoid product burn on. Typically the gas burners are at 1700°C and temperatures within the can are raised to 130°C for less than 10 min. The most recent review of flame sterilization is by Richardson (1987).

Direct heating systems often utilize the latent heat of steam to raise the temperature of the product as rapidly as possible. Whereas the temperature profile in an indirect system is largely exponential, in a direct system heating is virtually instantaneous. As there are no hot heat transfer surfaces to cause 'burn on', slightly higher temperatures (typically 140–145°C) are used in direct steam heating systems. The process is carried out by two methods: (1) steam injection into the product, or (2) steam infusion in which the product is injected into a steam atmosphere. For UHT milk processing the product is cooled in a vacuum chamber to remove the water which has condensed during heating.

Electrical heating methods such as the Elecster and Ohmic heating systems heat the food directly by passing an electric current through the electrically conductive product (Hersom 1985; Willhoft 1993). Temperatures up to 140°C can be achieved in such systems. The Atad equipment (Hersom 1985) utilizes the friction generated by a disc rotating at 5000 rev min^{-1} between two fixed discs spaced 0·3 mm apart. The heat generated raises the temperature of the product to 140°C. The APV Jupiter system (Hersom 1985) consists of a large conical vessel in which particulate food is mixed with steam in a tumbling action at temperatures in excess of 120°C. The Steriglen process (Hersom 1985) consists of a fluidized bed of superheated steam at 155°C which raises product temperature rapidly to 133°C before it passes into a second holding chamber. For further information on the various UHT food processing methods outlined above and in Table 1, readers are referred to recent publications by Hersom (1985), Richardson (1987), Rees and Bettison (1991), Holdsworth (1992), and Willhoft (1993).

2.2 Packaging

After UHT sterilization by one of the methods described in Section 2.1, the product is filled aseptically into presterilized packaging which is then sealed to prevent subsequent recontamination. There are several packaging systems which utilize high temperatures to sterilize the packaging prior to filling (Table 2). One of the oldest systems is the Dole aseptic canning system (Willhoft 1993). Superheated steam at temperatures up to 260°C is used to raise the temperature of can bodies and lids to approximately 220°C prior to filling and seaming. Spores exposed to superheated steam experience 'dry' heating conditions as opposed to 'wet' conditions in saturated steam, which explains the high temperatures required when superheated steam is used as the heat source.

Saturated steam is used in the Fran Rica and Bosch systems at 121 and 140–150°C to sterilize metal drums and bottle caps respectively (Hersom 1985; Holdsworth 1992). Some systems rely on the heat of extrusion of thermoplastics for sterility such as the Erca Neutral system (Willhoft 1993) which uses a temperature of 200°C in the extrusion die to destroy bacterial spores which may be present on the packaging material.

2.3 Equipment

An important aspect of any aseptic processing system is the pre-production sterilization of the equipment. In a product processing line it is vital that all sections of pipework, tanks, etc. downstream of the heat exchanger used for sterilization are free from micro-organisms which may grow in the product. Often UHT processing systems are run for several hours or even days between shut-down periods and survival of spores during pre-production sterilization must be avoided. It is customary to recirculate hot water under pressure through product processing lines (Hersom 1985) at temperatures from 130 to 140°C for 30 min or longer to achieve the desired sterility before switching over to the product (Table 3).

In the Ohmic system (Willhoft 1993) a solution of sodium sulphate or an alternative food-compatible mineral salt solution at a concentration that approximates to the electrical conductivity of the food material to be processed is recirculated through the Ohmic heater and downstream pipework at temperatures up to 140°C to sterilize the system.

The Dole aseptic canning line uses the same superheated steam generator to presterilize the equipment as that used for sterilization of cans and lids. Superheated steam at 260°C is used to raise all parts of the can and lid sterilizer sections to 177°C for a minimum of 20 min. The filling valves, can cooling and product filler assemblies are sterilized using hot water (Willhoft 1993).

Steam is used to sterilize several different pot and bag filler mechanisms (Table 3) at temperatures up to 150°C. The bags and pots will already have been sterilized by other means (e.g. irradiation, hydrogen peroxide or heat of extrusion) and the product lines upstream of the filler head will have been sterilized as described earlier. Several manufacturers also incorporate continuous steam bleeds on pipe couplings, valve spindles and scraped surface heat exchanger shafts to prevent contamination of the product via these routes.

Hot air at 250–300°C is also an effective machine sterilant as in the Erca Neutral system (Willhoft 1993). Dry heat sterilization tends to be used in situations where excessive moisture may have a deleterious effect on the equipment or packaging. One of the drawbacks of hot air as a sterilant is the low heat transfer and heat capacity, with the result that machinery may heat up very slowly. In the Erca Neutral system, additional electrical heating is required to provide additional heat to the filler mechanisms since hot air alone is insufficient to raise the temperature to the required 160°C.

Table 2 Examples of high temperature packaging sterilization systems

Sterilization system	Manufacturer	Packaging	Sterilization temp. (°C)
Superheated steam	Dole	Cans and lids	220
Heat of extrusion	Rommelag	Thermoplastic bottles	220
	Holopak	Thermoplastic bottles	164–234
	Erca/Conoffast	Thermoformed cups	>180
	Erca neutral	Coextruded plastic pots and lids	200
Steam	Fran Rica	Metal drums	121
	Bosch	Bottle caps	140–150

Table 3 Examples of high temperature equipment sterilization systems

Sterilization system	Manufacturer	Application	Sterilization temperature (°C)
Hot water	Alfa Laval Steritherm	Plate heat exchanger	137
	APV Ultramatic	Plate heat exchanger	130
Saturated steam	Metal Box Freshfill	Pot filler	130
	Scholle	Bag in box filler	121
Superheated steam	Dole	Can sterilizer	220
Steam	Bowater	Bag filler	145
	Intasept-Coloreed	Bag filler	150
	Gasti	Bag filler	143
Hot air	Bowater	Bag filler	275
	Erca Neutral	Container and filling machine	250–300 (equipment at 160)
Mineral salt solution	APV	Ohmic heater system	120–40

3. METHODS FOR DETERMINATION OF SPORE HEAT RESISTANCE

3.1 Wet heat resistance methods

There are several reviews of methods used to obtain wet heat resistance data of micro-organisms (Stumbo 1973; Pflug and Holcomb 1977; Hersom and Hulland 1980; Brown and Ayres 1982; Brown 1992). UHT methods may be grouped according to the type of heat transfer mechanism into four categories (Table 4).

3.1.1 Indirect heating. An indirect heating method is one where there is a physical barrier (e.g. wall of a capillary tube, can or heat exchanger) between the heating medium and the sample. Temperature will rise exponentially in the sample.

One of the most commonly used of the indirect heating methods is the capillary tube technique where a rapid rate of heating can be achieved: for example the aluminium thermal death time tubes used by Odlaug and Pflug (1977) which had a high heat transfer coefficient and large surface area to volume ratio. Unfortunately the sample size (1 ml) was large, resulting in a lag correction factor of 10·8–13·8 s at 121·1°C. An initial shoulder on a survivor curve could therefore be expected for exposure times below 13·8 s because the sample would not have reached temperature, and death rate would therefore be lower than at 121°C. Odlaug and Pflug (1977) used exposure times longer than 1 min. Larger tube systems such as the test tube or thermal death time tube methods suffer from longer heating lags of between 1 and 8 min at 120°C and are therefore unsuitable for high temperature short time studies.

Perkin *et al.* (1977), who used thin wire copper constantan thermocouples (0·34 mm o.d.), concluded that for heating times greater than 5 s, the 'come up' time would have a negligible effect on the determination of the slope of a thermal death time curve. In studies on spores of *Bacillus stearothermophilus* they found that for the heating times of less than 5 s that are involved at temperatures above 135°C, corrections would be necessary. Perkin *et al.* (1977) found that in 10% of their experiments a small kink appeared in the thermocouple trace, suggesting boiling of the tube contents. They calculated that 4% of the total heat input could be lost if the tube contents boiled. They also found that a difference of 0·45 s in 'come up' time could be introduced depending on where the tubes were placed in the oil bath. Various sizes of capillary tube are available and as the thermal lag calculations involve a radius squared term, slight variations in size may be critical at high temperatures where come up time would be a significant proportion of the total heating period.

Mechanical transfer systems for capillary tubes (Stern and Proctor 1954; Cerf *et al.* 1970; White *et al.* 1982) were designed to improve the reproducibility of the results. They were not intended to reduce the come up time, which was determined by heat transfer characteristics rather than speed of transfer in and out of the heating medium. This is contrary to the claims of White *et al.* (1982) that the technique minimized thermal lags.

In practice, capillary tubes can be used successfully up to 130°C for studies on *Clostridium botulinum* and *Cl. sporogenes*, and up to 140°C for *B. stearothermophilus*, *Cl. thermosaccharolyticum* and *Desulfotomaculum nigrificans* (provided sufficiently resistant spore crops are used), but above these temperatures the heating lag will be longer

Table 4 Examples of methods used for determination of high temperature wet heat resistance of bacterial spores

Method	Reference
Indirect heating methods	
Capillary tubes	Franklin *et al.* 1958b
	Resende *et al.* 1969
	Perkin *et al.* 1977
	Davies *et al.* 1977
	Neaves and Jarvis 1978a,b
Programmed heating of capillary tubes	Reichart 1979
	Ayres and Brown 1980
	Matsuda *et al.* 1983
Stern and Proctor capillary tube	Stern and Procter 1954
	Cerf *et al.* 1970
	Brown 1975
Aluminium thermal death time tubes	Odlaug and Pflug 1977
Thermal death time cans in miniature retorts	Townsend *et al.* 1938
	Sognefest *et al.* 1948
	Reynolds *et al.* 1952
	Schmidt 1950
Thermal tank method	Williams *et al.* 1937
Tubular heat exchangers	Martin 1950
	Galesloot 1956
	Cuncliffe *et al.* 1979
	Wadsworth and Bassette 1985
Plate heat exchangers	Williams *et al.* 1957
	Franklin *et al.* 1958
Thin film scraped surface heat exchanger	Dodeja *et al.* 1990
Mixing methods	
Injection of spores into preheated medium	Matsuda *et al.* 1973
	Kooiman 1974
	Cerny 1980
	Mikolajcik and Rajkowski 1980
Continuous flow mixing apparatus	Wang *et al.* 1964
	Oquendo *et al.* 1975
	Srimani *et al.* 1980
Particle methods	
Alginate beads	Dallyn *et al.* 1977
	Bean *et al.* 1979
Capillary bulbs in cubes of product	Hersom and Shore 1981
Alginate/food particles	Brown *et al.* 1984
	Gaze 1989
Inoculated thread in cubes of product	Mechura 1985
Ovalbumen particles	Cerny *et al.* 1989
Microporous plastic particles	Rönner 1990

Table 4 (*continued*)

Method	Reference
Direct steam heating methods	
Product-into-steam:	
Thermoresistometer	Stumbo 1948 Pflug and Esselen 1953 Pflug 1960
Computer linked thermoresistometer	Brown *et al.* 1988 David and Merson 1990
MISTRESS product-into-steam apparatus	Perkin 1974 Neaves and Jarvis 1978a,b
Steam-into-product:	
Dual plate/direct steam injection UHT sterilizer	Burton and Perkin 1970
Large scale direct steam injection	Edwards *et al.* 1965a,b Busta 1967
Alfa Laval Vacu-Therm instant sterilizer	Lindgren and Swartling 1963

than the total heating time. One of the features of some of the published data for capillary tubes (Davies *et al.* 1977; Neaves and Jarvis 1978b) was the appearance of survivor tails. Davies *et al.* (1977) attributed tailing to heterogeneity in the spore population. In their results, tailing occurred at plate counts of less than 10. Neaves and Jarvis (1978b) also obtained survivor tails at low survivor levels (less than 10 colonies per plate). Pflug (1987) observed that a survivor curve incorporating low levels of false positives would also produce an apparent survivor tail.

Bean *et al.* (1979) demonstrated that spores of *B. stearothermophilus* could survive longer than expected if they were dried on the inside of capillary tubes during heat sealing of the capillary. They obtained survivor tails from less than 10 to 500 per tube. Survivor tails were attributed to an artefact of the heating method. Survivor tails were not obtained when the same spore suspension was heated in alginate beads. One method of overcoming some of the heating lag problems with capillary tubes is to use a system where the tubes are heated in a rising temperature environment. Matsuda *et al.* (1983) pointed out that capillary tubes subjected to isothermal heating required correction for heating and cooling lags while tubes subjected to non-isothermal heating did not. However, with a temperature rise of 4°C min^{-1} for spores of *Cl. botulinum*, they reported a z-value of 18·4 C° compared with 10·3 C° from isothermal experiments. The difference was attributed to fluctuation of the results but Ayres and Brown (1980) also observed higher z-values from capillary tubes heated non-isothermally.

Thermal death time cans were introduced by Townsend *et al.* (1938). These were small cans, 2·5 in in diameter and 0·375 in high, which contained 13–16 ml of inoculum. Experimental results were expressed in the form of survival/destruction data. A process or F-value was then calculated for the destruction of the spores initially inoculated into the container. The cans were heated in special miniature retorts (Townsend *et al.* 1938; Sognefest *et al.* 1948; Reynolds *et al.* 1952) over the temperature range 100–142°C. Corrections for come up time ranged from 0·33 to 1·93 min. Thermal death time curves appeared to be invariably straight with z-values for *Cl. sporogenes* and *Cl. botulinum* close to 10 C°.

Williams *et al.* (1937) described a 900-ml capacity pressure vessel which could be heated to 121°C in 2·5 min and from which samples could be withdrawn at 2 min intervals. Survivor curves obtained for *B. stearothermophilus* spores were concave downwards with the greatest deviation from linearity at 115 and 110°C. No explanation was given for the non-linearity but some of the deviation could have been due to the chamber gradually achieving equilibrium temperature over a longer time period than that calculated by the investigators.

Other indirect heating methods include pilot scale systems which have been used principally to confirm process times and temperatures for factory scale equipment. Examples include tubular (Galesloot 1956; Cuncliffe *et al.* 1979; Wadsworth and Bassette 1985) and plate (Williams *et al.* 1957; Franklin *et al.* 1958a) heat exchangers. Galesloot (1956) reported that increasing residence times by 16·5 s at 120–135°C increased the sterilizing effect of the system by approximately two [sterilizing effect = log (initial concentration/final concentration)]. Martin (1950) used a tubular system and demonstrated that process efficiency in a continuous flow system can vary considerably from product to product. With three products, pea purée, bacon soup and liver soup, processed at temperatures between 130 and 148°C to F_0 6·2–6·7, spoilage levels in the three products were 98, 76 and 4% respectively.

Cuncliffe *et al.* (1979) used a stainless steel tubular system and was able to demonstrate that a temperature of 148°C was necessary to destroy foot-and-mouth disease virus in milk.

3.1.2 Mixing. In order to reduce the come up time, some investigators have mixed a small volume of a suspension of micro-organisms with a much larger volume of preheated substrate. The effectiveness of this approach depends on a number of factors including the quantities of suspension and substrate, their initial temperature and specific heats and the efficiency of mixing. The Kooiman (1974) method consisted of inoculation of a small volume (0·1 ml) of spore suspension into 10 ml of preheated heating medium in a screw capped Pyrex culture tube immersed in a heating bath. The temperature drops, when spore suspension at 0°C was added to heating medium at 110–125°C, were 0·90–1·87°C. An additional complication of the method was the time then taken to recover the temperature fall by heat transfer from the heating bath into the tube (calculated to be 47–112 s to recover 90% of the fall). In this respect the method suffers from some of the drawbacks of the capillary tube techniques, described earlier, and shoulders caused by heating lags may be observed.

Mikolajcik and Rajkowski (1980) modified the Kooiman (1974) method and used serum bottles with aluminium vial caps. They also increased the inoculum volume to 0·2 ml (initial temperature 4°C) and reduced the preheated volume to 4 ml. At 120°C the calculated temperature drop (5·5°C) would therefore be greater than that experienced by Kooiman (1974), but conversely the time to regain temperature equilibrium (64–76 s as opposed to 90 s) was shorter.

Cerny (1980) allowed 3 s for temperature equilibration when 1 ml of spore suspension was injected into 250 ml of heating medium in his small autoclave system. From calculation, 1 ml of spore suspension at 120°C injected into 250 ml of medium at 120–140°C would result in a temperature drop of 0·4–0·5°C. The design of Cerny's (1980) apparatus was very similar to that of Matsuda *et al.* (1973) in which 1 ml of spore suspension was injected into 400 ml of preheated buffer. The shortest heating time in their studies at 117·5–127·5°C was 30 s.

Continuous flow mixing devices have also been developed (Wang *et al.* 1964; Oquendo *et al.* 1975; Srimani *et al.* 1980) for mixing water, preheated to high temperatures, with spore suspension in appropriate volumes in a pressurized chamber. The flow was induced by compressed gases and flow rates determined by calibrated rotometers. Wang *et al.* (1964) reported a mixing time of 0·0006 s. Cooling was achieved by flash cooling in an expansion chamber. One of the main differences between the continuous flow mixing devices and the 'static' mixing devices was that the equilibrium temperature of the final mixture remained constant in the continuous system unlike the extra equilibration period of the Kooiman (1974) approach.

3.1.3 Particles. All of the systems listed in Table 4 have been used to determine processes at UHT processing temperatures in pilot plant and commercial scale product sterilization systems.

Particle systems are likely to produce shoulders on survivor curves as a result of the come up time. The alginate beads used by Bean *et al.* (1979) experienced come up times of 10–15 s as judged from the time the shoulder was present on their survivor curves. Calculation of integrated lethalities (Brown *et al.* 1984) which takes account of come up time is necessary to account for shoulders introduced by the particle method.

3.1.4 Direct steam heating. Direct steam heating methods are able to produce rapid rates of heating because of the latent heat energy released when steam condenses. There is sufficient latent heat energy in 0·22 g of condensing steam to raise the temperature of 1 g water from 20°C to 140°C. Steam heating profiles are either exponential or hyperbolic depending on whether the steam condensate is added to the volume of product being heated (hyperbolic) or not (exponential).

Direct steam methods may be divided into product-into-steam and steam-into-product methods. The steam-into-product methods (Lindgren and Swartling 1963; Edwards *et al.* 1965a,b; Busta 1967; Burton and Perkin 1970) reported generally involved pilot scale equipment where the investigators were using spore suspensions to measure the effectiveness of the thermal process rather than investigate death kinetics. Two product-into-steam techniques used principally for the study of the high temperature resistance of bacterial spores have been reported, the thermoresistometer developed originally by Stumbo (1948) and the MISTRESS (milk-into-steam treatment research equipment small scale), first developed by Perkin (1974). Both methods were capable of working at temperatures up to 140–150°C. The MISTRESS technique was designed to process sample volumes up to 500 ml whereas the thermoresistometer uses much smaller samples (0·01–0·02 ml). The shortest exposure time for the thermoresistometer is approximately 0·1 s while in the MISTRESS it is approximately 1 s, limited by both come up time and mechanical considerations.

3.2 Dry heat resistance methods

Pflug (1987) described dry heat sterilization as a situation where the relative humidity was any value between 0 and 100% where water was not present in the liquid state. The methods used for evaluation of dry heat resistance can be divided into open, closed, flow and non-flow systems depending on whether or not the dried spore suspension is open to the atmosphere and whether or not the heating medium is flowing over the spores (Table 5).

Dry heating methods often encounter problems with heat transfer and temperature control because of the low heat transfer coefficients and heat capacities of dry gases. In both open and closed systems, there may be variation from replicate to replicate while forced flow of heating medium across the dried spore suspension may alter the moisture content and resistance of the spores (Angelotti *et al.* 1968) and will affect the rate of heat transfer. As in wet heating, shoulders and tails have been reported for dry heat resistance survivor curves (Fox and Pflug 1968; Alderton and Snell 1969; Molin and Ostlund 1975). Shoulders may be caused by long heat up time. Tailing, however, may be caused by further drying of the spores during the experiment, or physiological differences between spores, with respect to water content, at the start of heating. The choice of heat transfer system (for example, superheated steam, hot air, oil bath, infrared, conduction through metal or plastic) is also likely to affect spore survival depending on the rate of heat transfer to the sample. If the experiments are being done to simulate what may happen to spores in packaging or machine sterilization procedures, then choice of experimental technique is very important. As a general observation it is often much more difficult to simulate commercial dry heating processes in the laboratory than to carry out wet heating of spores in food product.

4. RESISTANCE OF SPORES AT HIGH TEMPERATURES

4.1 Wet heat resistance

The most important bacterial spore-former with respect to heat processing is *Cl. botulinum* because of the potent neurotoxin produced by the organism. The most heat resistant spores (Table 6) are those produced by the proteolytic types A, B, F and G. Typically, for this group, the maximum D value at 121°C is between 0·1–0·2 min and the z-value is usually taken as 10 C° (Stumbo 1973; Hersom and Hulland 1980). Heat resistance data for the proteolytic A and B types exist between approximately 100 and 140°C. The most resistant type of *Cl. botulinum* at temperatures above 120°C is likely to be type G (Lynt *et al.* 1984; Table 6), and therefore the most likely to survive UHT processing. However, as proteolytic types A and B appear to be much more common than type G, which has never been implicated in food-borne botulism, the risk of type G causing an outbreak of botulism as a result of survival of UHT processing would appear to be remote.

The non-proteolytic types B, E and F are considerably less heat resistant than the proteolytic types. The non-proteolytic types are more important in pasteurized, chilled products because of their ability to grow down to 3·3°C whereas the proteolytic A, B and F strains only grow down to 10°C. The wet heat resistance of spores of type C is between that of proteolytic types A and B and non-

Table 5 Examples of methods used for determination of dry heat resistance of bacterial spores

Open/ closed	Flow/non-flow	Method: Carrier	Method: Heating device	Reference
Open	Flow	Tinplate	Superheated steam in Dole canning unit	Collier and Townsend 1956
		Tinplate	Superheated steam resistometer	Brown *et al.* 1981
		Tinplate	Dry heat resistometer	Pflug 1960
Open	Flow (natural convection)	Glass	Infrared heaters	Molin and Ostlund 1975
		Steel	Hot plate	Ruschmeyer *et al.* 1973
Open	Flow and non-flow	Tin, aluminium, glass, paper	Hot air oven	Fox and Pflug 1968
Open	Non-flow	Filter paper, glass, sand	Tubes in heated block	Bruch *et al.* 1963
Closed	Non-flow	Steel, paper, plastic	Tubes in oil bath	Angelotti *et al.* 1968
		Metal cups	In cans in a retort	Pheil *et al.* 1967
		Kapton, Teflon	Between aluminium blocks in hot air oven	Bruch and Smith 1968
		Perspex beads	Direct steam injection	Hunter 1972
		Aluminium pouches	Oil bath	Gurney and Quesnel 1980
		Plastic packaging	Between heated plates	Willhoft 1993

Table 6 Wet heat resistance of spores of *Clostridium botulinum*

Type	Typical *D*-values (min) 100°C	121°C	140°C	Typical *z*-value (C°)	Reference
A proteolytic	29·2*	0·05–0·13	0·001*	8·2–9·1	Stumbo *et al.* 1950
B proteolytic	10·5* 20·7–23·5	0·13*	0·002	11·0	Gaze and Brown 1988 Kaplan *et al.* 1954
B non-proteolytic	0·08*	0·0003*	—	8·6–9·8	Gaze and Brown 1990
C mildly or non-proteolytic					
terrestrial	3·4*	0·003*	—	10·0–11·5	
marine	0·6*	0·0002*	—	10·7–10·8	Segner and Schmidt 1971
D mildly or non-proteolytic†	—	—	—	—	Sakaguchi 1986
E non-proteolytic	0·03*	0·0002*	—	6·1–8·4	Lynt *et al.* 1977
F proteolytic	8·8–17·8*	0·14–0·22*	0·003–0·004	9·3–12·1	Lynt *et al.* 1981
F non-proteolytic	0·0001–0·002*	—	—	9·5–14·8	Lynt *et al.* 1979
G proteolytic	1·1–1·3*	0·14–0·19	0·02–0·04*	20·9–27·3	Lynt *et al.* 1984

* Calculated from published data.

† No data have been found for type D but it is reported (Sakaguchi 1986) to have similar resistance to type C.

proteolytic type B. Terrestrial strains of type C appear to be slightly more heat resistant than marine strains. No data have been published on the heat resistance of spores of type D although Sakaguchi (1986) reported that they were comparable in resistance to type C. *Clostridium sporogenes*, which is distinguished from the proteolytic strains of *Cl. botulinum* by its inability to produce neurotoxin, produces spores which are more heat resistant (Russell 1982). The D value of the spores at 121·1°C can be up to 1·7 min (Stumbo, 1948). Data are available (Table 7) for the wet heat resistance of *Cl. sporogenes* over a wide temperature range (104–144°C), much of it obtained by the thermoresistometer method. Because the spores are more heat resistant and non-toxigenic, *Cl. sporogenes* has often been used in place of *Cl. botulinum* in factory process challenge trials where it would be inadvisable to use *Cl. botulinum*.

The spores of the thermophilic spore-formers are the most resistant to wet heat (Table 8). *Bacillus stearothermophilus* spores have *D*-values as high as 16–17 min at 121°C and have been widely used in heat resistance studies (Holdsworth 1992) and as biological indicators of the efficacy of steam sterilization (Pflug 1987). For example, spores of *B. stearothermophilus* embedded in alginate particles were used (Willhoft 1993) to determine process lethality in the Ohmic heater. Provided care is taken to ensure complete recovery or removal of spores after a trial, spores of *B. stearothermophilus* can be used in commercial systems as a useful adjunct to other methods of process determination.

The most resistant spores are formed by *Cl. thermosaccharolyticum*, with *D*-values at 121°C of 68·0 and 295·0 reported by Brown (1983) and Xezones *et al.* (1965) respectively.

4.2 Dry heat resistance

Bacterial spores are much more resistant to dry heat than wet heat and mesophilic species, for example *B. subtilis*, are

Table 7 Wet heat resistance of spores of *Clostridium sporogenes* (PA 3679) obtained using the thermoresistometer

Suspending medium	*D*(121·1°C) (min)	*z* (C°)	Temperature range (°C)	Reference
Pea purée	1·68	9·44	104·4–132·2	Stumbo 1948
Several substrates	0·8–1·52	9·2–11·4	104·4–132·2	Stumbo *et al.* 1950
Phosphate buffer pH 7	1·06	9·3	112·8–148·9	Pflug and Esselen 1954
Several substrates	0·75–2·03	9·0–14·7	121·1–143·3	Esselen and Pflug 1956
Several substrates	0·24–0·58	9·4–10·4	110·0–132·2	Secrist and Stumbo 1956
Water	0·63–0·73	10·4	121·1–143·3	Secrist and Stumbo 1958
Strained peas	0·95–1·25	9·8	121·1–143·3	Secrist and Stumbo 1958

Table 8 Wet heat resistance of spores of some particularly resistant thermophilic spore-formers

Organism	Temperature (°C)	*D*-Value (min)	*z*-Value (C°)	Reference
Bacillus stearothermophilus	120	16·7	7·3	Davies *et al.* 1977
	121·1	16·0	7·7	Brown 1975
	128	2·2	7·8	Busta 1967
Clostridium thermosaccharolyticum	121	68·0	11·0	Brown 1983
	121	195·0	6·9	Xezones *et al.* 1965
Desulfotomaculum nigrificans	121	55·0	9·5	Donnelly and Busta 1980

often more resistant than thermophilic species, for example *B. stearothermophilus*. Typical dry heat resistance data from published sources are given in Table 9. A noticeable feature of dry heat resistance data is the magnitude of the *z*-value. Values for *z* between 13 and 31 C° are normally obtained compared with, on average, 10 C° for wet heating. This means that not only are much higher temperatures necessary for dry heat sterilization systems but to increase lethality by a factor of 10, a rise in temperature of up to 30 C° may be required. Angelotti *et al.* (1968) reported up to a 50-fold difference in heat resistance for spores of *B. subtilis* var. *niger* dried on different materials (Table 10). The differences in resistance of the spores on different surfaces was attributed to differences in the rate of reduction of moisture content of the spores in the different heating systems. The stainless steel strip and paper strips were 'open' test methods whereas the tests that used steel washers, lucite and epoxy were 'closed' methods.

Collier and Townsend (1956) showed that spores heated in superheated steam exhibit much greater resistance than those heated in saturated steam (Table 11). Systems that utilize superheated rather than saturated steam as the sterilizing medium require temperatures 40–50°C higher to effect similar destruction of bacterial spores. Spores of *Cl. botulinum* are much more resistant to dry heat than wet heat. Tanner and Dack (1922) reported destruction times from 140 min at 110°C to 20 min at 160°C under dry heat conditions. In a survey of seven type B strains and 9 type A

Table 9 Dry heat resistance of spores of *Bacillus subtilis*, *B. stearothermophilus*, *Clostridium sporogenes* and *Cl. botulinum*

Organism	Strain	Type of dry heat	Temperature (°C)	*D*-Value (min)	*z*-Value (C°)	Reference
B. subtilis	var. *niger*	Air dried on filter paper	160	1·8	27·2	Bruch *et al.* 1963
	5230	Air in TDT cans		1·46	18·3	Pheil *et al.* 1967
	var. *niger*	Dried on Kaplon or Teflon		0·84*	29·1	Bruch and Smith 1968
	5230	Hot air oven		2·80*	17	Fox and Pflug 1968
	5230	Sealed into cans		3·53*	17	Fox and Eder 1969
	ATCC 6633	Infrared heaters		0·38	23	Molin and Ostlund 1975
	var. *niger* (ATCC 9372)	Spores in organic substances heated in dry air		0·08–1·48	23–27	Molin 1977a
	var. *niger* (ATCC 9372)	Dry air		0·3	23	Molin 1977b
	var. *niger*	In aluminium pouches in hot oil		0·65	21·3	Gurney and Quesnel 1980
	786 (NIRD 1069)	Superheated steam		1·07–1·26	18·5	Brown *et al.* 1981
B. stearothermophilus	NCA 1518	Superheated steam	160	0·09–0·14	14·4	Collier and Townsend 1956
	NCA 1518	Air dried on filter paper		0·35	24·4	Bruch *et al.* 1963
	NCA 1518	Sealed in glass tube under vacuum		1·17–1·82*	14–22	Alderton and Snell 1969
	NCTC 10339	Dry air		0·16	23·29	Molin 1977b
	NCA 1518	Superheated steam		1·65*	17	Quast *et al.* 1977
	NCIB 8919	Superheated steam		0·73–1·35	27·8–31·3	Brown *et al.* 1981
Cl. sporogenes	PA 3679	Superheated steam	140	41–66·6	13·3	Collier and Townsend 1956
	PA 3679	Air in TDT cans		15	21·7	Pheil *et al.* 1967
	NCIB 8053	Superheated steam		1·2–2·4	—	Brown *et al.* 1981
Cl. botulinum	34B	Dry air	160	0·38	17·2	Denny and Matthys 1975

* Calculated from published data.

Table 10 *D*-Values (min) at different temperatures for spores of *Bacillus subtilis* var. *niger* heated under dry conditions in or on different materials (adapted from Angelotti *et al.* 1968)

Material	Temperature (°C) 125	135
Stainless steel strip	7·9–9·4	2·3–3·8
Paper strips	96·1–108·2	15·3–18·3
Between stainless steel washers	41·8–51·8	18·2–33·3
Lucite rods	150–234	72–90
Epoxy rods	300–342	102–126

Table 11 Comparison of resistance to saturated and superheated steam (adapted from Collier and Townsend 1956)

	Resistance to saturated steam		Resistance to superheated steam		
Organism	D_{121} (min)	z (C°)	D_{121} (min)	D_{177} (min)	z (C°)
Bacillus polymyxa	0·001	7·8	632	0·13	15·6
B. stearothermophilus (FS 1518)	1·9	10·1	936	0·14	14·4
Clostridium sporogenes (PA 3679)	0·4	7·6	6·4	0·13	33·3

strains, Denny and Matthys (1975) found that the type B strains were more resistant to dry heat at 154°C than the A strains. The most resistant (Strain 34B) had *D*-values at 148·9 and 165·6°C of 1·92 and 0·22 min respectively with a *z*-value of 17·18 C°.

5. COMMERCIAL PROCESS REQUIREMENTS

Stringent regulatory requirements in the United States apply to UHT and aseptic food processes. UHT processes for low acid foods in both the USA and UK must provide the same level of protection from *Cl. botulinum* spores as those used in conventional canning (Rees and Bettison 1991). The US Food and Drugs Administration also specify biological validation of thermal processes in addition to mathematical prediction of sterilizing value. Criteria established by the National Food Processors Association in the USA for low acid aseptic systems include requirements for data on bacteriological test procedures and results, and for on-site testing of each individual commercial installation for low acid foods.

For wet heat sterilization of processing equipment the minimum normally recommended in the USA (Rees and Bettison 1991) for low acid aseptic systems is the equivalent of 3·0 min at 121°C.

In the UK, UHT processing of milk and cream are subject to Statutory Instruments 1033 and 1509 under the Food and Drug Regulations, which specify the minimum processes for milk (not less than 132·2°C for not less than 1 s) and cream (140°C for not less than 2 s).

Sterilization of packaging equipment and machinery using dry heat is normally measured in terms of numbers of log reductions of spores of a test organism (usually *B. subtilis* or *B. stearothermophilus*). Figures from 4- to 8-log reductions (Holdsworth 1992; Willhoft 1993) of spores of these organisms are often quoted as the target set for sterilization performance. The Erca Neutral system, for example, achieves 8-log reductions of spores of *B. subtilis* var. *niger* (Willhoft 1993) using heat of extrusion at 200°C for 5 s, whereas the Dole aseptic canning system achieved 5-log reductions of spores of *B. stearothermophilus* with superheated steam at 177°C for 1 min or 160°C for 10 min (Collier and Townsend 1956).

6. CONCLUSION

The high temperature resistance of bacterial spores under both wet and dry heating conditions poses problems for food, packaging and machinery manufacturers in the production of safe, commercially sterile foodstuffs. Only through an understanding, not only of the heat resistance of the spores themselves but also the engineering aspects, including heat penetration, heat transfer and temperature distribution, can UHT processing be carried out successfully.

Pflug (1989) described the situation very succinctly when he stated that, 'Anyone who uses bacterial spores in a measurement mode has to proceed with the belief that the spores are always right. If we proceed in this manner it will keep us out of a lot of trouble.'

7. REFERENCES

Alderton, G. and Snell, N. (1969) Chemical states of bacterial spores: dry heat resistance. *Applied Microbiology* **17**, 745–749.

Angelotti, R., Maryanski, J.H., Butler, T.F., Peeler, J.T. and Campbell, J.F. (1968) Influence of spore moisture content on the dry heat resistance of *Bacillus subtilis* var *niger*. *Applied Microbiology* **16**, 735–745.

Ayres, C.A. and Brown, K.L. (1980) *Microbiological Aspects of Aseptic Packaging*. Technical Memorandum No. 250. Chipping Campden, Glos., UK: Campden Food and Drink Research Association.

Bean, P.G., Dallyn, H. and Ranjith, H.M.P. (1979) The use of alginate spore beads in the investigation of ultra high temperature processing. In *Food Microbiology and Technology* ed. Jarvis B., Christian, J.H.B. and Michener, H.D. Proceedings of the International Meeting on Food Microbiology and Technology 1978, Parma, pp. 281–294. Italy: Medicina Viva Servizio Congressi S.r.l.

Brown, K.L. (1975) *Aseptic Packaging of Vegetable Products.* Report 1975. Chipping Campden, Glos., UK: Campden Food and Drink Research Association.

Brown, K.L. (1983) Heat resistant thermophilic anaerobe isolated from composted forest bark. In *Fundamental and Applied Aspects of Spores*, Proceedings of Cambridge Spore Conference, pp. 387–394. London: Academic Press.

Brown, K.L. (1992) Heat resistance of bacterial spores. Ph.D. Thesis, University of Nottingham.

Brown, K.L. and Ayres, C.A. (1982) Thermobacteriology of UHT processed foods. In *Developments in Food Microbiology – 1* ed. Davies, R. Ch. 4, pp. 119–152. Barking, Essex: Applied Science Publishers.

Brown, K.L., Pitcher, J.E. and Mottishaw, J. (1981) *Microbiological Aspects of Aseptic Packaging.* Technical Memorandum No. 281. Chipping Campden, Glos., UK: Campden Food and Drink Research Association.

Brown, K.L., Ayres, C.A., Gaze, J.E. and Newman, M.E. (1984) Thermal destruction of bacterial spores immobilized in food/alginate particles. *Food Microbiology* **1**, 187–198.

Brown, K.L., Gaze, J.E., McClement, R.H. and Withers, P. (1988) Construction of a computer controlled thermoresistometer for the determination of the heat resistance of bacterial spores over the temperature range 100 to 150°C. *International Journal of Food Science and Technology* **23**, 361–371.

Bruch, C.W., Koesterer, M.G. and Bruch, M.K. (1963) Dry heat sterilisation: its development and application to components of exobiological space probes. *Developments in Industrial Microbiology* **4**, 334–342.

Bruch, M.K. and Smith, F.W. (1968) Dry heat resistance of spores of *Bacillus subtilis* var. *niger* on Kapton and Teflon film at high temperature. *Applied Microbiology* **16**, 1841–1846.

Burton, H. (1969) Ultra high temperature processed milk (Review article No. 151) *Dairy Science Abstracts* **31**, 287–297.

Burton, H. and Perkin, A.G. (1970) Comparison of milk processed by the direct and indirect methods of ultra-high temperature sterilisation. 1. An experimental UHT steriliser and its characteristics. *Journal of Dairy Research* **37**, 209–218.

Busta, F.F. (1967) Thermal inactivation characteristics of bacterial spores at ultra high temperatures. *Applied Microbiology* **15**, 640–645.

Cerf, O., Grosclaude, G. and Vermiere, D. (1970) Apparatus for the determination of heat resistance of spores. *Applied Microbiology* **19**, 696–697.

Cerny, G. (1980) Abhängigkeit der thermischen Abtötung von Mikroorganismem vom pH—Wert der Medien II Bakterien und Bakterien sporen. *Zeitschrift für Lebensmittel-Untersuchung und-Forschung* **170**, 180–186.

Cerny, G., Fink, A. and Pecher, A. (1989) Studies on thermal inactivation of bacterial spores in particulate foods. In *The First International Congress on Aseptic Processing Technologies: Innovations in Aseptic Processing of Particulates.* ed. Chambers, J.V. March 19–21, pp. 112–123. West Lafayette, IN: Purdue University.

Collier, C.P. and Townsend, C.T. (1956) The resistance of bacterial spores to superheated steam. *Food Technology* **10**, 477–481.

Cuncliffe, H.R., Blackwell, J.H., Dors, R. and Walker, J.S. (1979) Inactivation of milkborne foot and mouth disease virus at ultra high temperatures. *Journal of Food Protection* **42**, 135–137.

Dallyn, H., Falloon, W.C. and Bean, P.G. (1977) Method for the immobilisation of bacterial spores in alginate gel. *Laboratory Practice* **26**, 773–775.

David, J.R.D. and Merson, R.L. (1990) Kinetic parameters for inactivation of *Bacillus stearothermophilus* at high temperatures. *Journal of Food Science* **55**, 485–495, 515.

Davies, F.L., Underwood, H.M., Perkins, A.G. and Burton, H. (1977) Thermal death kinetics of *Bacillus stearothermophilus* spores at ultra high temperatures. I. Laboratory determination of temperature coefficients. *Journal of Food Technology* **12**, 115–129.

Denny, C.B. and Matthys, A.W. (1975) *NCA Tests on Dry Heat as a Means of Sterilization of Containers, Lids, and a Closing Unit for Aseptic Canning.* National Canners Association Research Report, RF4614, Washington, DC.

Dodeja, A.K., Sarma, S.C. and Abichandani, H. (1990) Thermal death kinetics of *B. stearothermophilus* in thin film scraped surface heat exchanger. *Journal of Food Processing and Preservation* **14**, 221–230.

Donnelly, L.S. and Busta, F.F. (1980) Heat resistance of *Desulfotomaculum nigrificans* spores in soy protein infant formula preparations. *Applied and Environmental Microbiology* **40**, 721–725.

Edwards, J.L. Jr., Busta, F.F. and Speck, M.L. (1965a) Thermal inactivation characteristics of *Bacillus subtilis* spores at ultra high temperatures. *Applied Microbiology* **13**, 851–857.

Edwards, J.L., Jr., Busta, F.F. and Speck, M.L. (1965b) Heat injury of *Bacillus subtilis* spores at ultra high temperatures. *Applied Microbiology* **13**, 858–864.

Esselen, W.B. and Pflug, I.J. (1956) Thermal resistance of putrefactive anaerobe No. 3679 spores in vegetables in the temperature range of 250–290°F. *Food Technology* **10**, 557–560.

Fox, K. and Eder, B.D. (1969) Comparison of survivor curves of *Bacillus subtilis* spores subjected to wet and dry heat. *Journal of Food Science* **34**, 518–521.

Fox, K. and Pflug, I.J. (1968) Effect of temperature and gas velocity on the dry heat destruction rate of bacterial spores. *Applied Microbiology* **16**, 343–348.

Franklin, J.G., Williams, D.J., Chapman, H.R. and Clegg, L.F.L. (1958a) Methods of assessing the sporicidal efficiency of an ultra-high-temperature milk sterilising plant. II. Experiments with suspensions of spores in milk. *Journal of Applied Bacteriology* **21**, 47–50.

Franklin, J.G., Williams, D.J. and Clegg, L.F.L. (1958b) Methods of assessing the sporicidal efficiency of an ultra-high-temperature milk sterilising plant. III. Laboratory determination of the heat resistance of spores of *Bacillus subtilis* in water and in milk. *Journal of Applied Bacteriology* **21**, 51–57.

Galesloot, Th.E. (1956) Een envoudige methode ter bepaling van het bacteriologisch effect van sterilisatie-processen voor milk toegepast op het steriliseren van melk in doorstroom en flessensterilisatoren. *Netherlands Milk and Dairy Journal* **10**, 79–100.

Gaze, J.E. (1989) The application of an alginate particle technique to the study of particle sterilisation. In *The First International Congress on Aseptic Processing Technologies: Innovations in Aseptic Processing of Particulates* ed. Chambers, J.V., March 19–21, pp. 105–111. West Lafayette, IN: Purdue University.

Gaze, J.E. and Brown, G.D. (1990) *Determination of the Heat Resistance of a Strain of Non-Proteolytic* Clostridium botulinum *Type B and a Strain of Type E, Heated in Cod and Carrot homogenate over the Temperature Range 70 to 92°C.* Technical Memorandum No. 592. Chipping Campden, Glos., UK: Campden Food and Drink Research Association.

Gaze, J.E. and Brown, K.L. (1988) The heat resistance of spores of *Clostridium botulinum* 213B over the temperatures range 120 to 140°C. *International Journal of Food Science and Technology* **23**, 373–378.

Gurney, T.R. and Quesnel, L.B. (1980) Thermal activation and dry heat inactivation of spores of *Bacillus subtilis* MD2 and *Bacillus subtilis* var. *niger*. *Journal of Applied Bacteriology* **48**, 231–247.

Hersom, A.C. (1985) Aseptic processing and packaging of food. *Food Reviews International* 1(2), 215–270.

Hersom, A.C. and Hulland, E.D. (1980) *Canned Foods, Thermal Processing and Microbiology* 7th edn. London: Churchill Livingstone.

Hersom, A.C. and Shore, D.T. (1981) Aseptic processing of foods comprising sauce and solids. *Food Technology* **35**(5), 53–62.

Holdsworth, S.D. (1992) *Aseptic Processing and Packaging of Food Products*. London: Elsevier Applied Science.

Hunter, G.M. (1972) Continuous sterilisation of liquid media containing suspended particles. *Food Technology in Australia* **24**, 158–165.

Kaplan, A.M., Reynolds, H. and Lichtenstein, H. (1954) Significance of variations in observed slopes of thermal death time curves for putrefactive anaerobes. *Food Research* **19**, 173–181.

Kooiman, W.J. (1974) The screw cap tube technique: a new and accurate technique for the determination of the wet heat resistance of bacterial spores. In *Spore Research 1973* ed. Barker, A.N., Gould, G.W. and Wolf, J. pp. 87–92. London: Academic Press.

Lindgren, B. and Swartling, P. (1963) *The Sterilizing Efficiency of the Alfa Laval Vacu-Therm Instant Sterilizer*. Milk and Dairy Research Report No. 69. Alnarp, Sweden: State Dairy Research Station.

Lynt, R.K., Solomon, H.M., Lilly, T., Jr and Kautter, D.A. (1977) Thermal death time of *Clostridium botulinum* type E in meat of the blue crab. *Journal of Food Science* **42**, 1022–1025, 1037.

Lynt, R.K., Kautter, D.A. and Solomon, H.M. (1979) Heat resistance of non-proteolytic *Clostridium botulinum* type F in phosphate buffer and crabmeat. *Journal of Food Science* **44**, 108–111.

Lynt, R.K., Kautter, D.A. and Solomon, H.M. (1981) Heat resistance of proteolytic *Clostridium botulinum* type F in phosphate buffer and crabmeat. *Journal of Food Science* **47**, 204–206, 230.

Lynt, R.K., Solomon, H.M. and Kautter, D.A. (1984) Heat resistance of *Clostridium botulinum* type G in phosphate buffer. *Journal of Food Protection* **47**(6), 463–466.

Martin, W. McK. (1950) *Martin Aseptic Canning System*, Decennial IFT Conference. Chicago: Institute of Food Technology.

Matsuda, N., Matsumoto, N. and Ushizawa, S. (1973) Mechanically simplified thermal death time tank. *The Canners Journal* 52(3), 255–261.

Matsuda, N., Komaki, M. and Matsunawa, K. (1983) Thermal death characteristics of spores of *Clostridium botulinum* 62A and *Bacillus stearothermophilus* subjected to programmed heat treatment. In *Heat Sterilization of Food* ed. Motohiro, T. and Hayakawa, K. pp. 51–62. Tokyo: Koseisha-Koseikaku.

Mechura, F.J. (1985) Shelf life of aseptically packaged food products in co-extruded plastic packages. In *Proceedings IUFoST Symposium on Aseptic Processing and Packaging of Foods*, pp. 290–299, Sept 9–12, Tylosand, Sweden, Lund University and SIK. Sweden: SIK.

Mikolajcik, E.M. and Rajkowski, K.T. (1980) Simple technique to determine heat resistance of *Bacillus stearothermophilus* spores in fluid systems. *Journal of Food Protection* **43**, 799–804.

Molin, G. (1977a) Dry heat resistance of *Bacillus subtilis* spores in contact with serum albumin, carbohydrates or lipids. *Journal of Applied Bacteriology* **42**, 111–116.

Molin, G. (1977b) Inactivation of *Bacillus* spores in dry systems at low and high temperatures. *Journal of General Microbiology* **101**, 227–231.

Molin, G. and Ostlund, K. (1975) Dry heat inactivation of *Bacillus subtilis* spores by means of infra red heating. *Antonie van Leeuwenhoek* **41**, 329–335.

Neaves, P. and Jarvis, B. (1978a) *Thermal Inactivation Kinetics of Bacterial Spores at Ultrahigh Temperatures with Particular Reference to* Clostridium botulinum. Research Report No. 280. Leatherhead, Surrey: Leatherhead Food RA.

Neaves, P. and Jarvis, B. (1978b) *Thermal Inactivation Kinetics of Bacterial Spores at Ultrahigh Temperatures with Particular Reference to* Clostridium botulinum*—second report*. Research Report No. 286. Leatherhead, Surrey: Leatherhead Food RA.

Odlaug, T.E. and Pflug, I.J. (1977) Thermal destruction of *Clostridium botulinum* spores suspended in tomato juice in aluminium thermal death time tubes. *Applied and Environmental Microbiology* **34**, 23–29.

Oquendo, R., Valdivieso, L., Stahl, R. and Loncin, M. (1975) Versuchsanlage zur thermischen Inaktivierung von Mikroorganismem bei Temperaturen über 120°C und vernachlässigbarer Aufheiz-und Abkühlphase. *Lebensmittel-Wissenschaft und Technologie* **8**, 181–182.

Perkin, A.G. (1974) A laboratory-scale ultra-high-temperature milk steriliser for batch operation. *Journal of Dairy Research* **41**, 55–63.

Perkin, A.G., Burton, H., Underwood, H.M. and Davies, F.L. (1977) Thermal death kinetics of *Bacillus stearothermophilus* spores at ultra high temperatures. II. Effect of heating period on experimental results. *Journal of Food Technology* **12**, 131–148.

Pflug, I.J. (1960) Thermal resistance of microorganisms to dry heat: design of apparatus, operational problems and preliminary results. *Food Technology* **14**, 483–487.

Pflug, I.J. (1987) *Textbook for an Introductory Course in Microbiology and Engineering of Sterilization Processes*. Union Street, Minneapolis: Environmental Sterilisation Laboratory.

Pflug, I.J. (1989) Biologically validating the sterilisation process delivered to particles in a heat-hold-cool system. In *The First International Congress on Aseptic Processing Technologies: Innovations in Aseptic Processing of Particulates* March 19–21 (1989) ed. Chambers, J.V. pp. 88-194. West Lafayette, IN: Purdue University.

Pflug, I.J. and Esselen, W.B. (1953) Development and application of an apparatus for study of thermal resistance of bacterial spores and thiamine at temperatures above 250°F. *Food Technology* 7, 237–241.

Pflug, I.J. and Esselen, W.B. (1954) Observations on the thermal resistance of putrefactive anaerobe No. 3679 spores in the temperature range of 250–300°F. *Food Research* 19, 92–97.

Pflug, I.J. and Holcomb, R.G. (1977) Thermal destruction of microorganisms. In *Disinfection, Sterilization and Preservation*, pp. 63-105. ed. Block, S.S. Philadelphia: Lea and Febiger.

Pheil, C.G., Pflug, I.J., Nicholas, R.C. and Augustin, J.A.L. (1967) Effect of various gas atmospheres on destruction of microorganisms in dry heat. *Applied Microbiology* 15, 120–124.

Quast, D.G., Leitão, M.F.F. and Kato, K. (1977) Death of *Bacillus stearothermophilus* 1518 spores on can covers exposed to superheated steam in a Dole aseptic canning system. *Lebensmittel-Wissenschaft und Technologie* 10, 198–202.

Rees, J.A.G. and Bettison, J. (1991) *Processing and Packaging of Heat Preserved Foods*. Glasgow: Blackie.

Reichart, O. (1979) A new experimental method for the determination of the heat destruction parameters of microorganisms. *Acta Alimentaria* 8, 131–155.

Resende, R., Stumbo, C.R. and Francis, F.J. (1969) Calculation of thermal processes for vegetable purée in capillary tubes at temperatures up to 350°F. *Food Technology* 23, 325–340.

Reynolds, H., Kaplan, A.M., Spencer, F.B. and Lichtenstein, H. (1952) Thermal destruction of Camerons putrefactive anaerobe 3679 in food substrates. *Food Research* 17, 153–167.

Richardson, P.S. (1987) Flame sterilisation—a review. *Journal of Food Technology* 22, 3–14.

Rönner, U. (1990) A new biological indicator for aseptic sterilisation. In *Food Technology International Europe 1990* ed. Turner, A. pp. 43–46 London: Sterling Publications International.

Ruschmeyer, O.R., Smith, G., Pflug, I.J., Gove, R. and Heisserer, Y. (1973) *Dry Heat Resistance of Selected Bacterial Spore Crops. Summary of Progress in the Project Environmental Microbiology as Related to Planetary Quarantine*. NASA NGL 24-005-160, pp. 3–32. Minnesota: University of Minneapolis.

Russell, A.D. (1982) *The Destruction of Bacterial Spores*. London: Academic Press.

Sakaguchi, G. (1986) Botulism. In *Progress in Food Safety*, Proceedings of symposium 'Progress in our Knowledge of Foodborne Disease during the Life of the Food Research Institute', held 28 May 1986 at University of Wisconsin-Madison, pp. 18–34. University of Wisconsin-Madison: Food Research Institute.

Schmidt, C.F. (1950) A method for the determination of the thermal resistance of bacterial spores. *Journal of Bacteriology* 59, 433–437.

Secrist, J.L. and Stumbo, C.R. (1956) Application of spore resistance in the newer methods of process evaluation. *Food Technology* 10, 543–545.

Secrist, J.L. and Stumbo, C.R. (1958) Some factors influencing thermal resistance values obtained by the thermoresistometer method. *Food Research* 23, 51–60.

Segner, W.P. and Schmidt, C.F. (1971) Heat resistance of spores of marine and terrestrial strains of *Clostridium botulinum* type C. *Applied Microbiology* 22, 1030–1033.

Sognefest, P., Hays, G.L., Wheaton, E. and Benjamin, H.A. (1948) Effect of pH on thermal process requirements of canned foods. *Food Research* 13, 400–416.

Srimani, B., Stahl, R. and Loncin, M. (1980) Death rates of bacterial spores at high temperatures. *Lebensmittel-Wissenschaft und-Technologie* 13, 186–189.

Stern, J.A. and Proctor, B.E. (1954) A micromethod and apparatus for the multiple determination of rates of destruction of bacteria and bacterial spores subjected to heat. *Food Technology* 8, 139–143.

Stumbo, C.R. (1948) A technique for studying resistance of bacterial spores to temperatures of the higher range. *Food Technology* 2, 228–240.

Stumbo, C.R. (1973) *Thermobacteriology in Food Processing*. New York: Academic Press.

Stumbo, C.R., Murphy, J.R. and Cochran, J. (1950) Nature of thermal death time curves for PA 3679 and *Clostridium botulinum*. *Food Technology* 4, 321–326.

Tanner, F.W. and Dack, G.M. (1922) *Clostridium botulinum*. *Journal of Infectious Diseases* 31, 92–100.

Townsend, C.T., Esty, J.R. and Baselt, F.C. (1938) Heat resistance studies on spores of putrefactive anaerobes in relation to determination of safe processes for canned foods. *Food Research* 3, 323–346.

Wadsworth, K.D. and Bassette, R. (1985) Laboratory scale system to process ultra high temperature milk. *Journal of Food Protection* 48(6), 530–531.

Wang, D.I.-C., Scharer, J. and Humphrey, A.E. (1964) Kinetics of death of bacterial spores at elevated temperatures. *Applied Microbiology* 12, 451–454.

White, T.C., Heidelbaugh, N.D. and McConnell, S. (1982) Survival of virus after thermoprocessing in capillary tubes. *Journal of Food Processing and Preservation* 6, 31–40.

Willhoft, E.M.A. (1993) *Aseptic Processing and Packaging of Particulate Foods*. Glasgow: Blackie Academic and Professional.

Williams, C.C., Merrill, C.M. and Cameron, E.J. (1937) Apparatus for determination of spore destruction rates. *Food Research* 2, 369–375.

Williams, D.J., Franklin, J.G., Chapman, H.R. and Clegg, L.F.L. (1957) Methods of assessing the sporicidal efficiency of an ultra-high-temperature milk sterilising plant. I. Experiments with suspensions of spores in water. *Journal of Applied Bacteriology* 20, 43–49.

Xezones, H., Segmiller, J.L. and Hutchings, I.J. (1965) Processing requirements for a heat tolerant anaerobe. *Food Technology* 19, 1001–1003

Journal of Applied Bacteriology Symposium Supplement 1994, **76**, 81S–90S

Tolerance of spores to ionizing radiation: mechanisms of inactivation, injury and repair

J. Farkas
Department of Refrigeration and Livestock Products Technology, University of Horticulture and Food Industry, Budapest, Hungary

1. Introduction, 81S
2. Tolerance of bacterial spores to ionizing radiation, 81S
3. Mechanism of radiation resistance of spores
   3.1 Chemical and structural factors, 82S
   3.2 Repair of radiation damage to spore DNA, 82S
4. Effect of environmental factors on radiation resistance of spores, 83S
5. Radiation injury of spores and its consequences
   5.1 Direct effects, 83S
   5.2 Effects of environmental factors on recovery of irradiated spores, 84S
   5.3 Increased heat sensitivity of irradiated spores, 84S
   5.4 Irradiation in combination with other adjuncts, 87S
6. Conclusions, 87S
7. Acknowledgements, 88S
8. References, 88S

1. INTRODUCTION

Radiation resistance of bacterial spores is of great practical importance both in radiation preservation of food and in radiation sterilization of medical products. Viruses and certain unusually radiation-resistant non-sporing bacteria are more resistant to radiation than are bacterial endospores. The latter, however, have high tolerance, not only to ionizing radiation but also to other antimicrobial factors. This makes them challenging subjects of investigation, although little information has been forthcoming in recent years. The action of ionizing radiation on micro-organisms and the repair of radiation damaged DNA in bacteria were reviewed by Moseley (1989). This paper attempts to review selected aspects of the effects of ionizing radiation on bacterial spores. It focuses on irradiation in the high-moisture environments that are the usual characteristic of food irradiation, with less emphasis on the equally important dry systems more common in radiation sterilization of medical products. The effects of ionizing radiation on dried spores and factors influencing their radiation resistance were reviewed by Tallentire (1970), and are detailed in the books of Russell (1982, 1992). Although ultraviolet (u.v.) radiation is also bactericidal, in view of the different practical connotations of its use, its effects are not considered in this paper. Medical and other uses of ionizing and u.v. radiation are considered by Dorpema (1990) and Russell (1992) respectively.

Correspondence to: J. Farkas, Department of Refrigeration and Livestock Products Technology, University of Horticulture and Food Industry, Ménesi ut 45, 1118 Budapest, Hungary.

2. TOLERANCE OF BACTERIAL SPORES TO IONIZING RADIATION

Radiation survival curves of bacterial spores have been reported. These show both semilogarithmically linear and sigmoid shapes, the latter with a more or less extensive 'shoulder' and, eventually, also a 'tail'. In some cases and in recovery media, low doses of irradiation may even result in an apparent increase in the number of colony forming units ('radiation activation') of a spore population (Levinson and Hyatt 1960; Roberts and Ingram 1965a; Gould and Ordal 1967). Population density does not seem to influence the kinetics of radiation inactivation of bacterial spores (Morgan and Reed 1954; Farkas *et al.* 1967).

Notwithstanding the occurrence of highly radiation-resistant specific non-sporing bacteria, spores are in general considerably more resistant to radiation (by a factor of about 5–15) than are vegetative cells of the same strain (Woese 1959; Farkas and Kiss 1964; Grecz 1965). This difference, however, is still much smaller than the difference between their heat resistance. Furthermore, e.g. in the genus *Bacillus*, there is a difference of almost 1000-fold between the heat resistance of spores of the most sensitive and the most resistant species, as measured and calculated by the *D*-values at 100°C (Murrell and Warth 1965), whereas studies of spores of a large number of species or strains have shown no more than fourfold differences in radiation resistance (expressed as *D*-values in the exponential part of the survival curve; Annellis and Koch 1962; Grecz 1965; Roberts and Ingram 1965a, b). It should be borne in mind, however, that there is often a considerable 'shoulder' (lag- or induction dose) with vegetative bacteria and spores before exponential loss of viability occurs.

It would be very difficult, and even misleading, to compare in a single list the radiation resistance characteristics of various spores as reported in the extensive literature, because experimental conditions during irradiation and during the counting of survivors vary from paper to paper. One should consider also the fact that a considerable variation may exist between the radiation resistance of different strains of some species. Nevertheless, in aqueous suspensions, the range of *D*-values given in previous reviews (Roberts and Ingram 1965a; Russell 1982; Grecz *et al.* 1983) seems to extend for various spores from *ca* 0·8 kGy to *ca* 2·5 kGy, and the width of the 'shoulder' varies between 0 and *ca* 6 kGy (Roberts and Ingram 1965a; Briggs 1966). It is unfortunate from the point of view of practical possibilities for the radiation sterilization of foods that spores of some strains of *Clostridium botulinum* types A and B are among the more highly radiation-resistant spores (Anellis and Koch 1962).

Pre-irradiation 'heat activation' does not seem to influence the radiation resistance of spores. Even lethal doses of radiation do not prevent the response of spores to physiological germinants which can evoke the first phase of germination, i.e. phase-darkening. Radiation inactivation of spores is expressed as the prevention of outgrowth and formation of vegetative cells. Thus, radiation injury is not associated with the germination lytic system but with the post-germination growth system. Similarly, the respiratory enzyme systems of spores are not inactivated even at radiation doses that reduce the count of viable spores by several log cycles (Levinson and Hyatt 1960; Farkas and Kiss 1964).

3. MECHANISM OF RADIATION RESISTANCE OF SPORES

3.1 Chemical and structural factors

Radiation inactivation of both spores and vegetative bacteria appears to be related primarily to radiation damage to DNA, particularly to the formation of single-strand and double-strand breaks (Moseley 1989).

Although both the heat resistance and radiation resistance of spores are high, structural and physico-chemical factors involved in heat resistance and radiation resistance are not necessarily the same. Indeed, there is no direct correlation between the radiation resistance and heat resistance of bacterial spores.

Vinter (1959) suggested that the high cyst(e)ine content of bacterial spores may play a role in tolerance to ionizing radiation. Hitchins *et al.* (1966), however, who investigated the effects of reducing disulphide bonds and used thiol-blocking agents suggest that the radiation resistance in spores cannot be due to a disulphide-rich structure.

While experimental evidence supports the premise that the high levels of dipicolinic acid (DPA) in bacterial spores governs to a certain extent their u.v. resistance (Berg and Grecz 1970; Kamat and Pradham 1987), their resistance to γ-radiation does not seem to be directly influenced by DPA content (Rowley and Newcomb 1964).

Other works point to the unusual conformation of DNA in the intact spore structure, which differs from that in the vegetative cells, as a possible cause of increased radiation resistance (Donellan and Setlow 1965; Tanooka and Sakakibara 1968).

It would appear that spore DNA is not intrinsically radiation resistant, but has resistance imposed on it by compositional and structural factors within the spore, i.e. the relatively 'dehydrated' state of the spore core and the specific conformation of the spore DNA may be responsible for its resistance (Gould 1989).

Neither α/β-type nor γ-type small acid-soluble spore proteins (SASPs) appear to be involved in γ-radiation resistance (Setlow 1988 and this Symposium, pp. 49S–60S).

3.2 Repair of radiation damage to spore DNA

It is probable that survival curves mirror not simply probabilities of 'hits' on static 'targets', but also distribution of resistance in a spore population. Differences in resistance of individual cells may be related to qualitative and quantitative differences in their chemical composition and fine structures as well as in their repair capacity. The same may be true for the overall average resistance of various strains and species. These considerations do not exclude the possibility that damage to DNA may be a major cause of radiation inactivation, but other components may also be involved in radiation damage. The regeneration processes can be considered also in the dynamic character of cells, i.e. the repair abilities of cells in a population may be mirrored for example in the length of the shoulder and the slope of the survival curve.

Because of the cryptobiotic nature of dormant spores, it is generally considered that metabolic DNA repair does not occur to any significant extent in them. Repair enzymes, however, may be present in an inactive state in dormant spores and may be activated during germination (Terano *et al.* 1969). Spores are well suited for maintenance of viability by the DNA-ligase mechanism. In comparative studies of the relatively radiation-sensitive spores of *Cl. botulinum* strain 51B and the radiation-resistant spores of *Cl. botulinum* strain 33A showing a shoulder of 2–6 kGy in the survival curve, Durban *et al.* (1974) observed that (1) after 3 kGy γ-radiation fewer single strand breaks (SSBs) were found in the DNA of the radiation-resistant strain than in that of the radiation-sensitive strain, and (2) in spores of the radiation-resistant strain, but not of the radiation-sensitive strain, rejoining of DNA SSBs occurred under anaerobic conditions during or immediately after irradiation.

These observations were attributed to the effect of the polynucleotide ligase enzyme, which does not require a substrate other than the damaged DNA. This type of spore recovery could be inhibited by various chelators, and the effect of chelators could be reversed by Mg^{2+} (Grecz *et al.* 1978). Such repair was almost completely eliminated by the presence of oxygen during irradiation of the spores. Recovery of spores of *Cl. botulinum* in the exponential portion of the survival curve seems to require excision–resynthesis DNA repair *after* germination, and may depend on the synthesis of new enzymes (Grecz *et al.* 1977).

4. EFFECT OF ENVIRONMENTAL FACTORS ON RADIATION RESISTANCE OF SPORES

Morgan and Reed (1954) evaluated the effect of environmental factors during spore formation, and reported that *B. coagulans* spores formed at 28°C were more resistant to γ-radiation than those produced at 55°C. The composition of the sporulation medium may also influence the radiation resistance of spores (Yamazaki *et al.* 1968).

The radiation resistance of bacterial spores is influenced much less by environmental factors during irradiation than that of vegetative cells (Powers *et al.* 1960; Ingram and Thornley 1961; Ma and Maxcy 1981). Nevertheless, over a wide range of temperatures, the effects of temperature and freezing may be considerable (Grecz *et al.* 1971; Anellis *et al.* 1977).

The maximum kill of irradiated spores has often been observed to occur at 0°C (Friedman and Grecz 1974). At lower temperatures survival increases as a result of the trapping of radiation-induced harmful free radicals by ice. Above this temperature, particularly at high but sublethal temperatures (50–70°C), radiation resistance may be higher than at or below room temperature. This is probably due to enhanced annealment of free radicals at higher temperatures ('thermal annealment'), preventing them from causing radiation damage ('thermorestoration effect', Webb *et al.* 1960).

Radiation resistance of *B. cereus* spores was not influenced by the pH of suspending citrate/phosphate buffer between pH 5·0 and 8·0, but decreased when the pH was reduced below pH 5·0 (Farkas *et al.* 1967). These observations are in agreement with those reported by Masokhina-Porshnyakova and Ladukhina (1967) with spores of *Cl. botulinum*, while Grecz and Upadhyay (1970) found that the effect of pH on the radiation-sensitivity of spores increased when the temperature was reduced to freezing point or lower. The addition of sodium nitrite to the medium sensitized *Cl. botulinum* spores to irradiation (Rowley *et al.* 1971).

Radiation resistance of spores increases with decreasing water activity, although this effect of a_w is less dramatic for radiation resistance than for heat resistance (Härnulv and Snygg 1973). Some media have a radioprotective effect which manifests itself mainly as an increase in the length of the shoulder of the survival curve (Wheaton and Pratt 1962; Grecz *et al.* 1977).

Grecz *et al.* (1977) analysed radiation survival curves of spores of fourteen strains of *Cl. botulinum* irradiated in poor and rich environments in the presence of radioprotective substances and found no strong correlation between the length of the shoulder part of the survival curve and the slope of the exponential part. The length of shoulder varied more between the different strains than the slope of the survival curve. Irradiation of spores in a radioprotective environment influenced the length of the shoulder more than the slope, and the slope of the exponential part decreased by a constant value, irrespective of the slope in a less protective environment. Spores that showed a long shoulder in a protection-free environment showed a greater increase of the shoulder in a protective environment than spores with a shorter shoulder in a non-protective environment.

5. RADIATION INJURY OF SPORES AND ITS CONSEQUENCES

5.1 Direct effects

In addition to inactivation of spores and damage to DNA, radiation can have other effects on spore properties. In some cases, sublethal doses of ionizing radiation may stimulate the germination of spores (Ando and Karashimada 1972; Farkas and Roberts 1976).

The smaller the fraction of a spore population that survives radiation treatment, the greater will be the radiation damage accumulated in less critical sites of spores. Increasing doses of ionizing radiation cause increasing losses of phase brightness of spores even under germinant-free conditions, which can be detected microscopically or turbidimetrically (Levinson and Hyatt 1960; Farkas and Kiss

1964). This 'pseudo-germination' is accompanied by a considerable loss of low molecular weight substances, including ninhydrin-positive compounds and DPA (Levinson and Hyatt 1960; Farkas and Kiss 1965) into the suspending media. This effect of high doses of irradiation appears to be similar to that of ultrasonic treatment (Palacios *et al.* 1991). Although heat inactivation of spores is also accompanied by loss of DPA, heat treatment caused much less exudation of ninhydrin-positive substances than radiation treatment that resulted in a similar lethality (Farkas and Kiss 1965). Thus, the exudates from irradiated spores are similar to those released during physiological germination, and there are also similarities between direct effects of high doses of ionizing radiation on spores and the effects of 'mechanical' germination (Rode and Foster 1960). These early observations on the increased permeability of irradiated spores indicated that it runs parallel with the damage of spore components that prevent hydration of the core.

5.2 Effects of environmental factors on recovery of irradiated spores

Demands on recovery media, and their modification by radiation damage, may differ widely, even for species belonging to the same genera (Roberts 1970; Farkas and Roberts 1982). As in heat-induced injury, radiation-injury is expressed as an increase in the sensitivity of the surviving spores to some secondary stress. Irradiation of spores increases their sensitivity to the pH of the recovery medium and this effect is enhanced when the incubation temperature deviates from the optimum temperature for growth (Farkas *et al.* 1967). Irradiation of spores also increases their sensitivity to sodium chloride in the plating medium (Roberts *et al.* 1965; Kiss *et al.* 1978). It is of interest that while $NaNO_2$ in the recovery medium reduced the recovery of untreated *Cl. botulinum* spores, it enhanced the recovery of irradiated spores (Rowley *et al.* 1971).

5.3 Increased heat sensitivity of irradiated spores

Kempe and his coworkers (1954) and Morgan and Reed (1954) first reported that bacterial spores that survived irradiation showed a decrease in heat resistance. This is illustrated by the results of some of Farkas and Robert's studies shown in Tables 1 and 2. The reverse order of combination treatment usually causes merely an additive effect.

Table 1 shows that synergistic reduction of colony-forming units by the combined treatment is enhanced if the incubation temperature of plates is lower than the optimum temperature for growth. The synergistic effect is increased by an increase in the radiation damage to spores as illustrated in Table 2. Turbidimetric measurements of spore suspensions and spectrophotometric investigations of their supernatant fluids strengthened the observation that synergistic combinations of pre-irradiation and heat treatment induce more drastic changes in spores than the reverse order of the same treatments (Farkas and Roberts 1976).

With pre-irradiated spores, heat treatment at a higher temperature resulted in greater synergism than heat treatment at a lower temperature (Table 3). Similar findings on the effect of radiation dose and heating temperature on the heat sensitization of *Cl. perfringens* spores by γ-irradiation were reported by Gombas and Gomez (1978).

The type of suspending medium and its pH during irradiation also have an influence on the radiation–heat interaction of spores (Shamsuzzaman and Lucht 1991).

Table 1 Combined effects of heat treatment, irradiation and recovery temperature on colony formation from spores of *Bacillus cereus* T (Farkas and Roberts, unpublished)

	Percentage apparent survival of spores at recovery temperatures of (°C)	
Treatment	30	20
Untreated	100 (10·2)*	100 (10·3)
90°C, 12 min	19·2	12·9
4 kGy (at 25°C)	8·3	7·6
90°C, 12 min + 4 kGy	1·6	0·9
4 kGy + 90°C, 12 min	0·0001	0·00002

* Initial log cfu on yeast–glucose agar are in brackets.

Table 2 The role of irradiation and heat in the combined effect of these treatments on spores of *Bacillus* 'F', a salt tolerant, denitrifying *Bacillus* isolated by Eddy and Ingram (1956), which probably belongs to the *B. circulans* group (Farkas and Roberts, unpublished)

Treatment	Survivors log cfu*	Synergy factor†
Untreated	9·7	—
0·5 kGy	9·4	—
4 kGy	7·7	—
90°C, 27 min	9·4	—
90°C, 95 min	7·7	—
0·5 kGy + 90°C, 27 min	8·3	1·9
0·5 kGy + 90°C, 95 min	6·4	10·8
4 kGy + 90°C, 27 min	4·5	772
4 kGy + 90°C, 95 min	1·5	15000

* Determined by plating on yeast–glucose agar.

† Synergy factor: number of cfu expected from an additive effect divided by cfu actually observed.

Table 3 The influence of heating temperature on the synergistic effect of irradiation plus heat treatment on spores of *Clostridium botulinum* type A no. 7272 (Farkas and Roberts, unpublished)

Treatment	Survivors* log cfu	Synergy factor†
Untreated	9·2	—
4 kGy	8·6	—
80°C, 30 min	8·3	—
85°C, 10 min	8·2	—
80°C, 30 min + 4 kGy	7·6	1·3
85°C, 10 min + 4 kGy	7·3	2·0
4 kGy + 80°C, 30 min	7·3	2·4
4 kGy + 85°C, 10 min	6·8	6·3

* Determined by plating in reinforced clostridial agar.
† See footnote to Table 2.

In studies on the effect of recovery medium on the synergistic inactivation, it was found that spores that survived irradiation followed by heat treatment were much more sensitive to inhibition by NaCl in the recovery medium than were untreated spores (Tables 4 and 5). If heating preceded irradiation, the effect of this triple combination remained approximately additive. The use of potassium nitrite instead of NaCl or a mixture of the two salts in these experiments showed that their effect in the recovery medium was due mainly to the reduction of a_w or increase of ionic strength, rather than to a specific effect of the salt.

With the nitrate-reducing spore-former, *Bacillus* 'F' (see Table 2), addition of $NaNO_2$ to the recovery medium increased the recovery of spores surviving radiation or heat treatment (Table 6). A combination of pre-irradiation and subsequent heat treatment, however, still resulted in a synergistic reduction of colony forming units.

This synergistic effect has been demonstrated not only with pure cultures, but with the native spore flora of spices (Farkas *et al.* 1973; Kiss and Farkas 1981; Farkas and Andrássy 1985) and flours (Farkas and Andrássy 1981).

The radiation-induced heat sensitivity of spores persists for a holding period of at least several weeks at room temperature in aqueous suspensions (Shamsuzzaman *et al.* 1990), and for several months in dry media such as flour and spices (Farkas and Andrássy 1985).

In some circumstances the simultaneous application of heat and radiation ('thermoradiation') has been shown to cause synergistic destruction of micro-organisms (Grecz *et al.* 1971). Fisher and Pflug (1977) reported that the thermoradiation synergistic mechanism shows a proportional dependency on radiation dose rate, an Arrhenius dependency on temperature, and that dry-heat thermoradiation is affected by relative humidity.

An explanation of the heat sensitizing effect of irradiation on spores was first proposed by Vinter (1959) as a result of mobilization of bound calcium, which may result in a reduced heat stability of protein structures. It may be due to heat inactivation of enzymes repairing DNA injuries (Grecz *et al.* 1981). Stegeman *et al.* (1977), who investigated the correlation between the ionic state of bacterial spores and the synergistic effect of combined radiation and heat treatment, concluded that radiation-induced breaks in, or decarboxylation of, cortex peptidoglycan may be responsible for the heat-sensitization of bacterial spores by ionizing radiation, including the weakening of osmoregulatory or core-dehydrating mechanisms. The

Table 4 The combined effect of irradiation, heat treatment and concentration of sodium chloride in the plating medium on formation of colonies from spores of *Clostridium botulinum* strains (Farkas and Roberts, unpublished)

Treatment	Log cfu in reinforced clostridial agar containing sodium chloride (% w/v)	
	0·5	6·0
Clostridium botulinum 7272A		
Untreated	9·2	7·8
4 kGy	8·6	4·8
80°C, 30 min	8·3	6·4
80°C, 30 min + 4 kGy	7·6 (7·7)*	3·2 (3·4)
4 kGy + 80°C, 30 min	7·3 (7·7)	<1·5 (3·4)
	0·5	2·5
Clostridium botulinum 1304E		
Untreated	7·2	6·0
4 kGy	5·4	3·3
70°C, 60 min	7·0	5·2
70°C, 60 min + 4 kGy	5·1 (5·4)	2·8 (2·5)
4 kGy + 70°C, 60 min	<1·8 (5·4)	<1·8 (2·5)

* Levels of assumed additive effect in brackets.

Table 5 The combined effect of irradiation, heat treatment and concentration of sodium chloride in the plating medium on formation of colonies from spores of *Bacillus* spp. (Farkas and Roberts, unpublished)

Treatment	Log cfu on yeast–glucose agar containing added sodium chloride of (%) 0		2	
Bacillus cereus T				
Untreated	10·2		10·3	
4 kGy	9·0		8·4	
90°C, 12 min	9·4		7·0	
90°C, 12 min + 4 kGy	8·3	(8·2)*	4·4	(5·1)
4 kGy + 90°C, 12 min	4·2	(8·2)	<2·5	(5·1)
	0		8	
Bacillus 'F'				
Untreated	9·5		9·3	
4 kGy	7·5		7·3	
90°C, 93 min	6·0		4·6	
90°C, 93 min + 4 kGy	3·9	(4·0)	2·8	(2·6)
4 kGy + 90°C, 93 min	<1·8	(4·0)	<1·8	(2·6)

* Levels of assumed additive effect in brackets.

latter was demonstrated by Gomez *et al.* (1980): when γ-irradiated spores were heated in the presence of increasing concentrations of glycerol and sucrose, the heat sensitivity induced by irradiation decreased progressively.

These hypotheses do not contradict, but rather complement each other. In a similar way to the trigger mechanism of spore germination, radiation damage to the cortex allows partial rehydration of the spore core. This may solubilize and release calcium dipicolinate and the partially rehydrated core results in an increased heat sensitivity of core components. This theory is supported also by the author's recent studies on the combined effects of ultrasonic treatment, γ-radiation and heat treatment on aqueous suspensions of aerobic bacterial spores (Farkas 1991). Lethalities of individual treatments and combinations on aqueous suspensions of *B. subtilis* spores are shown in Fig. 1. Although sonication alone had very little sporicidal effect, it enhanced significantly the synergistic inactivating effect of irradiation and heat treatment. Storage of treated suspensions for 1 d at room temperature prior to plating did not influence the lethalities observed.

According to the author's recent studies with the relatively heat-sensitive spores of *B. cereus* T (Farkas 1991), the increased heat sensitivity of irradiated spores can also be observed by differential scanning calorimetry which shows the temperature characteristics of endothermic transitions correlating with heat activation and spore death, respectively (Table 7). Apparently, by weakening the spore structures responsible for the refractory homeostasis (Gerhardt 1988), irradiation or ultrasonic treatment causes changes that decrease the heat stability of proteins, including those in the ribosomes which, it has been suggested, are the rate-

Table 6 The combined effect of irradiation, heat treatment and sodium nitrite in the plating medium on formation of colonies from spores of *Bacillus* 'F' (Farkas and Roberts, unpublished)

Treatment	Log cfu on yeast–glucose agar containing sodium nitrite ($\mu g\ g^{-1}$)† 0		150	
Untreated	9·5		9·6	
4 kGy	7·5		8·1	
90°C, 93 min	6·0	(4·0)*	8·4	
90°C, 93 min + 4 kGy	3·9	(4·0)	6·1	(6·9)
4 kGy + 90°C, 93 min	<1·8	(4·0)	<1·8	(6·9)

* Levels of assumed additive effect in brackets.

† The pH of the medium was adjusted to pH 6·0 and filter-sterilized sodium nitrite solution was given to the autoclave-sterilized medium directly before the plating.

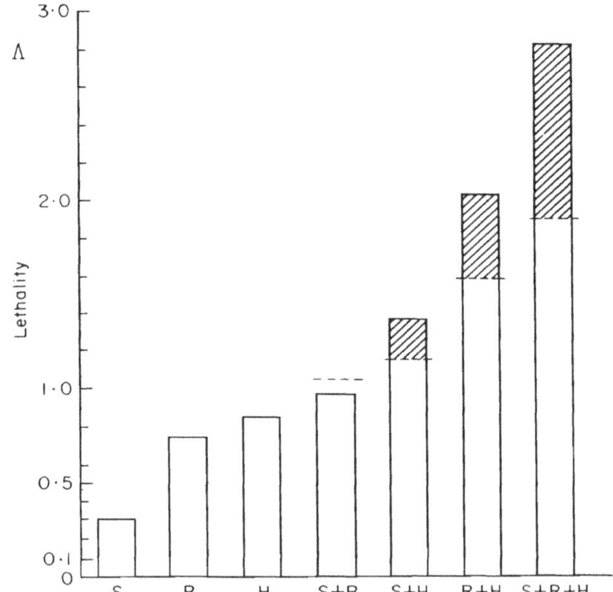

Fig. 1 Lethalities of ultrasonic, radiation and heat treatment alone and in combination on spores of *Bacillus subtilis* (Farkas 1991). Twin columns represent less than 1 h and 24 h standing, respectively, of treated aqueous spore suspensions at room temperature prior to plate counting. The shaded areas illustrate the synergistic lethality. S, Ultrasonic treatment (20 kHz, 120 W, 30 min); R, 2 kGy irradiation with γ-rays; H, 99°C, 60 min

limiting primary targets in the heat killing of microorganisms (Anderson *et al.* 1991; Mackey *et al.* 1991; Belliveau *et al.* 1992).

5.4 Irradiation in combination with other adjuncts

Under certain conditions, high hydrostatic pressures sensitize spores to radiation (Wills 1974). This may be due to partial rehydration of the core that occurs as a result of irradiation of the spore cortex peptidoglycan, and to the spore germinative effect of pressure (Gould and Jones 1989).

Effects of ionizing radiation on spores in combination with various inorganic and organic adjuncts have been also explored (Gould 1970; Gould and Jones 1989). However, the mechanisms of action of such synergistic effects are not known for certain, and many *in vitro* effects cannot be implemented in foodstuffs.

6. CONCLUSIONS

In principle, all physical and chemical factors that are able to damage the integrity of spore membranes and cortex structure can decrease the resistance of spores to antimicrobial treatments. From the practical point of view, the combination of irradiation, heat treatment and/or inhibitory factors offers methods for preventing the survival and growth of these resistant forms of bacteria. A better understanding of the synergism will require additional knowledge of the nature of sublethal injuries. Enhancement of injury, based on a better understanding of the mechanisms involved, may well contribute to the development of improved preservation and sterilization techniques (Gould 1989). With the wish of consumers to minimize chemical preservation, there is a clear potential to develop a process combining irradiation plus heat, reduced a_w and perhaps other preservative factors to make shelf-stable food products. Data in the literature are not organized adequately to permit this without some additional research. An approach similar to that used in developing predictive modelling would seem the most cost-effective and would lead to

Table 7 Characteristic endotherms in the DSC thermograms of *Bacillus cereus* T spores formed on Ca-poor (−Ca) and Ca-rich (+Ca) media as a function of the radiation treatment (Farkas 1991)*

		Temperature (°C)			
Sporulation medium†	Radiation dose (kGy)	'Heat activation' peak	Shoulder of main endotherm	Peak attributable to heat killing	Post-mortem peak
−Ca	0	—	76·4	86·0	—
−Ca	3	—	74·6	84·7	—
+Ca	0	68·8	80·7	87·2	98·0
+Ca	3	70·2	80·0	85·2	96·7

* Thermograms were established by SETARAM micro-DSC equipment with an upper temperature limit of 100°C. Table values are means of duplicate measurements.

† Nutrient agar containing 0·5% tryptone, 0·25% yeast extract, 0·15% glucose and 1·5% agar–agar containing 3·7 mmol l^{-1} Ca was considered as Ca-poor medium (−Ca); the medium rich in Ca (+Ca) was of the same composition, but 20 mmol l^{-1} $CaCl_2$ was added to it prior to sterilization. Spores formed on (−Ca)-medium contained 161 ± 14 μg Ca/10^{10} spores; spores produced on (+Ca)-medium had a Ca-content of 216 ± 2·5 μg Ca/10^{10} spores.

greater assurance of safety, e.g. with respect to *Cl. botulinum* (Roberts 1989).

7. ACKNOWLEDGEMENTS

The author thanks Dr Barbara M. Lund and Dr T.A. Roberts for valuable advice and discussions.

8. REFERENCES

Anderson, W.A., Hedges, N.D., Jones, M.V. and Cole, M.B. (1991) Thermal inactivation of *Listeria monocytogenes* studied by differential scanning calorimetry. *Journal of General Microbiology* **137**, 1419–1424.

Ando, Y. and Karashimada, T. (1972) Radiation resistance of spores of *Clostridium botulinum* type E. III. Effect of exposure of spores to a low dose level of gamma-irradiation on subsequent growth and toxin production (in Japanese). *Food Irradiation [Shokuhin-Shosha]* 7(1), 1–5.

Anellis, A. and Koch, R.B. (1962) Comparative resistance of strains of *Clostridium botulinum* to gamma rays. *Applied Microbiology* **10**, 326–330.

Anellis, A., Berkowitz, D. and Kemper, D. (1977) Comparative radiation death kinetics of *Clostridium botulinum* spores at low-temperature gamma irradiation. *Journal of Food Protection* **40**, 313–316.

Belliveau, B.M., Beaman, T.C., Pankratz, H.S. and Gerhardt, P. (1992) Heat killing of bacterial spores analyzed by differential scanning calorimetry. *Journal of Bacteriology* **174**, 4463–4474.

Berg, P.E. and Grecz, N. (1970) Relationship of dipicolinic acid content in spores of *Bacillus cereus* T to ultraviolet gamma radiation resistance. *Journal of Bacteriology* **103**, 517–519.

Briggs, A. (1966) The resistance of spores of the genus *Bacillus* to phenol, heat and radiation. *Journal of Applied Bacteriology* **29**, 490–504.

Donellan, F.J. and Setlow, R.B. (1965) Thymine photoproducts but not thymine dimers found in ultraviolet-irradiated bacterial spores. *Science* **149**, 303–310.

Dorpema, J.W. (1990) Review and state of the art on radiation sterilization of medical devices. *Radiation Physics and Chemistry* **35**, 357–360.

Durban, E., Grecz, N. and Farkas, J. (1974) Direct enzymatic repair of DNA single strand breaks in dormant spores. *Journal of Bacteriology* **118**, 129–138.

Eddy, B.P. and Ingram, M. (1956) A salt-tolerant denitrifying *Bacillus* strain which 'blows' canned bacon. *Journal of Applied Bacteriology* **19**, 62–70.

Farkas, J. (1991) *Studies on the Resistance of Bacterial Spores and the Mechanism of Combination of Factors Inactivating Spores* (in Hungarian). Report to the National Foundation for Scientific Research, OTKA 153. Budapest: University of Horticulture and Food Industry.

Farkas, J. and Andrássy, É. (1981) Decrease of bacterial spoilage of bread by low dose irradiation of its flour. In *Combination Processes in Food Irradiation*, pp. 81–94. Vienna: International Atomic Energy Agency.

Farkas, J. and Andrássy, É. (1985) Increased sensitivity of surviving bacterial spores in irradiated spices. In *Fundamental and Applied Aspects of Spores* ed. Dring, G.J., Ellar, D.J. and Gould, G.W. pp. 397–407. London: Academic Press.

Farkas, J. and Kiss, I. (1964) Observations on radiation-induced damage in bacterial spores (in Hungarian). *Communications of the Central Food Research Institute, Budapest* **1**, 13–19.

Farkas, J. and Kiss, I. (1965) Observations on biochemical changes in irradiated spores of *Bacillus cereus*. *Acta Microbiologica*, **12**, 15–28.

Farkas, J. and Roberts, T.A. (1976) The effect of sodium chloride, gamma irradiation and/or heat on germination and development of spores of *Bacillus cereus* T in single germinants and complex media. *Acta Alimentaria* **5**, 289–302.

Farkas, J. and Roberts, T.A. (1982) The effect of the composition of the recovery medium upon the colony-forming capacity of clostridial spores damaged by gamma-radiation. *Acta Alimentaria* **10**, 393–398.

Farkas, J., Kiss, I. and Andrássy, É. (1967) The survival and recovery of irradiated bacterial spores as affected by population density and some external factors. In *Radiosterilization of Medical Products*, pp. 343–354. Vienna: International Atomic Energy Agency.

Farkas, J., Beczner, J. and Incze, K. (1973) Feasibility of irradiation of spices with special reference to paprika. In *Radiation Preservation of Food*, pp. 383–402. Vienna: International Atomic Energy Agency.

Fisher, D.A. and Pflug, I.J. (1977) Effect of combined heat and radiation on microbial destruction. *Applied and Environmental Microbiology* **33**, 1170–1176.

Friedman, Y.S. and Grecz, N. (1974) The role of water in thermorestoration of bacterial spores. *Acta Alimentaria* **3**, 251–265.

Gerhardt, P. (1988) The refractory homeostasis of bacterial spores. In *Homeostatic Mechanism in Micro-organism*, FEMS Symposium No. 44, pp. 41–49. Bath: Bath University Press.

Gombas, D.E. and Gomez, R.F. (1978) Sensitization of *Clostridium perfringens* spores to heat by gamma radiation. *Applied and Environmental Microbiology* **36**, 403–407.

Gomez, R.F., Gombas, D.E. and Herrero, A. (1980) Reversal of radiation dependent heat sensitization of *Clostridium perfringens* spores. *Applied and Environmental Microbiology* **39**, 525–529.

Gould, G.W. (1970) Potentiation by halogen compounds of the lethal action of gamma-radiation on spores of *Bacillus cereus*. *Journal of General Microbiology* **64**, 289–300.

Gould, G.W. (ed.) (1989) *Mechanism of Action of Food Preservation Procedures*. London and New York: Elsevier Applied Science.

Gould, G.W. and Jones, M.V. (1989) Combination and synergistic effects. In *Mechanisms of Action of Food Preservation Procedures* ed. Gould, G.W. pp. 401–421. London and New York: Elsevier Applied Science.

Gould, G.W. and Ordal, R.J. (1967) Activation of spores of *Bacillus cereus* by gamma-radiation. *Journal of General Microbiology* **50**, 77–84.

Grecz, N. (1965) Biophysical aspects of *Clostridia*. *Journal of Applied Bacteriology* **28**, 17–35.

Grecz, N. and Upadhyay, J. (1970) Radiation survival of bacterial spores in neutral and acid ice. *Canadian Journal of Microbiology* **16**, 1045–1049.

Grecz, N., Walker, A.A., Anellis, A. and Berkovitz, D. (1971) Effect of irradiation temperature in the range −196 to 95 degree on the resistance of spores of *Clostridium botulinum* 33A in cooked beef. *Canadian Journal of Microbiology* **17**(2), 135–142.

Grecz, N., Lo, H., Kang, T. and Farkas, J. (1977) Characteristics of radiation survival curves of spores of *Clostridium botulinum* strains. In *Spore Research 1976* ed. Barker, A.N., Wolf, L.J., Ellar, D.J., Dring, G.J. and Gould, G.W. Vol. II, pp. 603–630. London: Academic Press.

Grecz, N., Wiatr, C. and Farkas, J. (1978) *In vivo* evidence for the role of DNA-ligase in radiation resistance of *Clostridium botulinum* 33A. In *Food Preservation by Irradiation*, Vol. II, pp. 135–143. Vienna: International Atomic Energy Agency.

Grecz, N., Bruszer, G. and Amin, I. (1981) Effect of radiation and heat on bacterial spore DNA. In *Combination Processes in Food Irradiation*, pp. 3–20. Vienna: International Atomic Energy Agency.

Grecz, N., Rowley, D.B. and Matsuyama, A. (1983) The action of radiation on bacteria and viruses. In *Preservation of Food by Ionizing Radiation* ed. Josephson, E.S. and Peterson, M.S. Vol. II, pp. 167–218. Boca Raton, Fl.: CRC Press.

Härnulv, B.G. and Snygg, B.G. (1973) Radiation resistance of spores of *Bacillus subtilis* and *B. stearothermophilus* at various water activities. *Journal of Applied Bacteriology* **36**, 677–682.

Hitchins, A.D., King, W.L. and Gould, G.W. (1966) Role of disulphide bonds in the resistance of *Bacillus cereus* spores to gamma irradiation and heat. *Journal of Applied Bacteriology* **20**, 505–511.

Ingram, M. and Thornley, M.J. (1961) The effect of low temperatures on the inactivation by ionising radiations of *Clostridium botulinum* spores in meat. *Journal of Applied Bacteriology* **24**, 94–103.

Kamat, A.S. and Pradhan, D.S. (1987) Involvement of calcium and dipicolinic acid in the resistance of *Bacillus cereus* BIS-59 spores to u.v. and gamma radiations. *International Journal of Radiation Biology* **31**(1), 7–18.

Kempe, L.L., Graikoski, J.T. and Gillies, R.A. (1954) Gamma ray sterilization of canned meat previously inoculated with anaerobic spores. *Applied Microbiology* **2**, 330–352.

Kiss, I. and Farkas, J. (1981) Combined effect of gamma irradiation and heat treatment on microflora of spices. In *Combination Processes in Food Irradiation*, pp. 107–115. Vienna: International Atomic Energy Agency.

Kiss, I., Rhee, C.P., Grecz, N., Roberts, T.A. and Farkas, J. (1978) Relation between radiation resistance and salt sensitivity of spores of five strains of *Clostridium botulinum* types A, B and E. *Applied and Environmental Microbiology* **35**, 533–539.

Levinson, H.S. and Hyatt, M.T. (1960) Some effects of heat and ionizing radiation on spores of *Bacillus megaterium*. *Journal of Bacteriology* **80**, 441–451.

Ma, K. and Maxcy, R.B. (1981) Factors influencing radiation resistance of vegetative bacteria and spores associated with radappertization of meat. *Journal of Food Science* **46**(2), 612–616.

Mackey, B.M., Miles, C.A., Parsons, S.E. and Seymour D.A. (1991) Thermal denaturation of whole cells and cell components of *Escherichia coli* examined by differential scanning calorimetry. *Journal of General Microbiology* **137**, 2361–2374.

Masokhina-Porshnyakova, N.N. and Ladukhina, G.V. (1967) The effect of ionizing radiation on *Clostridium botulinum* spores. In *Microbiological Problems in Food Preservation by Irradiation*, pp. 89–98. Vienna: International Atomic Energy Agency.

Morgan, B.H. and Reed, J.M. (1954) Resistance of bacterial spores to gamma irradiation. *Food Research* **19**, 357–366.

Moseley, B.E.B. (1989) Ionizing radiation: action and repair. In *Mechanism of Action of Food Preservation Procedures* ed. Gould, G.W. pp. 43–70. London and New York: Elsevier Applied Science.

Murrell, W.G. and Warth, A.D. (1965) Composition and heat resistance of bacterial spores. In *Spores III* ed. Campbell, L.L. and Halvorson, H.O. pp. 1–24. Ann Arbor: American Society for Microbiology.

Palacios, P., Burgos, J., Hoz, L., Saviz, B. and Ordónez, J.A. (1991) Study of substances released by ultrasonic treatment from *Bacillus stearothermophilus* spores. *Journal of Applied Bacteriology* **71**, 445–451.

Powers, E.L., Webb, R.B. and Ehret, C.F. (1960) Storage transfer and utilization of energy from X-rays in dry bacterial spores. *Radiation Research, Supplement* **2**, 94–121.

Roberts, T.A. (1970) Recovering spores damaged by heat, ionizing radiation or ethylene oxide. *Journal of Applied Bacteriology* **33**, 74–94.

Roberts, T.A. (1989) Combinations of antimicrobials and processings methods. *Food Technology* **43**(1), 156–163.

Roberts, T.A. and Ingram, M. (1965a) Radiation resistance of spores of *Clostridium* species in aqueous suspension. *Journal of Food Science* **30**, 879–885.

Roberts, T.A. and Ingram, M. (1965b) The resistance of spores of *Clostridium botulinum* type E to heat and radiation. *Journal of Applied Bacteriology* **28**, 125–139.

Roberts, T.A., Ditchett, P.J. and Ingram, M. (1965) Effect of sodium chloride on radiation resistance and recovery of irradiated anaerobic spores. *Journal of Applied Bacteriology* **28**, 336–348.

Rode, L.J. and Foster, J.W. (1960) Mechanical germination of bacterial spores. *Proceedings of the National Academy of Sciences* **46**, 118–128.

Rowley, D.B. and Newcomb, H.R. (1964) Radiosensitivity of several dehydrogenases and transaminases during sporogenesis of *Bacillus subtilis*. *Journal of Bacteriology* **87**, 701–709.

Rowley, D.B., Feeherry, F. and Powers, E. (1971) Effect of curing salt on the gamma radiation resistance and recovery of *Clostridium botulinum* type 62A spores. *Bacteriological Proceedings*, 64.

Russell, A.D. (1982) *The Destruction of Bacterial Spores*. London: Academic Press.

Russell, A.D. (1992) Ultraviolet radiation. In *Principles and Practice of Disinfection, Preservation and Sterilization* ed. Russell, A.D., Hugo, W.B. and Ayliffe, G.A.J. 2nd edn, pp. 544–556. Oxford: Blackwell Scientific Publications.

Setlow, P. (1988) Small acid-soluble, spore proteins of *Bacillus* species: structure, synthesis, function and degradation. *Annual Review of Microbiology* **42**, 319–338.

Shamsuzzaman, K. and Lucht, L. (1991) Effect of freezing, freeze drying and pH on radiation induced heat sensitivity of *Clostridium sporogenes*. In *Proceedings of the XVIIIth International Congress of Refrigeration*, Aug. 10–17, 1991. Vol. IV, pp. 1652–1655. Montreal.

Shamsuzzaman, K., Payne, B., Cole, L. and Borsa, J. (1990) Radiation induced heat sensitivity and its persistence in *Clostridium sporogenes* spores in various media. *Canadian Institute of Food Science and Technology Journal* 23(2/3), 114–120.

Stegeman, H., Mossel, D.A.A. and Pilnik, W. (1977) Studies on the sensitizing mechanism of preirradiation to a subsequent heat treatment on bacterial spores. In *Spore Research 1976* ed. Barker, A.W., Wolf, J., Ellar, D.J., Dring, G.J. and Gould, G.W. Vol. 2, pp. 565–587. London: Academic Press.

Tallentire, A. (1970) Radiation resistance of spores. *Journal of Applied Bacteriology* 35, 141–146.

Tanooka, H. and Sakakibara, Y. (1968) Radioresistant nature of the transforming activity of DNA in bacterial spores. *Biochimica and Biophysica Acta* 155, 130–142.

Terano, H., Tanooka, H. and Kadota, H. (1969) Germination induced repair of single strand breaks of DNA in irradiated *Bacillus subtilis* spores. *Biochemical and Biophysical Research Communications* 37, 66–71.

Vinter, V. (1959) Differences in cyst(e)ine content between vegetative cells and spores of *Bac. cereus* and *Bac. megaterium*. *Nature* 183, 998–999.

Webb, R.B., Ehret, C.F. and Powers, E.L. (1958) A study of the temperature dependence of radiation sensitivity of dry spores of *Bacillus megaterium* between 5°K and 309°K. *Experimenta* 14, 324–326.

Webb, R.B., Powers, E.L. and Ehret, C.F. (1960) Thermorestoration of radiation damage in dry bacterial spores. *Radiation Research* 12, 682–693.

Wheaton, E. and Pratt, G.B. (1962) Radiation survival curves of *Clostridium botulinum* spores. *Journal of Food Science* 27, 327–334.

Wills, P.A. (1974) Effects of hydrostatic pressure and ionising radiation on bacterial spores. *Experimental Cell Research* 170, 64–79.

Woese, C.R. (1959) Further studies on the ionizing radiation inactivation of bacterial spores. *Journal of Bacteriology* 77, 38–42.

Yamazaki, K., Ito, M., Sato, K. and Oka, M. (1968) Effects of growth media composition on radioresistance of bacterial spores (in Japanese). *Food Irradiation* [*Shokuhin-Shosha*] 3(1), 13–19.

Journal of Applied Bacteriology Symposium Supplement 1994, **76**, 91S–104S

Mechanisms of inactivation and resistance of spores to chemical biocides

Sally F. Bloomfield and M. Arthur
Department of Pharmacy, King's College London, UK

1. Introduction, 91S
2. The role of the spore coat and cortex in biocide resistance
   2.1 The spore coat, 91S
   2.2 The spore cortex, 92S
3. Interaction of sporicides with the spore coat, 94S
4. Interaction of sporicides with the spore cortex
   4.1 Action of halogen-releasing agents and hydrogen, peroxide, 95S
   4.2 Glutaraldehyde, 97S
5. Osmoregulation and resistance to biocides, 97S
6. Action of sporicides on the spore protoplast
   6.1 Leakage of dipicolinic acid, 97S
   6.2 Leakage of phosphate, 98S
7. Inhibition of spore germination and outgrowth, 99S
8. Sublethal injury of spores by chemical agents, 99S
9. Conclusions, 101S
10. Future development of sporicidal agents
   10.1 Use of potentiators, 101S
   10.2 Use of combinations of biocides, 102S
11. References, 102S

1. INTRODUCTION

Bacterial spores are the most resistant life forms known. Although a number of chemical compounds effective as bactericidal agents are also sporicidal, for the most part much higher concentrations and longer contact times are required for sporicidal compared with bactericidal action. Chemical agents which also have a sporicidal action include glutaraldehyde and formaldehyde, chlorine and iodine, acids and alkalis, hydrogen peroxide and the peroxy acids, ethylene oxide, β-propionolactone and ozone. By contrast, some highly active bactericidal agents such as the phenolics, quaternary ammonium compounds, alcohols, bisguanides, organic acids and esters, and the mercurials have little or no sporicidal activity, although these compounds may be effective sporistatic agents which prevent germination and/or outgrowth of spores.

A considerable amount of information is available about the mechanisms of resistance of bacterial spores to biocides. This evidence has been previously reviewed by Russell (1983, 1990), Gould (1984, 1985), Waites (1985) and Bloomfield (1992). By contrast, the precise mechanisms of sporicidal action have been relatively little studied and may involve interaction at one or more target sites within the spore coat, cortex or protoplast. In general it is assumed that the primary lethal action of chemical biocides is the same for spores as for vegetative cells. This supposition is supported by the fact that where the effects of biocides on isolated spore protoplasts have been investigated their sensitivity is found to be similar to that of the corresponding vegetative cells. In some cases, however, e.g. glutaraldehyde, experimental evidence indicates that the mechanism of sporicidal action may be different from its bactericidal action.

Experimental evidence indicates that resistance to chemical agents is associated with the spore coat and cortex which act as permeability barriers preventing access to the underlying spore protoplast. Increasingly, the evidence also suggests that an integral part of the action of some sporicidal agents involves their ability to disrupt the outer spore layers thereby facilitating access to their sites of action on or in the spore protoplast. These various aspects of the mechanisms of action and resistance of spores to chemical biocides are reviewed below.

Correspondence to: S.F. Bloomfield, Department of Pharmacy, King's College London, Manresa Road, London, UK.

2. THE ROLE OF THE SPORE COAT AND CORTEX IN BIOCIDE RESISTANCE

2.1 The spore coat

The results of various studies described in this section indicate that the spore coat plays an important part in biocide resistance by limiting penetration to the underlying protoplast. It is apparent, however, that the nature and extent of this effect varies from one sporicide to another and between different spore species.

The role of the spore coat in biocide resistance has generally been studied with coatless or coat-defective spore forms. These are either coat-deficient mutant strains or coat-extracted spores prepared by treatment with sodium hydroxide or with combinations of thioglycolic acid, 2-mercaptoethanol or dithiothreitol, with urea and/or sodium dodecyl (lauryl) sulphate (SLS). Electron microscopy and chemical analysis of spores indicate that the spore coat consists of two layers, an inner coat which is alkali soluble and an outer coat which is alkali resistant (Kulikovsky *et al.* 1975; Labbé *et al.* 1978; Nishihara *et al.* 1981). Aronson and Fitz-James (1971), Vary (1973) and Gorman *et al.* (1983b) have demonstrated that treatment with disulphide-reducing agents extracts both inner and outer spore coat protein ranging from 38% of total coat protein for urea–mercaptoethanol (UME) at pH 7·0 to 70% or more for dithiothreitol with urea and SLS (UDS).

Studies with a number of sporicidal agents indicate, that for the most part, progressive increases in inner and outer coat removal are associated with progressive increases in sporicidal action although some conflicting results have been obtained. These effects are illustrated by recent studies of the action of a range of chlorine-releasing agents (CRAs) and iodine on various coat-defective forms of *Bacillus subtilis*. Preliminary studies with CRAs (Bloomfield and Arthur 1989; Arthur 1991) showed that although sodium hypochlorite (NaOCl), sodium dichloroisocyanurate (NaDCC) and chloramine-T solutions buffered to pH 7·4 have similar activity against vegetative cells of *B. subtilis*, being effective at 50–100 ppm available chlorine $avCl_2$ (5 min contact), there was not only increased resistance but also variation in activity of these agents against *B. subtilis* spores; although NaOCl (200 ppm, 5 min contact) showed good activity, NaDCC showed little activity at this concentration and contact time, 1000 ppm being required to produce a significant effect. Chloramine-T was least effective and had little sporicidal action at concentrations even up to 5000 ppm. Bloomfield and Megid (1994) found that with buffered solutions of iodine (pH 7·0) a concentration of 5000 ppm available iodine (avI_2) (15–45 min contact) was required to produce a significant effect against *B. subtilis* spores.

Chemical analysis of *B. subtilis* spores indicated, as shown in Table 1, that, with a number of agents, progressive increases in coat protein extraction could be achieved ranging from 10 to 24% with NaOH (which presumably represents inner coat protein) up to 67·5% for UDS (which represents both outer and inner coat protein). Evaluations with halogen-releasing agents (HRAs) (Bloomfield and Arthur 1989, 1992; Bloomfield and Megid 1994), as summarized in Fig. 1, show that treatment with CRAs in the presence of increasing concentrations of NaOH was associated with increased sensitivity to NaOCl and NaDCC—although little or no effects were observed for chloramine-T. When spores pretreated with UDS or UDS/lysozyme which achieved further coat protein extraction were used, further substantial increases in sensitivity to all three agents were obtained. For example, for iodine, Bloomfield and Megid (1994) showed that concentrations of NaOH which extracted alkali-soluble outer coat protein caused some increase in sensitivity. With spores pretreated with UDS or UDS/lysozyme to achieve both spore coat and cortex degradation, further substantial increases in sensitivity to iodine were obtained.

Progressive increases in the sensitivity of *Clostridium bifermentans*, *Bacillus cereus* and *B. subtilis* spores to HRAs, and also to chloroform, following treatment with combinations of reagents causing extraction of spore coat protein have also been demonstrated by Cousins and Allan (1967), Wyatt and Waites (1975), Sousa *et al.* (1978), Waites and Bayliss (1979) and Gorman *et al.* (1983b, 1984c, 1985). Similar results have also been obtained with other sporicidal agents, although in some instances conflicting results were reported. Marletta and Stumbo (1970) showed that *B. subtilis* spores treated with urea/thioglycollic acid were more sensitive to ethylene oxide, although resistance was regained on storage. Dadd and Daley (1982) showed that the resistance to ethylene oxide of *B. cereus* spores pretreated with dithiothreitol remained unchanged whilst spores of a coat-defective mutant of *B. subtilis* were actually more resistant. Investigations by Bayliss and Waites (1976) with spores of *Cl. bifermentens* showed that intact spores were 500 times more resistant to hydrogen peroxide than spores pretreated with dithiothreitol. Waites and Bayliss (1979) also showed, however, that *B. cereus* spores made sensitive to lysosyme by UME treatment were only slightly more sensitive to hydrogen peroxide, sodium hydroxide and ethanol, than spores with intact coats, suggesting that the intact coat had little effect in this situation. Gorman *et al.* (1984b) and McErlean *et al.* (1980) demonstrated that pretreatment of *B. subtilis* spores with UME or UDS (causing 38 and 70% extraction of total coat protein respectively) caused a progressive increase in sensitivity to glutaraldehyde. By contrast Thomas and Russell (1974) showed that pretreatment with mercaptoethanol or thioglycolic acid had little or no effect on the resistance of *Bacillus pumilis* spores to glutaraldehyde.

2.2 The spore cortex

Although the role of the spore coat in resistance to biocides is fairly well established, increasingly there is evidence to suggest that the spore cortex also plays a part in determining biocide sensitivity. Recent investigations in this labor-

Table 1 Extraction of spore coat protein and cortex peptidoglycan from *Bacillus subtilis* spores by various treatments

Treatment	Total coat protein extracted (%)	Cortex hexosamine extracted (%)
0·2% NaOH*	10·0	0
0·4% NaOH	17·5	0
1·0% NaOH	24·5	0
UDS†	67·5	50
UDS/lysozyme	—	75

* 5 min contact.
† UDS = Urea/dithiothreitol/sodium lauryl sulphate.
For *B.subtilis* spores, total coat protein = 60–62 μg/3 × 10^8 spores, total cortex hexosamines = μg/3 × 10^8 spores.
From Bloomfield and Arthur (1992); Bloomfield and Megid (1994).

Table 2 Resistance of spores to antimicrobial agents associated with various stages of sporulation

Agent	Resistance develops at sporulation stage
Xylene, toluene	Forespore engulfment (stage III)
↓	↓
Benzene, butanol, methanol, ethanol	Cortex and coat formation (stages IV and V)
↓	↓
Chlorhexidine	
↓	
Chloroform, phenol	Spore maturation (stages VI and VII)
↓	
Glutaraldehyde	

From Shaker *et al.* (1988a); Russell (1990).

atory indicate that UDS treatment of *B. subtilis* spores produced substantial extraction of cortex peptidoglycan as well as coat protein. Chemical analysis of extracted spores as shown in Table 1 confirmed that alkali treatment, whilst producing significant extraction of coat protein, had no measurable effect on cortex peptidoglycan. By contrast, treatment with UDS and UDS/lysozyme caused progressive extraction of cortex material (50 and 75% of total cortex material respectively) as well as coat protein.

In correlating these results with the evaluations of sporicidal activity shown in Fig. 1, it is seen that pretreatment of spores with NaOH to extract alkali soluble coat protein caused increased sensitivity to NaOCl and particularly to

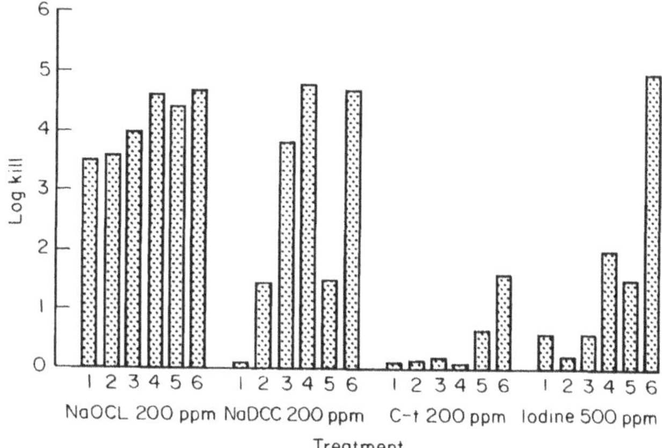

Fig. 1 Sporicidal activity of chlorine-releasing agents and iodine against *Bacillus subtilis* spore forms (5 min contact): 1, intact spores; 2, 0·2% NaOH; 3, 0·4% NaOH; 4, 1·0% NaOH; 5, UDS; 6, UDS/lys. From Bloomfield and Arthur (1992); Bloomfield and Megid (1994)

NaDCC but had only limited effect on iodine and no effect on chloramine-T, whereas pretreatment of spores with UDS, and UDS followed by lysozyme produced successive increases in sensitivity to all four agents: i.e. whereas extraction of alkali soluble coat protein alone increases sensitivity to NaOCl and NaDCC, degradation of both coat and cortex material is required to achieve significant activity with chloramine-T. The inference from these results is that the spore coat as well as the cortex plays a major role in resistance. In line with this it was found that the sensitivity of UDS/lysozyme-treated spore protoplasts to NaOCl, NaDCC, chloramine-T and iodine was of the same order as that achieved with vegetative cells, 100–200 ppm available halogen producing 5 log reduction within 5 min.

Evidence for the role of the cortex in resistance to octanol and xylene was reported by Imae and Strominger (1976). They used mutants of *Bacillus sphaericus* which were defective in the synthesis of *meso*-diaminopimelate and showed that muramic lactam in the spore cortex increased linearly with an increase in the concentration of *meso*-diaminopimelic acid in the sporulation medium. When 25% of maximum cortex concentration was produced the spores became resistant to both octanol and xylene, but 90% of the maximum cortex development was required for heat resistance. The time at which cortex synthesis occurred during spore formation also coincided with the ordered appearance of resistance to chloroform, methanol and octanol (Sousa *et al.* 1978). Thus it would seem that a complete cortex is important in conferring resistance to these chemicals. However an incomplete cortex may affect development of other spore components. Thus, Imae and Strominger showed that spores without cortices were irregularly shaped, suggesting that coats may also have been defective thereby reducing resistance.

Some further insights into the role of the spore coat and cortex in biocide resistance have been gained by correlating the emergence of resistance to the stage of spore development. In these experiments the development of resistance to biocides was evaluated with spore strains which were blocked at different stages of sporulation. The results of these studies, as shown in Table 2, are in general agreement with the results described above. Investigations by Sousa *et al.* (1978), Gorman *et al.* (1984a) and Shaker *et al.* (1988) suggest that resistance to toluene and xylene is an early event whilst resistance to heat, chlorhexidine, alcohols and halogens is associated with coat and cortex formation. Resistance to phenols, chloroform, lysozyme and finally glutaraldehyde (Power *et al.* 1988) appears only late in the sporulation process.

3. INTERACTION OF SPORICIDES WITH THE SPORE COAT

Experimental evidence from studies with sporicidal agents which exert an oxidative action, such as hydrogen peroxide (Bayliss and Waites 1976) and hypochlorite (Rode and Williams 1966; Kulikovsky *et al.* 1975; Wyatt and Waites 1975; Foegeding and Busta 1983) suggests that these agents may themselves cause disruption and extraction of spore coat material thereby facilitating penetration to the underlying cortex and protoplast. Wyatt and Waites (1975) and Bayliss and Waites (1976) demonstrated that treatment of *Cl. bifermentans*, *B. cereus* and *B. subtilis* spores with chlorine, and of *Cl. bifermentans* with hydrogen peroxide, caused extraction of coat protein; gel electrophoresis of protein solubilized with chlorine or hydrogen peroxide produced an identical electrophoretic band to that demonstrated by Vary (1973) from alkali extracts of *Cl. bifermentans*, but other less dense bands similar to those obtained from UD extracts (corresponding to outer coat protein material) were absent. Disruption of spore coat material by treatment with chlorine is also indicated by studies of Wyatt and Waites (1975) and Foegeding and Busta (1983). These workers also showed that chlorine treatment of *B. subtilis*, *B. cereus*, *Cl. bifermentans* and *Cl. botulinum* spores increased their sensitivity to lysozyme.

Results from this laboratory (Bloomfield and Arthur 1992) show that CRAs solubilized spore coat protein and that the extent of this extraction correlated with sporicidal activity. Figure 2 shows that, with buffered solutions of CRAs (pH 7·4), NaOCl, which was sporicidal at relatively low concentrations (100–200 ppm avCl$_2$, 5 min contact) produced most extensive extraction of coat protein. With NaDCC, for which the concentration required to produce 99% kill in 5 min was 1000 ppm or more, there was relatively less extraction of protein. Chloramine-T caused some

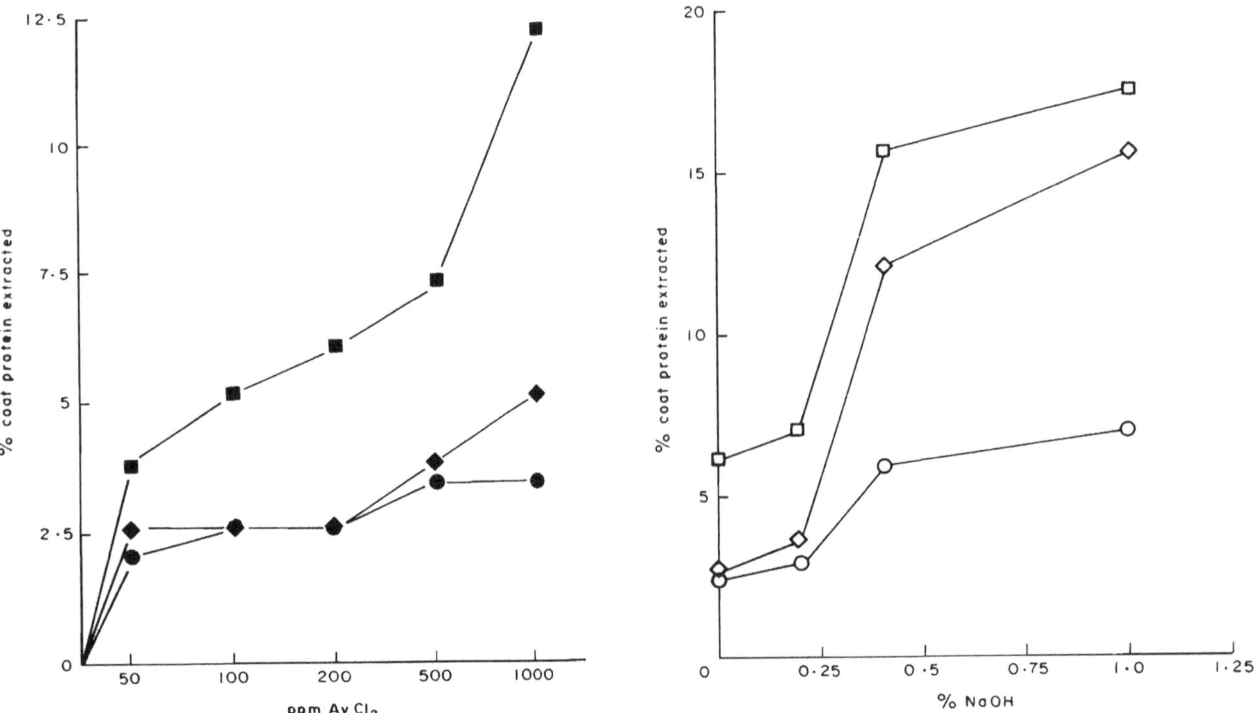

Fig. 2 Release of coat protein from *Bacillus subtilis* spores (1 × 10^8 cfu ml^{-1}) by treatment (5 min contact) with buffered solutions (pH 7·4) of chlorine-releasing agents in the presence and absence of NaOH: ■, NaOCl, ◆, NaDCC, ●, chloramine-T, □, NaOCl 200 ppm avCl$_2$, ◇, NaDCC 200 ppm avCl$_2$, ○, chloramine-T 200 ppm avCl$_2$. From Bloomfield and Arthur (1992)

coat extraction but was sporicidal only at very high concentrations (5000 ppm for 2–3 h).

When the effect of NaOH on coat protein extraction by CRAs was investigated, it was found that there was a significant increase in extraction with increasing NaOH concentration for NaDCC and NaOCl. For NaDCC it was found that both the level of coat protein extraction and the sporicidal activity achieved in the presence of 0·4 and 1·0% NaOH was similar to that produced by NaOCl. For chloramine-T on the other hand, there was relatively little increase in coat protein extraction in the presence of NaOH and no observed increase in activity. Further studies by Bloomfield and Megid (1994) showed that buffered solutions of iodine at pH 7·0, were effective against intact spores only at relatively high concentrations and produced only limited coat protein extraction under these conditions. The addition of 1 and 2% NaOH produced a significant increase in coat protein extraction and was associated with an increase in activity against *B. subtilis* spores.

For agents other than the HRAs and hydrogen peroxide there is little evidence available to indicate whether any direct interaction with coat protein occurs. Some studies by Munton and Russell (1970) and King *et al.* (1974) showed that, whereas under acid conditions, glutaraldehyde resides at the spore surface, treatment of spores with glutaraldehyde under alkaline conditions (in the presence of bicarbonate) facilitated penetration of glutaraldehyde into the spores with increased activity. From this they suggested that alkaline but not acid glutaraldehyde may cause disruption of spore coats.

4. INTERACTION OF SPORICIDES WITH THE SPORE CORTEX

Experimental evidence increasingly suggests that agents such as glutaraldehyde, CRAs and hydrogen peroxide also interact with the spore cortex, although the nature of this interaction varies considerably from one agent to another.

4.1 Action of halogen-releasing agents and hydrogen peroxide

The relationship between cortex degradation and activity of HRAs was investigated by determining the total hexosamine content of *B. subtilis* spores before and after biocide treatment. The results of studies with CRAs (Bloomfield and Arthur 1992) as illustrated in Fig. 3 show that a buffered solution of NaOCl, which was the most effective sporicide, produced substantial release of cortex hexosamine, the extent of this release increasing with increasing avCl$_2$ concentration. Buffered solutions of NaDCC, which were relatively less effective, also produced relatively less cortex

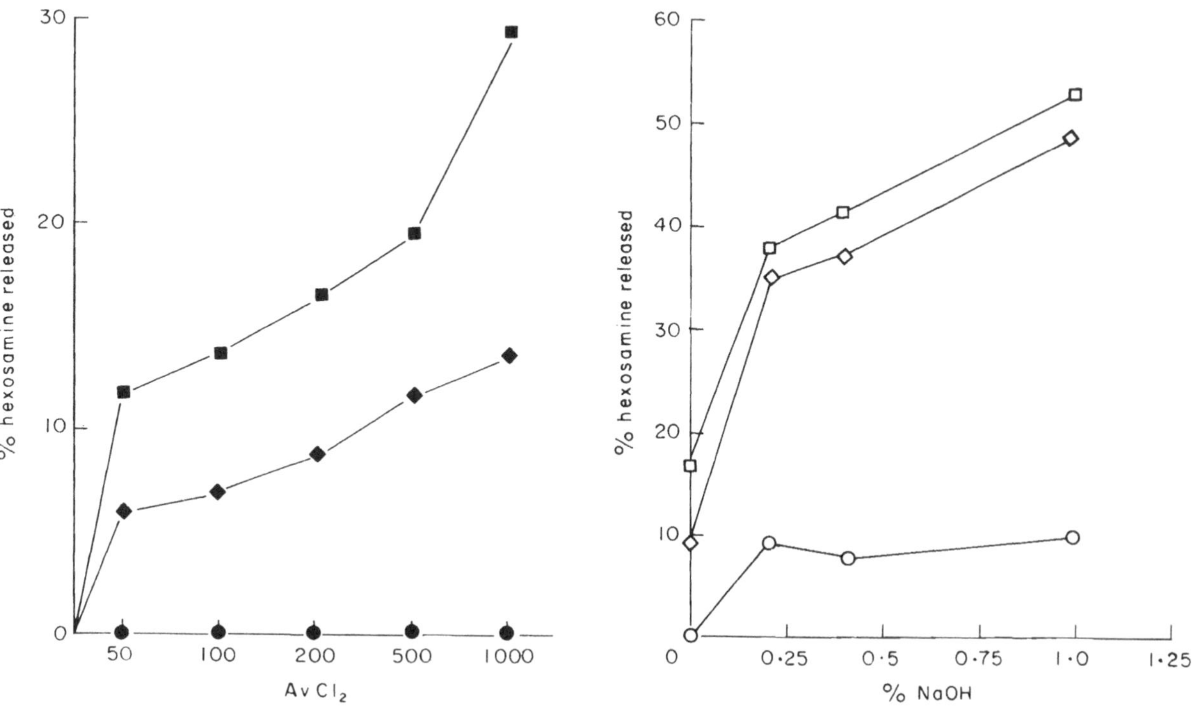

Fig. 3 Release of cortex hexosamines from *Bacillus subtilis* spores (1 × 10^8 cfu ml^{-1}) by treatment (5 min contact) with buffered solutions (pH 7·4) of chlorine-releasing agents in the presence and absence of NaOH: ■, NaOCl, ◆, NaDCC, ●, chloramine-T, □, NaOCl 200 ppm avCl$_2$, ◇, NaDCC 200 ppm avCl$_2$, ○, chloramine-T 200 ppm avCl$_2$. From Bloomfield and Arthur (1992)

degradation. In the presence of NaOH, under conditions in which the activity of NaDCC was similar to that of NaOCl, the extent of release of hexosamine was significantly increased to a level similar to that produced by NaOCl. By contrast chloramine-T, which has relatively little sporicidal activity, had little or no effect on cortex hexosamine either in the presence or absence of NaOH.

Similar studies with iodine (Bloomfield and Megid 1994) indicated some inconsistencies. Whereas for NaDCC it was found that NaOH potentiated both activity and cortex degradation, for iodine there was some potentiation of activity in the presence of NaOH but no evidence of cortex degradation at any of the concentrations tested, either in the presence or absence of NaOH.

It could be argued that resistance of spores to cortex extraction by NaDCC, chloramine-T and iodine resulted from the impermeability of the spore coat, but further investigations showed that these two agents also produced little or no extraction of hexosamine from coat-extracted UDS spores.

Overall, the results of these investigations suggest that the relative sensitivity of spores to agents which exert an oxidizing action is correlated with their ability to cause coat and cortex degradation. If the results for HRAs are summarized (Table 3) it is seen that for NaOCl, and NaDCC in the presence of NaOH, i.e. conditions under which they produce significant degradation of both coat and cortex, these formulations are effective at relatively low concentrations (100–200 ppm). By contrast for iodine, which produces only coat protein degradation, and chloramine-T, which causes little coat and no cortex degradation, concentrations and contact times required to achieve adequate penetration by diffusion of spore outer layers as required for destruction of spores are relatively much higher (500–5000 ppm).

Degradation of cortex material has also been shown with hydrogen peroxide as well as hypochlorites, although no attempt has been made to correlate these effects with sporicidal activity. Waites *et al.* (1976, 1977) and Gorman *et al.* (1984c) showed that treatment of *Cl. bifermentans* spores with hydrogen peroxide and *B. subtilis* spores with hypochlorite degraded cortex material.

4.2 Glutaraldehyde

For glutaraldehyde on the other hand, current evidence suggests that interaction with the cortex may be directly responsible for spore death. Thomas and Russell (1974) and Hughes and Thurman (1970) showed that low concentrations of glutaraldehyde inhibit spore germination and suggested that this occurred as a result of interaction with and cross-linking of amino groups in the cortex peptidolycan. Thomas and Russell (1974) also showed that pretreatment of spores with glutaraldehyde caused a reduction in the release of dipicolinic acid (DPA) caused by subsequent treatment with heat. Further investigations (Thomas 1977) suggest that, at higher concentrations (10% alkaline glutaraldehyde), extensive irreversible interaction with

Table 3 Comparison between coat protein extraction, cortex degradation and sporicidal activity of halogen-releasing agents

Agent	Extraction by treatment with HRAS (5 min contact) Conc. ppm	Coat protein (%)	Cortex hexosamine (%)	Concentration required to produce 2- or more log reduction (5 min contact) (ppm available halogen)
NaOCl	200	6·1	16·7	200
	500	7·4	19·6	
NaDCC + 1% NaOH	200	15·7	49·0	200
NaDCC	200	2·6	8·9	1000
	500	3·8	12·0	
Iodine + 1% NaOH	500	40·8	0	1000
Iodine	500	4·5	0	5000
Chloramine-T + 1% NaOH	200	7·0	8·9	—
Chloramine-T	200	2·6	0	>5000
	500	2·0	0	

From Bloomfield and Arthur (1992); Bloomfield and Megid (1994).

spore outer layers produces a tough sealed structure such that germination can no longer occur and spore death results. Gorman *et al.* (1984b) showed that pretreatment with glutaraldehyde reduced solubilization of hexosamines from cortical fragments and also protected coatless spores of *B. subtilis* from the lytic action of lysozyme and sodium nitrite.

In view of the observed differences between the action of glutaraldehyde and that of CRAs and hydrogen peroxide, it is interesting to note that, whereas studies with glutaraldehyde suggest that interaction with spore outer layers produces sealing of spores thereby preventing germination, the series of changes which occur following treatment with hypochlorite and hydrogen peroxide resemble those associated with spore germination, namely decrease in refractility whereby phase bright spores become phase dark (Wyatt and Waites 1973, 1975; Waites *et al.* 1976), degradation of cortex peptidoglycan (Gorman *et al.* 1984c; Bloomfield and Arthur 1992), decrease in dry weight and optical density (Wyatt and Waites 1973, 1975; Waites *et al.* 1976).

5. OSMOREGULATION AND RESISTANCE TO BIOCIDES

A possible explanation of the role of the cortex in spore resistance to biocides is provided by the expanded osmoregulatory cortex theory of Gould (1977). This theory suggests that, as well as having a rigid protective function, the spore cortex is also responsible for maintaining dehydration of the spore protoplast. This theory suggests that dehydration of the spore protoplast is brought about and maintained by the osmotic activity of expanded electronegative peptidoglycan in the spore cortex, the stability of the peptidoglycan being maintained by the presence of mobile positively-charged counterions within the cortex. Gould (1977) estimated that the expanded cortex may exert an osmotic potential of as much as 2 MPa, causing water to pass from the core to the cortex until the osmotic pressure is equalized. Further evidence suggests that the spore protoplast contains mainly macromolecules and insolubilized salts such as DPA or calcium salts of other weak acids such as glutamic and phosphoglyceric acids (Nelson *et al.* 1969) which make only a small osmotic contribution thereby maintaining an equilibrium in the direction of a high water content cortex surrounding a dehydrated protoplast.

From this it could be expected that in the presence of a very low water level and a consequent high viscosity in the protoplast, the diffusion of water-soluble antimicrobial agents into the protoplast is restricted. It may be therefore that rehydration of the spore protoplast following cortex degradation by the action of agents such as CRAs causes rehydration of the protoplast thereby facilitating diffusion of these agents to their site of action on the underlying protoplast.

Gould and Dring (1974, 1975) have proposed that spore heat resistance also probably results from dehydration of the spore protoplast. In line with this it was found that coat-defective spores treated to allow penetration of multivalent cations were more sensitive to heat indicating that, in this situation, the multivalent cations interact with and cross-link the carboxyl groups of the peptidoglycan causing it to contract, allowing partial rehydration of the protoplast and loss of heat resistance. Sacks and Alderton (1961) showed that titration with acid, producing H-form spores, also reduced heat resistance, but this was recovered by titration with Ca^{2+} to produce Ca-form spores. Gould and Dring (1975) suggested that, in this situation, acid treatment displaces positively charged cations from the cortex by protonating peptidoglycan carboxyl groups, thus reducing cortex expansion and lowering osmotic pressure. They proposed that re-equilibration at high pH values reimposes expansion and increases the osmotic pressure exerted by the cortex, thereby accounting for the observed increase in heat resistance.

There is also some evidence to suggest that the presence of cations may affect spore resistance to chemical biocides. Although further investigation is required, Thomas and Russell (1975), McErlean *et al.* (1980) and Gorman *et al.* (1984b) showed that stable Ca^{2+} forms of *B. subtilis* spores were more sensitive to glutaraldehyde which, as described above, appears to act by preventing cortical breakdown and spore germination, whereas NaOCl and NaOCl/methanol, which cause cortical breakdown and destabilization of spores, are more effective against unstable H-forms. Tawasatini and Shibasaki (1973) have shown that H-form spores were also more sensitive to propylene oxide than normal or Ca-form spores.

6. ACTION OF SPORICIDES ON THE SPORE PROTOPLAST

Although the evidence available increasingly suggests that for oxidizing agents such as HRAs and peroxide, degradation of the spore coat and cortex is an integral part of their sporicidal action facilitating penetration to the underlying spore protoplast, it seems unlikely that degradation of spore cortex is responsible for spore death *per se* since lysozyme treatment alone under conditions producing cortex degradation does not result in significant spore death (Fitz-James 1971; Bloomfield and Arthur 1989). Rather it would seem that cortex degradation facilitates penetration of these agents to their site of action on the underlying spore protoplast.

6.1 Leakage of dipicolinic acid

From a number of observations in which leakage of intracellular materials from spores treated with agents which exert an oxidizing action was reported, it has been suggested that the primary lethal effect of these agents involves disruption of spore protoplast permeability barriers. Investigations by Alderton and Halbrook (1971), Dye and Mead (1972) and Gorman *et al.* (1984c) showed that hydrogen peroxide and CRAs produce leakage of Ca^{2+} and/or DPA from spores whilst glutaraldehyde which appears to act by stabilization of the spore cortex, does not (Thomas and Russell 1975).

To evaluate biocide damage to spore permeability barriers, the leakage of DPA from intact and coat- and cortex-deficient spores treated with HRAs was evaluated (Bloomfield and Arthur 1992; Bloomfield and Megid 1994). The results as shown in Fig. 4 indicate that treatment of UDS/lysozyme-treated spores with CRAs produced leakage of DPA to varying extents. It was found that NaOCl caused release of DPA which was significantly greater than that produced by NaDCC. For spore protoplasts treated with chloramine-T and also iodine, no release of DPA was detectable at concentrations up to 5000 ppm. Thus, whereas observations as described in Section 3 above indicate that all four agents were active at relatively low concentrations (50–300 ppm $avCl_2$) against UDS and UDS/lysozyme-treated spores, the results of these investigations together with those in Section 4.2 indicate that, under comparable conditions, NaOCl, produced both significant DPA release and solubilization of hexosamines, but NaDCC, chloramine-T and iodine did not. From this it is concluded that DPA leakage resulting from damage to spore permeability barriers is not a primary lethal effect but is more closely associated with degradation of spore cortex. Additionally it was noted that cortex degradation alone does not cause DPA release since the total DPA content of normal spores was found to be the same as that of UDS/lysozyme-treated spores (11–11·5 $\mu g/3 \times 10^8$ spores). From this it is concluded that cortex destabilization caused by CRAs is different from that associated with lysozyme action or spore germination and that, in addition to destabilizing effects on the spore cortex, formulations of CRAs such as buffered NaOCl and, to a lesser extent, NaDCC in the presence of NaOH, produce damage of the spore protoplast membrane which is responsible for the leakage of DPA from the cortex-destabilized spores.

6.2 Leakage of phosphate

Some further evidence of the effects of HRAs on the spore protoplast permeability barriers has been obtained by studying the effects of iodine on uptake of radiolabelled phosphate by intact spores (Bloomfield and Megid 1994).

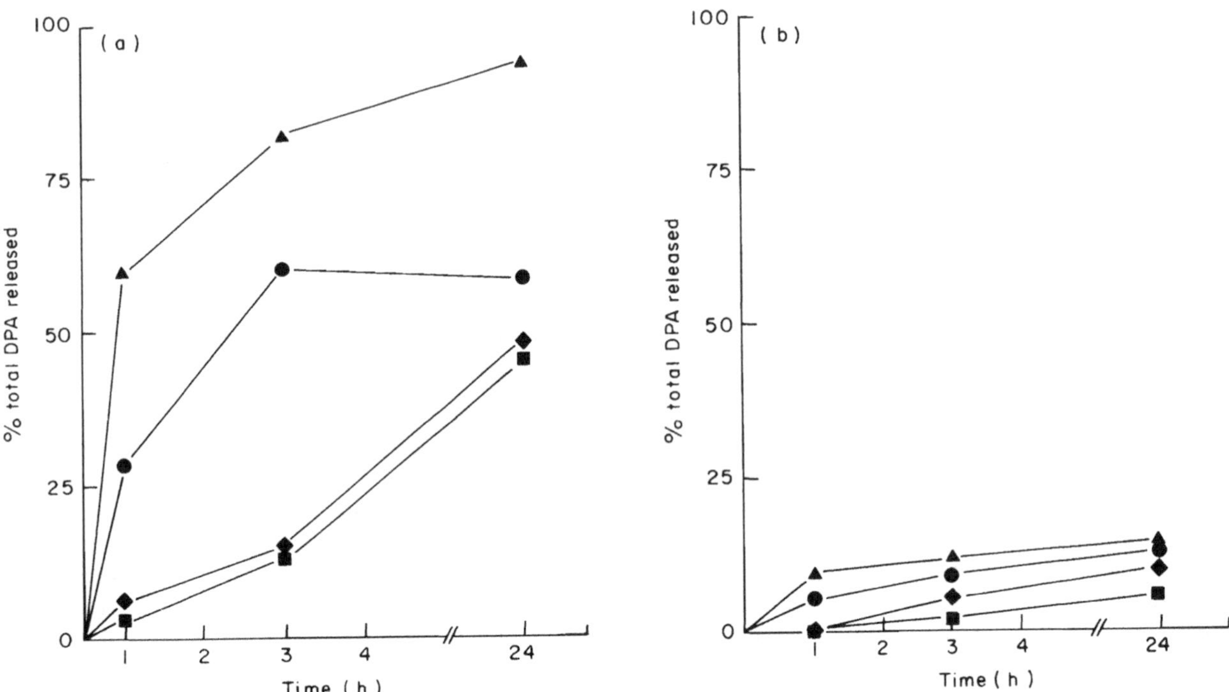

Fig. 4 Release of dipicolinic acid from UDS/lysozyme-treated *Bacillus subtilis* spores (1×10^8 cfu ml^{-1}) by treatment with buffered solutions (pH 7·4) of (a) NaOCl and (b) NaDCC: ■, 100 ppm $avCl_2$, ◆, 200 ppm $avCl_2$, ●, 500 ppm $avCl_2$, ▲, 1000 ppm $avCl_2$. From Bloomfield and Arthur (1992)

Results indicate a marked difference between uptake of phosphate by untreated and iodine-treated spores. Whereas, with untreated spores, uptake of phosphate occurred at a steady rate over a 60 min contact period, treatment with sporicidal concentrations of iodine (500/1000 ppm) stimulated phosphate uptake over the initial 30–40 min period, but following this, phosphate then leaked from cells, a concentration of 1000 ppm producing total loss within 60 min. Although further investigations are required, the results suggest a possible correlation between damage to the spore protoplast membrane and sporicidal action.

In any evaluation of the mechanisms of action of biocides against spores there are two other aspects as in the following, which require consideration, both of which have been the subject of investigation and both of which have a significant bearing on the practical use of sporicides.

7. INHIBITION OF SPORE GERMINATION AND OUTGROWTH

As stated previously, although a whole range of agents are active bactericidal agents with little or no sporicidal action, many of these compounds have a sporistatic effect. For these compounds it is generally found that the concentrations which are sporistatic are very much lower than those which are sporicidal and are similar to those which inhibit vegetative cell growth.

Studies with these compounds indicate two quite separate mechanisms of action; whereas compounds such as the phenols, cresols, glutaraldehyde, etc., inhibit spore germination processes, experimental evidence suggests that agents such as QACs, ethylene oxide, chlorhexidine and CRAs at sporistatic concentrations act on the processes leading to outgrowth of the spore. What is not clear, however, is why these different agents have different actions; there is little or no evidence to suggest why some compounds should inhibit the degradative process associated with germination, whereas other compounds have no effect on this process, but inhibit later stages of outgrowth.

Although this review is primarily concerned with biocidal effects of sporicidal agents, the potential value of antibacterial agents which are not sporicidal but which inhibit outgrowth/germination processes should not be overlooked. In general, such agents are of value as preservatives in situations, such as in foods, where spores, especially those that are capable of surviving food-processing treatments, are a potential problem and where it is also necessary to restrict biocide concentrations in order to avoid the risk of toxic side effects.

Inhibition of spore germination and outgrowth is reviewed in more detail by Russell (1990).

8. SUBLETHAL INJURY OF SPORES BY CHEMICAL AGENTS

The other important aspect of sporicidal action relates to the sublethal effects induced by the action of biocides on spores. This subject has been previously reviewed by Gould (1984), Waites and Bayliss (1984) and Russell (1990).

Quantitative estimates of sporicidal action are used for a number of purposes. Primarily they are used in the development of sporicidal formulations, for establishing use concentrations and for comparing the activity of sporicidal formulations. Sporicidal concentrations are normally derived by carrying out suspension tests in which a sample of the product is inoculated with a spore suspension and, after the prescribed contact time, a sample is taken. The biocide is neutralized by suitable means and a sample plated to assess numbers of colony forming units. The log reduction is then computed by reference to the numbers of cfu's in the untreated suspension. In general, for practical purposes, the assumption is then made that those spores which form colonies are viable whilst those spores which do not form colonies are non-viable. Increasingly, however, there are data to suggest that, as with all microbicidal processes, within any population of biocide-treated spores there will be a proportion of cells that are sublethally injured but not killed. Within that population also, there will be different levels of sublethal injury, ranging from spores which are extensively damaged to those which are little affected. As suggested by Gould (1984), a consideration of the spore cell cycle indicates that the opportunity for sublethal injury and expression of this sublethal injury in response to subsequent stress is potentially much greater than for vegetative cells. In order to form a vegetative cell which undergoes cell division to form a colony, spores cells must go through a complex development process in which sublethal injury may be expressed at any stage.

Within a biocide-damaged population, the number of cells which actually outgrow to form colonies will depend on two factors. Firstly, it will depend on the method of neutralization of the biocide and secondly on the subsequent method of treatment and plating of the biocide-treated spores. One practical implication of this is that, according to the method of sporicidal testing adopted, it may be possible to overestimate sporicidal efficacy of a product, i.e. the numbers of damaged spores which recover to form colonies under laboratory conditions may be less than those which might recover under in-use situations. A recent example where this has been demonstrated is with glutaraldehyde. Investigations by Dancer *et al.* (1989) showed that, for spores treated with glutaraldehyde, recovery could be increased by post-treatment with NaOH at concentrations up to 50 mmol l^{-1}. The results as shown in Table 4 indicate that for a challenge inoculum of 10^8 cfu ml^{-1} of *Bacillus* spp. spore forms treated with 2% glutaraldehyde followed by neutralization with glycine, the recovery of viable spores was $<10^1$. By contrast, where spores were treated with NaOH after neutralization and before plating, the recovery was increased up to 10^2–10^3.

If these results are considered in conjunction with findings previously discussed, which suggest that glutaraldehyde treatment of spores produces a tough sealed structure, it may be suggested that glutaraldehyde treatment produces a spore form which is superdormant, but not killed, which could under certain specific conditions germinate and outgrow. These findings, if correct, could have serious practical implications for the use of glutaraldehyde as a chemosterilant.

In an earlier study, Spicher and Peters (1981) showed that exposure of formaldehyde-treated spores to post-treatment heat shock at 60–80°C enabled most of the apparently killed spores to revive. Some revival of glutaraldehyde-treated spores was also achieved by post-treatment heating (Gorman *et al.* 1983a; Dancer *et al.* 1989; Power *et al.* 1989) although the extent of this effect was much less than with formaldehyde or with glutaraldehyde-treated/NaOH post-treated spores.

More recent evaluations of the effects of different post treatment conditions on the resuscitation of *B. subtilis* spores treated with glutaraldehyde, formaldehyde, iodophors and CRAs were reported by Williams and Russell (1993a). These workers showed that although post-treatment with NaOH 25 $mmol^{-1}$ revived a small portion of glutaraldehyde-treated spores, there was no effect on recovery of spores treated with the other biocides. In a second study (Williams and Russell 1993b) biocide-treated spores were enumerated on nutrient agar containing potential revival agents (subtilisin, lysozyme, calcium dipicolinate, calcium lactate). Of these, only calcium lactate had any significant enhancing effect and then only with iodine-treated spores. The effects of incubation temperature were also investigated (Williams and Russell 1993c) and it was found that the recovery of biocide-treated spores was more sensitive to changes in incubation temperature than the recovery of control spores.

On the beneficial side, the phenomenon of sublethal injury also gives the potential for potentiation of sporicidal activity by the application of combined or sequential treatment with combinations of agents. The effects of agents causing damage to the spore coat and cortex on resistance to biocides has already been discussed. In a number of recent studies by Williams and Russell (1992, 1993d) the effects of stressing agents on the recovery of biocide-treated spores were evaluated. In these studies *B. subtilis* spores were treated with a range of biocides at concentrations which produced a 2 log reduction in spore count. The treated spores were then plated on several media containing a range of different stressing agents including NaOH, sodium lauryl sulphate, cetylpyridinium chloride and polymyxin B sulphate. The numbers of cfu recovered were compared with the numbers of cfu recovered on plain agar. The results, as shown in Table 5, indicate that spores pretreated with various sporicidal agents such as alkaline glutaraldehyde, iodophors, NaOCl and NaDCC were more susceptible to the sporistatic effects of the various stressing

Table 4 NaOH-induced revival of *Bacillus* spores after treatment with 2% w/v glutaraldehyde (15 min contact)

Bacterial strain	Viable count after glutaraldehyde treatment: Control	Viable count after glutaraldehyde treatment: 1 mol l^{-1} NaOH post-treatment ($\times 10^2$)
B. subtilis 168	$<10^1$	8
B. subtilis var. *niger*	$<10^1$	6
B. subtilis NCTC 8236	$<10^1$	52
B. licheniformis NCIB 6346	$<10^1$	7·4
B. licheniformis NCIB 11107	$<10^1$	1·2
B. cereus NCIB 8933	$<10^1$	2
B. thuringiensis BGSG 4D1	$<10^1$	1
B. megaterium NCIB 9376	$<10^1$	8·2

Initial count 10^7–10^8 cfu ml^{-1}.
From Dancer *et al.* (1989).

Table 5 The effects of stressing agents on biocide-treated spore suspensions of *Bacillus subtilis* (3×10^8 cfu ml^{-1})

Stressing agent added to agar plating medium	Reduction (%) in cfu following exposure to: Glut	PVPI	NaDCC	NaOCl
None	0	0	0	0
NaOH 5 mmol l^{-1}	68	38	74	78
SLS 20 $\mu g\ ml^{-1}$	3	62	73	80
CPC 0·1 $\mu g\ ml^{-1}$	13	4	13	41
PBS	86	76	81	79

SLS = sodium lauryl sulphate; CPC = cetyl pyridinium chloride; PBS = polymixin B sulphate; Glut = glutaraldehyde: 2% alkaline; PVPI = polyvinyl–pyrrolidone–iodine 45 ppm avI_2; NaDCC = sodium dichloroisocyanurate 25 ppm $avCl_2$.
From Williams and Russell (1992).

agents. Further studies suggested that the increased sensitivity to the stressing agents was manifested at the outgrowth stage of the spore cell cycle.

Another example which illustrates the potential for use of combined treatments is provided by evaluations of the sporicidal action of alcohol/hypochlorite mixtures by Coates and Death (1978). They used spores of *B. subtilis*, and showed that significant potentiation of the action of buffered NaOCl against spores could be achieved by the addition of methanol, ethanol or propanol, offering the potential for utilizing NaOCl at concentrations which are significantly less corrosive but still highly effective as a sporicidal formulation. From their observations Death and Coates (1979) suggested that alcohols potentiate the action of NaOCl by softening of spore coats, allowing penetration to occur. The results described by Bloomfield and Arthur (1989, 1992) suggest rather that the action of chlorine is to produce coat and cortex degradation thereby facilitating diffusion of methanol to its site of action on the underlying protoplast.

9. CONCLUSIONS

The review of the literature as presented in this paper indicates that the mechanisms of action of sporicides, and the mechanisms which account for the resistance of spores relative to that of vegetative bacterial forms are still relatively poorly understood. Most certainly it would appear that the spore coat plays an important part in determining resistance, by limiting penetration to the underlying cortex and protoplast. There is increasing evidence that the spore cortex also plays a vital role in the resistance of spores to heat and chemical agents. It is suggested that this derives from the stable expanded spore cortex, which maintains a stable dehydrated state and a consequent high viscosity within the spore protoplast thereby limiting diffusion into the protoplast.

As far as mechanisms of sporicidal action are concerned, investigations currently suggest that sporicidal formulations such as hydrogen peroxide, buffered NaOCl and NaDCC in the presence of NaOH are effective at relatively low concentrations because of their ability to degrade both spore coat and cortex, which contributes to their action by allowing access to the underlying sensitive spore protoplast where irreversible damage occurs. By contrast, agents such as iodine, chloramine-T and buffered NaDCC produce some coat protein but no cortex extraction, such that concentrations and contact times required to achieve adequate penetration as required for destruction of the spore protoplast are higher.

For sporistatic agents such as chlorhexidine which have little or no sporicidal action it is suggested that these agents may be almost completely excluded from the spore. This is demonstrated by the investigations of Shaker *et al.* (1988b) which showed that UDS extraction of *B. subtilis* spores decreased resistance to chlorhexidine; whereas 25 $\mu g\ ml^{-1}$ chlorhexidine produced only a 0·5-log reduction of intact spores within 2 h, the same concentration produces a 2-log reduction with UDS-treated spores.

For glutaraldehyde, the indications are that it may interact to prevent cortex degradation to produce a so called 'sealed structure' thereby preventing rehydration, germination and outgrowth of the spore.

10. FUTURE DEVELOPMENT OF SPORICIDAL AGENTS

One of the major problems in achieving sporicidal action in practice results from their resistance to chemical biocides which necessitates the use of high concentrations of these materials for situations where a sporicidal action is required. As many of the most active sporicidal agents are compounds with high chemical reactivity and as a result also tend to have corrosive and irritant effects, this is obviously a far from ideal situation.

The results of the various investigations described in this paper suggest a number of aspects of sporicidal action which offer the possibility of optimizing sporicidal performance of individual formulations with the possibility of reducing the 'use' concentrations.

10.1 Use of potentiators

The well-defined role of the spore coat and cortex in the resistance of spores to chemical biocides suggests the possibility of using potentiating agents. The results described in this paper suggest that many of the most active sporicidal agents owe this activity to their ability to interact with and damage spore outer layers. From this, the possibility arises that other reagents—not necessarily sporicidal in themselves—which cause coat and cortex degradation could be used to potentiate the action of known sporicidal agents. To achieve this, however, a more detailed understanding of the mechanisms of coat and cortex degradation is required.

A consideration of the relationship between coat extraction and sporicidal action for different agents and combinations of agents indicates certain anomalies which require further investigation. A summary of the results of investigations with HRAs (Bloomfield and Arthur 1992; Bloomfield and Megid 1994) in Table 6 shows that the amounts of protein extracted by CRAs in the presence of NaOH (up to about 18%) were significantly less than those achieved by treatment with NaOH alone (up to 67·5%) or by treatment with UDS (67·5%). From this it might have been expected that spores pretreated with NaOH would be

Table 6 Relationship, between release of coat protein and sporicidal activity for *Bacillus subtilis* spores (log count 8·1) following various treatments (5 min contact)

	Release of coat protein (%)		
NaOH concentration % w/v	NaOH alone	NaOH + NaOCl 200 ppm avCl_2	NaOH + iodine 500 ppm avI_2
0	10·0	6·1	4·5
	(0)	(3·7)	(0·6)
0·2	17·5	7·0	—
	(0)	(3·2)	(0·7)
0·4	24·5	15·7	20·0
	(0)	(4·0)	(0·0)
1·0	67·5	17·5	40·8
	(0)	(4·6)	(2·0)

From Bloomfield and Arthur (1992); Bloomfield and Megid (1994).
Sporicidal activity (as log kill) in parentheses.

more sensitive to subsequent treatment with CRAs than spores treated with combinations of CRAs with NaOH, but the opposite effect occurred. Similarly, it was found that the amounts of protein extracted by iodine in the presence of NaOH were much higher than for NaOCl in the presence of NaOH, but the sporicidal action of the iodine formulation was relatively less—although it is possible that this may relate to the ability of NaOCl to interact with cortex material, as compared with iodine which does not.

Investigations of the nature of spore coat protein indicates that this material is rich in cystine disulphide bonds. It might therefore be expected that whereas disulphide-reducing agents such as DTT cause solubilization by reduction of insoluble disulphide bonds to soluble SH groups, oxidizing agents such as CRAs cause oxidation of disulphide bonds, which produces a disruption of coat structure and increased permeability without causing extensive solubilization, thereby accounting for the lower levels of coat protein extraction by CRAs. Overall however the results suggest that the disruption of coat protein produced by dithiothreitol or NaOH treatment is different from that caused by CRAs, which in turn is different from that caused by iodine.

As far as cortex degradation is concerned, investigations with the oxidizing agent sodium nitrite suggest that it behaves differently from lysozyme. Investigations by Ando (1980) have indicated that whereas nitrite appears to degrade cortex at the 'spore unique muramic lactam residue' giving free hexosamine in nitrite digests of cortical fragments, lysozyme on the other hand hydrolyses glycosidic β-1,4-linkages between *N*-acetyl glucosamine and *N*-acetylmuramic acid.

10.2 Use of combinations of biocides

The complexity of the spore cell structure and the spore cell cycle compared with that of vegetative cells obviously offers the potential for much greater variability in mechanisms of action of sporicides compared with bactericidal action. It is reasonable to suggest that this complexity and variability in turn offers significant potential for using combinations of sporicidal agents either sequentially or in combination to achieve enhanced sporicidal action at lower biocide concentrations. The evaluation of chlorine/alcohol mixtures by Coates and Death (1978) and Death and Coates (1979) as described in Section 6 is an example of this type of approach.

In looking for potential combinations of sporicidal agents or combinations of potentiators with sporicidal agents however, a more detailed and more focused evaluation of the mechanisms of action and the mechanisms of resistance of biocides on bacterial spores is required.

11. REFERENCES

Alderton, G. and Halbrook, W.V. (1971) Action of chlorine on bacterial spores. *Bacteriological Proceedings* **11**, 12.

Ando, Y. (1980) Mechanism of nitrite induced germination of *Clostridium perfringens* spores. *Journal of Applied Bacteriology* **49**, 527–535.

Arthur, M. (1991) Studies of the mode of action of chlorine-releasing antibacterials. Ph.D. Thesis, London.

Aronson, A.I. and Fitz-James, P.C. (1971) Reconstitution of bacterial spore coat layers *in vitro*. *Journal of Bacteriology* **108**, 571–578.

Bayliss, C.E. and Waites, W.M. (1976) The effect of hydrogen peroxide on spores of *Clostridium bifermentans*. *Journal of General Microbiology* **96**, 401–407.

Bloomfield, S.F. (1992) Resistance of bacterial spores to chemical agents. In : *Principles and Practice of Disinfection, Preservation and Sterilisation* 2nd edn ed. Russell, A.D., Hugo, W.B. and Ayliffe, G.A.J. pp. 688–702. Oxford: Blackwell Scientific Publications.

Bloomfield, S.F. and Arthur, M. (1989) Effect of chlorine-releasing agents on *Bacillus subtilis* vegetative cells and spores. *Letters in Applied Microbiology* **8**, 101–104.

Bloomfield, S.F. and Arthur, M. (1992) Interaction of *Bacillus subtilis* spores with sodium hypochlorite, sodium dichloroisocyanurate and chloramine-T. *Journal of Applied Bacteriology* **72**, 166–172.

Bloomfield, S.F. and Megid, R. (1994) Interaction of iodine with *Bacillus subtilis* spores and spore forms. *Journal of Applied Bacteriology* **76**, 492–499.

Coates, D. and Death, J.E. (1978) Sporicidal activity of mixtures of alcohol and hypochlorite. *Journal of Clinical Pathology* **31**, 148–152

Cousins, C.M. and Allan, C.D. (1967) Sporicidal properties of some halogens. *Journal of Applied Bacteriology* **30**, 168–174.

Dadd, A.H. and Daley, G.M. (1982) Role of the coat in resistance of bacterial spores to inactivation by ethylene oxide. *Journal of Applied Bacteriology* **53**, 109–116.

Dancer, B.N., Power, E.G.M. and Russell, A.D. (1989) Alkali-induced revival of *Bacillus* spores after inactivation by glutaraldehyde. *FEMS Microbiology Letters* **57**, 345–348.

Death, J.E. and Coates, D. (1979) Effect of pH on sporicidal and microbicidal activity of buffered mixtures of alcohol and sodium hypochlorite. *Journal of Clinical Pathology* **32**, 148–153.

Dye, M. and Mead, G.C. (1972) The effect of chlorine on the viability of clostridial spores. *Journal of Food Technology* **7**, 173–181.

Fitz-James, P.C. (1971) Formation of protoplasts from resting spores. *Journal of Applied Bacteriology* **105**, 1119–1136.

Foegeding, P.M. and Busta, F.F. (1983) Proposed mechanism for sensitization by hypochlorite treatment of *Clostridium botulinium* spores. *Applied and Environmental Microbiology* **45**, 1374–1379.

Gorman, S.P., Hutchinson, E.P., Scott, E.M. and McDermott, L.M. (1983a) Death, injury and revival of chemically-treated *Bacillus subtilis* spores. *Journal of Applied Bacteriology* **54**, 91–99.

Gorman, S.P., Scott, E.M. and Hutchinson, E.P. (1983b) The effect of sodium hypochlorite–methanol combinations on spores and spore forms of *Bacillus subtilis*. *International Journal of Pharmaceutics* **17**, 291–298.

Gorman, S.P., Scott, E.M. and Hutchinson, E.P. (1984a) Emergence and development of resistance to antimicrobial chemicals and heat in spores of *Bacillus subtilis*. *Journal of Applied Bacteriology* **57**, 153–163.

Gorman, S.P., Scott, E.M. and Hutchinson, E.P. (1984b) Interaction of *Bacillus subtilis* spore protoplast, cortex, ion-exchange and coatless forms with glutaraldehyde. *Journal of Applied Bacteriology* **56**, 95–102.

Gorman, S.P., Scott, E.M. and Hutchinson, E.P. (1984c) Hypochlorite effects on spores and spore forms of *Bacillus subtilis* and on a spore lytic enzyme. *Journal of Applied Bacteriology* **56**, 295–303.

Gorman, S.P., Scott, E.M. and Hutchinson, E.P. (1985) Effects of aqueous and alcoholic povidone-iodine on spores of *Bacillus subtilis*. *Journal of Applied Bacteriology* **59**, 99–105.

Gould, G.W. (1977) Recent advances in the understanding of resistance and dormancy in bacterial spores. *Journal of Applied Bacteriology* **42**, 297–309.

Gould, G.W. (1984) Injury and repair mechanisms in bacterial spores. In *The Revival of Injured Microbes* ed. Andrew, M.H.E. and Russell, A.D. pp. 199–220. Society for Applied Bacteriology Symposium Series No. 12. London: Academic Press.

Gould, G.W. (1985) Modifications of resistance and dormancy. In *Fundamental and Applied Aspects of Bacterial Spores* ed. Dring, G.J., Ellar, D.J and Gould, G.W. London: Academic Press.

Gould, G.W. and Dring, G.J. (1974) Mechanisms of spore heat resistance. *Advances in Microbial Physiology* **11**, 137–164.

Gould, G.W. and Dring, G.J. (1975) Heat resistance of bacterial endospores and concept of an expanded osmoregulatory cortex. *Nature, London* **258**, 402–405.

Hughes, R.C. and Thurman, P.F. (1970) Cross-linking of bacterial cell walls with glutaraldehyde. *Biochemical Journal* **119**, 925–926.

Imae, Y. and Strominger, J.L. (1976) Relationship between cortex content and properties of *Bacillus sphaericus* spores. *Journal of Bacteriology* **126**, 907–913.

King, J.A., Woodside, W. and McGucken, P.V. (1974) Relationship between pH and antibacterial activity of glutaraldehyde. *Journal of Pharmaceutical Sciences* **63**, 804–805.

Kulikovsky, A., Pankratz, H.S. and Sadoff, H.L. (1975) Ultrastructural and chemical changes in spores of *Bacillus cereus* after action of disinfectants. *Journal of Applied Bacteriology* **38**, 39–46.

Labbé, R.G., Reich, R.R. and Duncan, C.L. (1978) Alteration in ultrastructure and germination of *Clostridium perfringens* type A spores following extraction of spore coats. *Canadian Journal of Microbiology* **24**, 1526–1536.

Marletta, J. and Stumbo, C.R. (1970) Some effects of ethylene oxide on *Bacillus subtilis*. *Journal of Food Science* **35**, 627–631.

McErlean, E.P., Gorman, S.P. and Scott, E.M. (1980) Physical and chemical resistance of ion–exchange and coat-defective spores of *Bacillus subtilis*. *Journal of Pharmacy and Pharmacology* **32**, 32P.

Munton, T.J. and Russell, A.D. (1970) Aspects of the action of glutaraldehyde on *Escherichia coli*. *Journal of Applied Bacteriology* **33**, 410–419.

Nelson, D.L., Spudich, J.A., Donsen, P.P.M., Bertsch, L. and Kornberg, A. (1969) Biochemical studies of bacterial sporulation and germination. XVI Small molecules in spores. In *Spores IV* ed. Campbell, L.L. Washington, DC: American Society for Microbiology.

Nishihara, T., Yutsudo, T., Ichikawa, T. and Kondo, M. (1981) Studies on the bacterial spore coat on the SDS-DTT extract from *Bacillus megaterium* spores. *Microbiology and Immunology* **25**, 327–331.

Power, E.G.M., Dancer, B.N. and Russell, A.D. (1988) Emergence of resistance to glutaraldehyde in spores of *Bacillus subtilis* 168. *FEMS Microbiology Letters* **50**, 223–226.

Power, E.G.M., Dancer, B.N. and Russell, A.D. (1989) Possible mechanisms for the revival of glutaraldehyde-treated spores of *Bacillus subtilis* NCTC 8236. *Journal of Applied Bacteriology* **67**, 91–98.

Power, E.G.M., Dancer, B.N. and Russell, A.D. (1990) Effect of sodium hydroxide and two proteases on the revival of aldehyde-treated spores. *Letters in Applied Microbiology* **10**, 9–13.

Rode, L.J. and Williams, M.G. (1966) Utility of sodium hypochorite for ultrastructure study of bacterial spore integuments. *Journal of Bacteriology* **92**, 1772–1778.

Russell, A.D. (1983) Mechanisms of action of chemical sporicidal and sporistatic agents. *International Journal of Pharmaceutics* **10**, 127–140.

Russell, A.D. (1990) The bacterial spore and chemical sporicidal agents. *Clinical Microbiological Reviews* **3**, 99–111.

Sacks, L.E. and Alderton, G. (1961) Behavior of bacterial spores in aqueous polymer two- phase systems. *Journal of Bacteriology* **82**, 331–341.

Shaker, L.A., Dancer, B.N., Russell, A.D. and Furr, J.R. (1988a) Emergence and development of chlorhexidine resistance during sporulation of *Bacillus subtilis* 168. *FEMS Microbiology Letters* **51**, 73–76.

Shaker, L.A., Furr, J.R. and Russell, A.D. (1988b) Mechanism of resistance of *Bacillus subtilis* spores to chlorhexidine. *Journal of Applied Bacteriology* **64**, 531–539.

Sousa, J.C.F., Silva, M.T. and Balassa, G. (1978) Ultrastructural effects of chemical agents and moist heat on *Bacillus subtilis*. II. Effects on sporulating cells. *Annales de Microbiologie* **129B**, 377–390.

Spicher, G. and Peters, J. (1981) Heat activation of bacterial spores after inactivation by formaldehyde. Dependence of heat activation on temperature and duration of action. *Zentralblatt für Bakteriologie, Parasitenkunde, Infektionskrankheiten und Hygiene, I. Abteilung originale, Reihe B* **173**, 188–196.

Tawasatini, T. and Shibasaki, I. (1973) Change in the chemical resistance of heat sensitive and heat resistant bacterial spores against propylene oxide. *Journal of Fermentation Technology* **51**, 824–891.

Thomas, S. (1977) Effect of high concentrations of glutaraldehyde upon bacterial spores. *Microbios Letters* **4**, 199–204.

Thomas, S. and Russell, A.D. (1974) Studies on the mechanism of the sporicidal action of glutaraldehyde. *Journal of Applied Bacteriology* **37**, 83–92.

Thomas, S. and Russell, A.D. (1975) Sensitivity and resistance to glutaraldehyde of the hydrogen and calcium forms of *Bacillus pumilus* spores. *Journal of Applied Bacteriology* **38**, 315–317.

Vary, J.C. (1973) Germination of *Bacillus megaterium* spores after various extraction procedures. *Journal of Bacteriology* **116**, 797–803.

Waites, W.M. (1985) Inactivation of spores with chemical agents. In *Fundamental and Applied Aspects of Bacterial Spores* ed. Dring, G.J., Ellar, D.J. and Gould, G.W. pp. 383–396. London: Academic Press.

Waites, W.M. and Bayliss, C.E. (1979) The effect of changes in spore coat on the destruction of *Bacillus cereus* spores by heat and chemical treatment. *Journal of Applied Biochemistry* **1**, 71–76.

Waites, W.M. and Bayliss, C.E. (1984) Damage to bacterial spores by combined treatments and possible revival and repair processes. In *The Revival of Injured Microbes* ed. Andrew, M.H.E. and Russell, A.D. pp. 221–240. Society for Applied Bacteriology Symposium Series No. 12. London: Academic Press.

Waites, W.M., Wyatt, L.R., King, N.R. and Bayliss, C.E. (1976) Changes in spores of *Clostridium bifermentans* caused by treatment with hydrogen peroxide and cations. *Journal of General Microbiology* **93**, 388–396.

Waites, W.M., King, N.R. and Bayliss, C.E. (1977) The effect of chlorine and heat on spores of *Clostridium bifermentans*. *Journal of General Microbiology* **102**, 211–213.

Williams, N.D. and Russell, A.D. (1992) The nature and site of biocide-induced sublethal injury in *Bacillus subtilis* spores. *FEMS Microbiology Letters* **99**, 277–280.

Williams, N.D. and Russell, A.D. (1993a) Revival of biocide-treated spores of *Bacillus subtilis*. *Journal Applied of Bacteriology* **75**, 69–75.

Williams, N.D. and Russell, A.D. (1993b) Revival of *Bacillus subtilis* spores from biocide-induced injury in the germination process. *Journal Applied of Bacteriology* **75**, 74–81.

Williams, N.D. and Russell, A.D. (1993c) Conditions suitable for the recovery of biocide-treated spores of *Bacillus subtilis*. *Microbios* **74**, 121–129.

Williams, N.D. and Rusell, A.D. (1993d) Injury and repair in biocide-treated spores of *Bacillus subtilis*. *FEMS Microbiology Letters* **106**, 183–186.

Wyatt, L.R. and Waites, W.M. (1973) The effect of hypochlorite on the germination of spores of *Clostridium bifermentans*. *Journal of General Microbiology* **70**, 383–385.

Wyatt, L.R. and Waites, W.M. (1975) The effect of chlorine on spores of *Clostridium bifermentans*, *Bacillus subtilis* and *Bacillus cereus*. *Journal of General Microbiology* **89**, 337–344.

Journal of Applied Bacteriology Symposium Supplement 1994, **76**, 105S–114S

Effects of water activity and pH on growth of *Clostridium botulinum*

P.J. McClure, M.B. Cole and J.P.P.M. Smelt[1]
Unilever Research, Colworth Laboratory, Sharnbrook, Bedford, UK and [1]Unilever Research, Vlaardingen Laboratory, Olivier van Noortlaan, Vlaardingen, The Netherlands

1. Introduction, 105S
2. Distribution of *Clostridium botulinum*, 105S
3. Outbreaks, 106S
4. Outbreaks from commercially produced foods, 107S
5. Trends affecting safety, 108S
6. Factors affecting growth and toxin production, 108S
7. Effect of pH, 108S
8. Effect of water activity, 109S
9. Combination effects, 110S
10. Predictive models, 110S
11. Future aspects, 111S
12. References, 112S

1. INTRODUCTION

Although a relatively rare disease, there have been more than 16 000 cases of botulism, and more than 2 700 fatalities recorded since it was recognized as being associated with food poisoning by van Ermingen in the 1890s. The number of botulism cases and the mortality rate have decreased in recent years, due to the availability of antisera, but botulism still remains a significant public health hazard. Many of the traditional methods of food processing and preservation used by the food industry today aim to ensure the control of *Clostridium botulinum*, e.g. canning, pickling. *Clostridium botulinum* spores are widely distributed in nature, and may find their way into processed foods through raw materials, whilst a product is growing or being harvested, or by post-process contamination. The main objectives of food processors in controlling *Cl. botulinum* are to destroy all spores in the product or to prevent growth and subsequent production of toxin if destruction is not possible. In order to appreciate the potential botulinum hazard from a food, it is important to understand the possible sources of contamination, which depend on the incidence and distribution of different types of *Cl. botulinum* in the environment.

2. DISTRIBUTION OF *CLOSTRIDIUM BOTULINUM*

Despite the large number of surveys describing the incidence of *Cl. botulinum* spores in the environment, the factors influencing the distribution of spores are poorly understood. Different types of *Cl. botulinum* prevail in different geographical locations. In the western United States, Brazil, Argentina and China, type A spores predominate, whilst proteolytic type B spores tend to predominate in soils from the eastern United States, and non-proteolytic type B in Britain and Europe. Type E is often associated with freshwater and marine sediments, and northern regions such as northern Europe, the former Soviet Union, Alaska, northern Canada and Japan. The higher incidence of type E in colder regions may be explained by its psychrotolerant nature.

Correspondence to: P.J. McClure, Unilever Research, Colworth Laboratory, Sharnbrook, Bedford MK44 1LQ, UK.

Surveys describing the distribution of *Cl. botulinum* spores in foods are less common than environmental surveys. Although fresh fruits and vegetables are in contact with the soil, there are few studies which have looked at the occurrence of spores on these products. Spores have been found in a variety of fruits and vegetables such as asparagus, beans, carrots, celery, corn, olives, potatoes, turnips, apricots, cherries and peaches (Meyer and Dubovsky 1922a,b; Rozanova *et al.* 1972), frozen spinach (Insalata *et al.* 1969), mushrooms (Hauschild *et al.* 1975), onion and garlic skins (Rozanova *et al.* 1972; Solomon and Kautter 1986, 1988), cabbage (Solomon *et al.* 1990). Other surveys (Kautter *et al.* 1978; Vergieva and Incze 1979; Notermans *et al.* 1989) carried out have failed to detect any spores of *Cl. botulinum* in fresh produce.

Most food surveys have concentrated on meats, fish and products fed to infants, such as honey. Although the incidence in meat is generally low, it is higher in Europe (Roberts and Smart 1976, 1977; Klarmann 1989) than in the United States (Greenberg *et al.* 1966; Taclindo *et al.* 1967; Abrahamsson and Riemann 1971) or Canada (Hauschild and Hilsheimer 1980, 1983). A relatively large number of articles describe the incidence of *Cl. botulinum* spores in fish and have been reviewed by Hobbs (1976) and Huss (1980). A number of these studies have revealed the presence of *Cl. botulinum* spores in a high percentage of fish tested from the United States (Bott *et al.* 1966; Eklund and Poysky 1970; Lindroth and Genigeorgis 1986; Garcia and Genigeorgis 1987) and Europe (Cann *et al.* 1965; Johannsen 1965; Huss *et al.* 1974a, b; Rouhbakhsh-Khaleghdoust 1975; Eyles and Warth 1981). Other foods, such as honey (Sugiyama *et al.* 1978; Arnon *et al.* 1979; Midura *et al.* 1979; Kautter *et al.* 1982) and corn syrup (Kautter *et al.* 1982) have also been shown to contain spores. In general, fish contain the highest level of contamination due to the presence of high numbers of spores in aquatic environments.

3. OUTBREAKS

Countries with the largest numbers of outbreaks of botulism are shown in Table 1. An accurate estimate of the total number of cases/outbreaks is difficult, and is dependent on proper diagnoses, and efficient, up-to-date reporting. Countries with the largest total numbers of cases include the United States, Poland, China and Japan. The excellent recording and reporting in countries such as Poland undoubtedly contributes significantly to the total number of known outbreaks. Countries with the highest number of cases per year are Poland, Iran and China. Table 2 shows the different types of foods involved in outbreaks in different countries. These data show that vegetables are the main source of botulism in the United States and China, meats are the main source in Poland and continental Europe (apart from Spain and Italy), and fish is the main source in Japan, Iran and the former Soviet Union. The table also shows that a large percentage of outbreaks were from home-preserved products.

As a result of the distribution of different types of spores

Table 1 Recorded food-borne botulism in countries with relatively frequent occurrences

			Cases		Averages/year	
Country	Period	Outbreaks	Total	Fatal	Outbreaks	Cases
United States	1971–89	272	597	63	14	31
Canada	1971–89	79	202	28	4	11
Poland	1984–87	1301	1791	46	325	448
Hungary	1979–84	31	57	1	6	11
W. Germany	1983–89	63	154	0	10	26
France	1978–89	175	304	7	15	25
Italy	1979–87		310			34
Spain	1969–88	63	198	11	3	10
USSR	1958–64	95	328	95	14	47
Iran	1972–74		314			105
China	1958–83	986	4377	548	38	168
Japan	1951–87	97	479	110	3	13

Data from Hauschild (1992) by courtesy of Marcel Dekker Inc.

Table 2 Food involved in botulism outbreaks

		Food (%)				Food source (%)	
Country	Outbreaks with food identified	Meats	Fish	Fruits and vegetables	Other	Home	Commerce
United States	222	16	17	59	9	92	8
Canada	75	72	20	8	0	96	4
Poland	1500	83	12	5	0	75	25
Hungary	28	89	0	4	7	100	0
W. Germany	55	78	13	9	0	100	0
France	123	89	3	6	2	88	12
Italy	13	8	8	77	8		
Spain	48	38	2	60	0	90	10
USSR	83	17	67	16	0	97	3
Iran	63	3	97	0	0		
China	958	10	0	86	4		
Japan	95	0	99	1	0	98	2

Data from Hauschild (1992) by courtesy of Marcel Dekker Inc.

in the environment, the indigenous spores in different regions dictate the type of botulism most commonly found. The influence of regional preferences for different types of food is also an important factor when considering the numbers of cases of botulism. Local delicacies which rely on traditional methods of preservation have often been identified as common sources of botulism. For example, Canadian Inuits use raw meats from marine animals, which are often contaminated with type E spores, and rely on fermentation in a food which does not contain sufficient fermentable carbohydrate to reduce the pH quickly enough to prevent toxin production. Poorly controlled fermentation is also responsible for the large number of outbreaks in China, but the products in these cases tend to be fermented bean curd or sauce. Reliance on salting as a preservation system can also result in potentially hazardous products. Salt has to diffuse into the product from the surface, and the time taken for this to happen can often allow growth and toxin production in the product. Examples here include home cured hams in France and Portugal and fish eggs (ashbal) in Iran.

The majority of outbreaks reported occur in the northern hemisphere. In the southern hemisphere, only Argentina has reported a substantial number of outbreaks. Countries with frequent occurrences show annual rises and falls (apart from the United States). Cases in Alaska, Canada, Poland, the former Soviet Union and Iran peak between May and October. Chinese outbreaks occur in winter and early spring.

4. OUTBREAKS FROM COMMERCIALLY PRODUCED FOODS

Although most of the botulism outbreaks around the world have been caused by home-processed meat, fish and vegetable products (Table 2), a number have been caused by commercially manufactured foods. A summary of recent outbreaks from commercially produced foods is shown in Table 3. The most recent outbreak, in Egypt, is unusual since no previous reports have been received from this region. It has been suggested (Weber *et al.* 1993) that this may be due to modification of traditional fermentation methods, or the fact that *Cl. botulinum* spores have only recently been introduced to the area.

In simple terms, the control of food-borne botulism from commercial products can be achieved either by decontamination (e.g. heat sterilization) and prevention of recontamination or by preventing growth of the organism. Therefore the most common errors responsible for commercial outbreaks are failure of required time/temperatures during cooking, cooling and storage, post-process contamination from raw foods or unclean equipment, and/or inadequate preservation. In the first case, contamination due to

Table 3 Recent outbreaks of food-borne botulism from commercially produced foods including restaurant associated outbreaks

		Outbreak formation					
Product	Cause	Year	Location	Type	No. of cases	No. of deaths	Reference
Canned peppers	Underprocessing	1977	Michigan, USA	B	59	0	Terranova *et al.* 1978
Bottled peanuts	Underprocessing	1986	Taiwan	A	9	2	Chou *et al.* 1988; Tsai *et al.* 1990
Kosher airline meal	Underprocessing/temperature abuse	1987	UK	A	1	0	Colebatch *et al.* 1989
Canned salmon	Post-process leakage	1978	Lancashire, UK	E	4	2	Ball *et al.* 1979; Gilbert and Willis 1980
Canned salmon	Post-process leakage	1982	Belgium	E	2	1	Marchal *et al.* 1985
Sautéed onions	Temperature abuse	1983	Illinois, USA	A	28	1	MacDonald *et al.* 1985; Solomon and Kautter 1986
Chopped garlic in oil	Temperature abuse	1985	Vancouver, Canada	B	36	0	Blatherwick and Peck 1985; Hauschild and Simonsen 1986; St. Louis *et al.* 1988
Hazelnut yoghurt	Inadequately reduced a_w	1989	Lancashire and North Wales, UK	B	27	1	Critchley *et al.* 1989; O'Mahony *et al.* 1990
Bologna sausage	Inadequate preservation	1982	Madagascar	E	60	30	Vicens *et al.* 1985
Kapchunka (salt-cured, air-dried uneviscerated whitefish)	Poorly controlled salting process	1981	California, USA	B (np)	1	0	California State Department of Health Services 1981
		1985	New York, USA	E	2	2	CDC 1985
		1987	New York, USA and Israel	E	8	1	CDC 1987; Slater *et al.* 1989; Telzak *et al.* 1990
Karashi-renkon (vacuum-packed, deep-fried, lotus root)	Underprocessing/temperature abuse	1984	Japan	A	36	11	Otofuji *et al.* 1987; Hayashi *et al.* 1986
Bottled mushrooms	Inadequate acidification	1973	Canada	B	1	0	Todd *et al.* 1974
Faseikh (uneviscerated fish)	Poorly controlled fermentation/salting	1991	Egypt	E	91	18	Weber *et* al. 1993

np, Non-proteolytic.

the presence of *Cl. botulinum* where food is produced, harvested, processed or stored results in cells or spores surviving the processing treatment. The organism must be able to grow and produce toxin in the food and the food must be consumed without sufficient cooking to destroy toxin.

5. TRENDS AFFECTING SAFETY

In recent years there has been an increasing consumer demand for healthier, less heavily preserved, more natural products. Hence there have been moves to eliminate $NaNO_2$ and reduce NaCl and sugar levels in foods (to reduce the calorific value). There has also been a change in consumer lifestyle, with a wider range of refrigerated convenience foods becoming available. This demand has led to an increase in minimally or partially processed foods. Such products are not commercially sterile and rely on good refrigeration to prevent spoilage and ensure product safety. They are often packaged in high-barrier packaging material under vacuum or modified atmosphere to extend the shelf-life.

Lower levels of preservatives and alternative means of packaging may affect the stability of products with respect to growth and toxin production by *Cl. botulinum*. Equally such foods may allow conditions that do not permit spoilage to occur prior to toxin formation, which is an additional safety factor in some circumstances. Hazards in new generation foods can be reduced by using combination preservation systems, e.g. acidification, lowered water activity (a_w) and preservatives in addition to refrigeration. One way of optimizing and assessing the reliability of multiple barriers or combination preservation systems is to use predictive models which allow the interactive effects between different factors influencing the growth and/or toxin production in foods to be accurately quantified.

6. FACTORS AFFECTING GROWTH AND TOXIN PRODUCTION

Control of *Cl. botulinum* in most foods is achieved by inhibition of growth and toxin production by controlling factors such as temperature, pH, a_w (reduced by adding NaCl, sugars, etc.), redox potential, gaseous atmosphere and preservatives such as $NaNO_2$, sorbic acid, parabens, etc. The inhibitory effects of these factors acting alone on growth and toxin production of *Cl. botulinum* are well characterized in the literature.

7. EFFECT OF pH

Acidification (low pH) is used widely to control growth and toxin production by *Cl. botulinum* in food products. It has been generally accepted, for some time, that *Cl. botulinum* will not grow and produce toxin in food products with an equilibrium pH of 4·6 or below. There is evidence from a substantial body of research that the minimum pH for growth is in the range 4·6–4·8. This is supported by work carried out using laboratory media (Ingram and Robinson 1951) and food products (Townsend *et al.* 1954), and also by the excellent safety record of acid foods. Examples of the minimum pH values allowing toxin production in canned foods are shown in Table 4. These data have been used for the current US Government regulations, which state that a canned food at pH 4·6 or less is safe without resorting to conventional commercial sterilization.

There are, however, examples which have demonstrated that *Cl. botulinum* can grow and produce toxin at pH values below 4·6. These data have been obtained in laboratory media or under laboratory conditions where strict anaerobiosis prevails and there is a high concentration of protein present and various acidulants used. It has been suggested that the protein acts as a reducing agent, a source of essential metabolites or as a buffer, slowing down the decrease in the internal pH of the cell (Young-Perkins and Merson 1987; Wong *et al.* 1988). Tanaka (1982) and Raatjes and Smelt (1979) have also shown growth at low pH in the presence of high levels of precipitated protein, postulating that microenvironments within the matrices of the protein were at higher pH levels, allowing growth to occur.

It has also been demonstrated that growth of organisms, other than clostridia, such as moulds (*Aspergillus*, *Cladosporium*, or *Penicillium* spp.; Huhtanen *et al.* 1976; Odlaug and Pflug 1979) and bacilli (*Bacillus licheniformis* and *B. subtilis*; Fields *et al.* 1977; Montville 1982) will increase the pH in the substrate, allowing growth and toxin production to occur. All of the 35 outbreaks of botulism attributed to normally 'high' acid foods (Sperber 1982), such as fruit juice (14), pickles (3) and tomato products (18), had been spoiled by yeasts or moulds.

There is also an example of growth and toxin production below pH 4·6, with no protein or other micro-organisms

Table 4 Minimum pH values for toxin production by *Clostridium botulinum* types A and B in canned foods

Food	Minimum pH for toxin production
Prune pudding	5·44
Pears	5·42
Pimientos	5·25
Pineapple rice pudding	4·94
Pork and beans	4·93
Zucchini	4·86
Vegetable juice	4·84

From Townsend *et al.* (1954).

Table 5 Minimum levels of water activity (a_w) permitting growth of *Clostridium botulinum* at temperatures near optimum in sodium chloride-containing substrates

Cl. botulinum type	Minimum a_w	% NaCl (w/w)	Reference
A and B (prot.)	0·96	6·6	Baird-Parker and Freame 1967
	0·955	7·5	Denny *et al.* 1969
	0·95	8·0	Scott 1957
	0·94	9·4	Ohye and Christian 1966
	0·94	9·4	Marshall *et al.* 1971
E	0·972	4·5	Emodi and Lechowich 1969
	0·97	5·0	Baird-Parker and Freame (1967)
	0·97	5·0	Ohye and Christian (1966)

present. Tsang *et al.* (1985) demonstrated no growth of types A and B below pH 4·6 in media acidified with acetic acid or citric acid, but showed growth and toxin production by type E in media acidified with citric acid down to pH 4·2. This observation could not be repeated in food products prepared to the same formulation and has not been repeated in laboratory media. The inhibitory effects of acids depend on a number factors such as their physical properties (pK_a, molecular weight and number of carboxyl groups), pH, acid concentration and the buffer capacity of the medium. In general, the inhibitory effects of weak acids are due to the undissociated acid form, and not to the anion. Citric acid and HCl tend to be less inhibitory than lactic acid which is less inhibitory than acetic acid (Smelt *et al.* 1982). Overall, it is still correct to assume that *Cl. botulinum* will not grow at or below pH 4·6, unless there is concomitant growth of other micro-organisms such as bacilli, yeasts or moulds, or aggregates of precipitated protein present, perhaps causing localized pH increases.

Table 6 Influence of solute on minimum levels of water activity permitting growth of *Clostridium botulinum*

	Minimum a_w in:	
Growth parameter	Glycerol	NaCl
Spore germination	0·89	0·93
Growth of types A and B	0·93	0·96
Growth of type E	0·94	0·97

From Baird-Parker and Freame (1967).

Table 7 Combined effect of water activity and pH on the growth of *Clostridium botulinum* type B

	Growth of *Cl. botulinum* type B at pH:				
a_w	7·0	6·0	5·5	5·3	5·0
0·997	+	+	+	+	+
0·99	+	+	+	+	−
0·98	+	+	+	+	−
0·97	+	+	+	−	−
0·96	+	−	−	−	−
0·95	−	−	−	−	−

From Baird-Parker and Freame (1967).

Table 8 Interaction of water activity and pH on toxin production by *Clostridium botulinum* types A and B (proteolytic) in cooked, vacuum-packed potatoes

a_w	pH	Days to toxin detection
0·980	6·10	7
0·981	5·45	7
0·977	4·83	35
0·972	6·07	7
0·973	5·50	14
0·969	4·96	35
0·959	5·74	35
0·960	5·46	>35
0·964	4·95	>35

Data of Dodds (1989).

8. EFFECT OF WATER ACTIVITY

Water activity (a_w) in foods can be adjusted with a number of different solutes. The most common humectant which is used in foods is NaCl and it is generally accepted that the NaCl–water activity minima allowing growth are representative of those found in foods. Although the effects of NaCl on growth of *Cl. botulinum* have received considerable attention over the past 60 years, there are relatively few definitive studies since much of the earlier work determined moisture content as opposed to water activity. The minimum a_w values allowing growth of *Cl. botulinum* are summarized in Table 5. From these data, it is generally accepted that, under otherwise optimum conditions for growth, 5% NaCl is required to prevent growth of non-proteolytic strains and 10 % NaCl is necessary to prevent growth of proteolytic strains (Segner *et al.* 1966; Abrahamsson 1967; Baird-Parker and Freame 1967; Spencer 1967; Emodi and Lechowich 1969; Boyd and Southcott 1971; Roberts and Ingram 1973). Glycerol, a compatible solute, will allow growth at a_w values lower than those permitted by other solutes such as NaCl, as illustrated in Table 6. As with pH, if conditions are not optimum for growth (low pH values or low temperatures) then less NaCl is required to inhibit growth. Other factors important when considering the minimum a_w for growth are incubation temperature, strain(s) tested, solute type and presence of other preservatives, such as $NaNO_2$.

9. COMBINATION EFFECTS

An example of an early study examining the effects of combinations of antibotulinal factors in a laboratory medium (Baird-Parker and Freame 1967) is shown in Table 7. As well as describing the effects of different solutes on the minimum a_w allowing germination and growth of *Cl. botulinum* spores, this study also identified the combinations of NaCl and pH necessary to prevent growth of *Cl. botulinum*. Similar studies (Hauschild and Hilsheimer 1975; Dodds 1989) have been carried out in food products, where toxin production has also been noted. Table 8 shows the combined effects of NaCl and pH on toxin production of *Cl. botulinum* in vacuum packed potatoes (Dodds 1989).

A number of studies have examined the effects of combinations of factors, such as heat treatment/smoking/brine concentration (Christiansen *et al.* 1968), pH/E_h/a_w (Smoot and Pierson 1979), nisin/pH/heating temperature (Scott and Taylor 1981) against growth and/or toxin production of *Cl. botulinum*. Unfortunately, many of these studies are only relevant to the product or conditions tested, and sometimes only toxin formation is considered. An alternative approach, which has been taken up in recent years, is to develop a systematic database of responses of micro-organisms to factors most important in foods. The purpose of this exercise is to construct and validate empirical predictive models which can describe accurately those responses. This allows interactions between factors to be described, and predictions of the response under conditions not tested and new/modified formulations to be considered.

10. PREDICTIVE MODELS

A number of different approaches have been taken to describe the effects of combinations of preservatives on growth or toxin production of *Cl. botulinum*. Although predictive modelling is regarded as novel, food technologists have been applying models for microbial death for many years, e.g. controlling spore inactivation via the 12-D concept, the minimum heat process for low acid canned foods based on the heat resistance of spores of *Cl. botulinum* (Esty and Meyer 1922). This is an example of a probabilistic model, which predicts that if there were a single spore of *Cl. botulinum* in each of 10^{12} cans, after thermal processing there would be one can containing a viable spore or, expressed in another way, there would be a one in 10^{12} chance of a single spore surviving. There are other examples of probabilistic models, which predict the probability of growth or toxin production. Riemann (1967) suggested that the effectiveness of a preservation system to prevent growth of *Cl. botulinum* depends on the probability (*P*) that not even a single spore will be able to grow and produce toxin. A continuous probability scale can be used to analyse qualitative plus/minus data and allows quantitative comparisons to be made:

$$P = \frac{MPN \text{ spores outgrowing}}{MPN \text{ spores inoculated}}$$

Probability models usually take the form of linear equations with the overall probability of growth equal to the summation of products of each controlling factor (e.g. pH, NaCl, temperature) being considered multiplied by a constant derived from multiple regression analysis. Published models which predict the effects of storage temperature and NaCl concentration (Roberts *et al.* 1982) or pH (Lund *et al.* 1987), NaCl and pH in cheese spreads (Tanaka *et al.* 1986), pH, sorbate, isoascorbate, nitrite and NaCl in pork slurry (Roberts *et al.* 1981a,b, 1982) serve to identify the most important single factors and interactions influencing growth and toxin production of *Cl. botulinum*. Models that predict the probability of growth can be used quickly to give an indication of the likely effect of changes in product formulation, or storage conditions.

Other studies have used a two-step approach which first uses regression analysis to model the effects of the response (e.g. lag time, or time to toxin production) and these are then incorporated into secondary equations based on logistic regression, relating the probability of one spore initiating toxigenesis. Such studies have been developed for predicting the probability of toxin production in fish homogenates as a function of temperature and inoculum size (Lindroth and Genigeorgis 1986), in fish fillets stored under modified atmospheres (Ikawa and Genigeorgis 1987), in laboratory media (Jensen *et al.* 1987), vacuum packed processed turkey roll (Genigeorgis *et al.* 1991) and vacuum packed mash potatoes (Dodds 1989).

More recently, predictive models have concentrated on establishing a quantitative relationship between the controlling factors used in foods and the kinetic parameters of growth, e.g. lag time and growth rate, of food poisoning micro-organisms. For instance, kinetic models have been constructed to predict the lag time and growth rate of *Cl. botulinum* in raw minced turkey (Adair *et al.* 1989) and vacuum packed minced meat (Baird-Parker and Kilsby 1987). These studies describe the effect of temperature on growth of *Cl. botulinum* and compare different types of models. Studies in the authors' laboratory (Lawson and Adair, unpublished) have used this approach to investigate the effects of temperature and a_w (adjusted with NaCl) on time to toxin production by *Cl. botulinum* in soya peptone broth. These data (Fig. 1a) were used to generate a response surface describing the effects of varying temperature and a_w on time to toxin production by a cocktail of non-proteolytic *Cl. botulinum* strains (shown in Fig. 1b). Comparisons of predictions from the model developed with observed growth in vacuum packed trout are shown in Fig. 2.

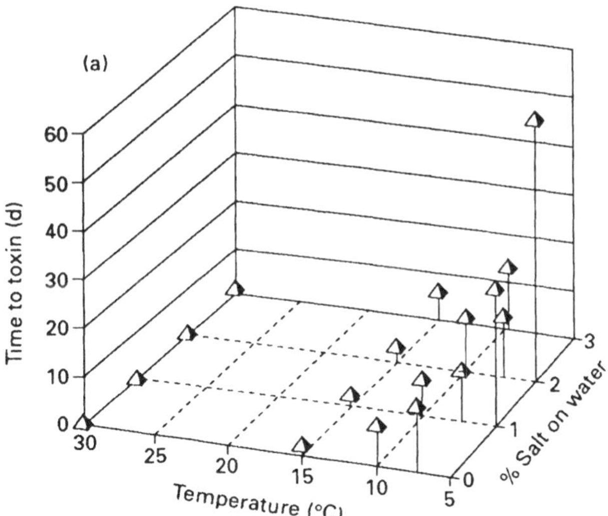

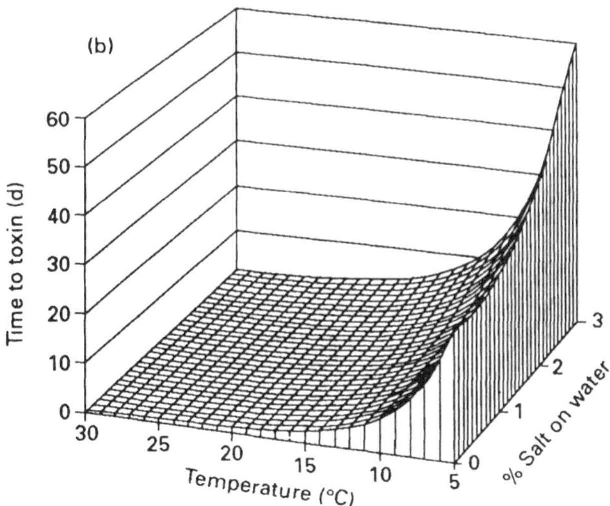

Fig. 1 (a) Data used to generate a model predicting the time to toxin for a cocktail of non-proteolytic *Clostridium botulinum* strains in a laboratory medium in response to a_w and temperature (Lawson and Adair, unpublished). (b) Response surface showing the effects of a_w and temperature on the predicted time to toxin for a cocktail of non-proteolytic *Cl. botulinum* strains in a laboratory medium (Lawson and Adair, unpublished)

Models generated using otherwise optimum conditions for growth, i.e. in laboratory media, should predict faster growth than is observed in food products. For example, in Fig. 2, although the predicted times to toxin obtained in broth are similar to those in trout, the predicted times to toxin are always faster than in the real food. In other words, where the model deviates from growth in food, it will 'fail safe'.

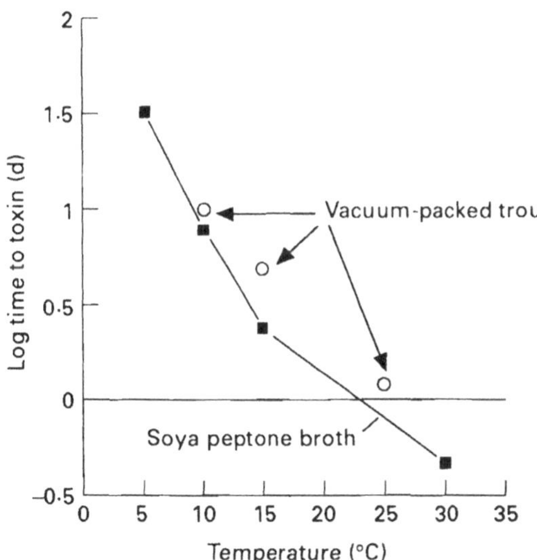

Fig. 2 Comparisons of the predicted time to toxin for a cocktail of non-proteolytic *Clostridium botulinum* strains with observed growth in vacuum-packed trout (Lawson and Adair, unpublished)

11. FUTURE ASPECTS

Despite the abundance of studies which have examined the effects of preservatives and their interactions on growth and/or toxin production by *Cl. botulinum* there is very little understanding of the fundamental mechanisms of inhibition involved. Comparison of data from different sources has been made difficult by a lack of standard experimental conditions, common qualitative approach or terminology. Although the effects of the main controlling factors in foods have been studied, interactions may not be limited to two or three factors as has already been mentioned in this article. Factors such as pH, acid type, a_w, solute type, E_h, presence of other preservatives and presence of other micro-organisms will influence the growth of *Cl. botulinum* in foods. In order to further develop and refine existing predictive models we need to complete the knowledge base detailing the growth/survival/death responses of *Cl. botulinum* to likely influencing factors. This will aid food microbiologists to gain a better understanding of the relative risks associated with particular food products.

We also need a better understanding of the mechanisms of action of preservative systems, particularly those factors commonly used to control growth of pathogens in food. These studies will include identification of sites of action, different stages of the spore/vegetative cell cycle, biovariability, distributions of germination rates, lag times and rates of outgrowth. Accurate, statistically based studies and predictive modelling are the only reliable ways to reduce preservative levels by using novel combinations of factors without compromising food safety.

12. REFERENCES

Abrahamsson, K. (1967) Occurrence of type E *Clostridium botulinum* in smoked eel. In *Botulism 1966*, Proceedings of the Fifth International Symposium on Food Microbiology, Moscow, July 1966, ed. Ingram, M. and Roberts, T.A., pp. 73–75. London: Chapman and Hall.

Abrahamsson, K., and Riemann, H. (1971) Prevalence of *Clostridium botulinum* in semipreserved meat products. *Applied Microbiology* **21**, 543–544.

Adair, C., Kilsby, D.C. and Whittall, P.T. (1989) Comparison of the Schoolfield (non-linear Arrhenius) model and the square root model for predicting bacterial growth in foods. *Food Microbiology* **6**, 7–18.

Arnon, S.S., Midura, T.F., Damus, K., Thompson, B., Wood, R.M. and Chin, J. (1979) Honey and other environmental risk factors for infant botulism. *Journal of Paediatrics* **94**, 331–336.

Baird-Parker, A.C. and Freame, B. (1967) Combined effect of water activity, pH and temperature on the growth of *Clostridium botulinum* from spore inocula. *Journal of Applied Bacteriology* **30**, 420–429.

Baird-Parker, A.C. and Kilsby, D.C. (1987) Principles of predictive food microbiology. *Journal of Applied Bacteriology, Symposium Supplement*, 43S–49S.

Ball, A.P., Hopkinson, R.B., Farrell, I.D. Hutchison, J.G.P., Paul, R., Watson, R.D.S. *et al.* (1979) Human botulism caused by *Clostridium botulinum* type E: the Birmingham outbreak. *Quarterly Journal of Medicine* **191**, 473–475.

Blatherwick, F.J. and Peck, S.H. (1985) An international outbreak of botulism associated with a restaurant in Vancouver, British Columbia. *Canadian Diseases Weekly Report* **11**(42), 177.

Bott, T.L., Defner, J.S., McCoy, E. and Foster, E.M. (1966) Clostridium botulinum type E in fish from the Great Lakes. *Journal of Bacteriology* **91**, 919- 924.

Boyd, J.W. and Southcott, B.A. (1971) Effects of sodium chloride on outgrowth and toxin production of *Clostridium botulinum* type E in cod homogenates. *Journal of Fisheries Research of the Board of Canada* **28**, 1071–1075.

California State Department of Health Services (1981) Alert: botulism associated with commercially produced, dried salted whitefish. *Californian Morbity Weekly Reports* Nov. 6, Supplement 43.

Cann, D.C., Wilson, B.B. and Shewan, J.M. (1965) The Incidence of *Clostridium botulinum* type E in fish and bottom deposits in the North Sea and off the coast of Scandinavia. *Journal of Applied Bacteriology* **28**, 426–430.

CDC (Centers for Disease Control) (1985) Botulism associated with commercially distributed kapchunka – New York City. *Californian Morbity Weekly Reports* **34**, 546–547.

CDC (Centers for Disease Control) (1987) International outbreak of type E botulism associated with ungutted, salted whitefish. *Californian Morbity Weekly Reports* **36**, 812–813.

Chou, J.H., Hwang, P.H. and Malison, M.D. (1988) An outbreak of type A foodborne botulism due to commercially preserved peanuts. *International Journal of Epidemiology* **17**, 899–902.

Christiansen, L.N., Deffner, J., Foster, E.M. and Sugiyama, H. (1968) Survival and outgrowth of Clostridium botulinum type E spores in smoked fish. *Applied Microbiology* **16**, 133–137.

Colebatch, J.G., Wolff, A.H., Gilbert, R.J., Mathias, C.J., Smith, S.E., Hirsch, N. and Wiles, C.M. (1989) Slow recovery from severe foodborne botulism. *Lancet* **II**, 1216–1217.

Critchley, E.M.R., Hayes, P.J. and Isaacs, P.E.T. (1989) Outbreak of botulism in North West England and Wales, June 1989. *Lancet* **II**, 849–853.

Denny, C.B., Goeke, D.J. and Sternberg, R. (1969) *Inoculation tests of* Clostridium botulinum *in Canned Breads with Specific Reference to Water Activity*. Research Report No. 4–69, Washington, DC: National Canners Association.

Dodds, K.L. (1989) Combined effect of water activity and pH on inhibition of toxin production by *Clostridium botulinum* in cooked, vacuum-packed potatoes. *Applied and Environmental Microbiology* **55**, 656–660.

Eklund, M.W. and Poysky, F. (1970) Distribution of *Clostridium botulinum* on the Pacific coast of the United States. In *Toxic Microorganisms* ed. Herzburg, M. pp. 304–308. Washington, DC: *US Department of Interior*.

Emodi, A.S. and Lechowich, R.V. (1969) Low temperature growth of type E *Clostridium botulinum* spores. I. Effects of sodium chloride, sodium nitrite and pH. *Journal of Food Science* **34**, 78–81.

Esty, J.R. and Meyer, K.F. (1922) The heat resistance of spores of *Clostridium botulinum* and allied anaerobes. *Journal of Infectious Diseases* **31**, 650–663.

Eyles, M.J. and Warth, A.D. (1981) Assessment of the risk of botulism from vacuum-packed raw fish: a review. *Food Technology Australia*, **33**, 574–580.

Fields, M.L., Zamora, A.F. and Bradsher, M. (1977) Microbiological analysis of home canned tomatoes and green beans. *Journal of Food Science* **42**, 931–934.

Garcia, G. and Genigeorgis, C. (1987) Quantitative evaluation of *Clostridium botulinum* nonproteolytic types B, E, and F growth risk in fresh salmon tissue homogenates stored under modified atmospheres. *Journal of Food Protection* **50**, 390–397.

Genigeorgis, C., Meng, J. and Baker, D.A. (1991) Behaviour of non-proteolytic *Clostridium botulinum* type B and E spores in cooked turkey and modelling lag phase and probability of toxigenesis. *Journal of Food Science* **56**, 373–379.

Gilbert, R.J. and Willis, A.T. (1980) Botulism. *Communications in Medicine* **2**, 25–26.

Greenberg, R.A., Tompkin, R.B., Bladel, B.O., Kittaka, R.S. and Anellis, A. (1966) Incidence of mesophilic *Clostridium* spores in raw pork, beef and chicken in processing plants in the United States and Canada. *Applied Microbiology* **14**, 789–793.

Hauschild, A.H.W. (1992) Epidemiology of human foodborne botulism. In Clostridium botulinum: *Ecology and Control in Foods* ed. Hauschild, A.H.W. and Dodds, K.L. pp. 69–104. New York: Marcel Dekker.

Hauschild, A.H.W. and Hilsheimer, R. (1980) Incidence of *Clostridium botulinum* in commercial bacon. *Journal of Food Protection* **43**, 564–565.

Hauschild, A.H.W. and Hilsheimer, R. (1983) Prevalence of *Clostridium botulinum* in commercial liver sausage. *Journal of Food Protection* **46**, 242–244.

Hauschild, A.H.W. and Simonsen, B. (1986) Safety assessment for shelf-stable canned cured meats—an unconventional approach. *Food Technology* **40**(4), 155- 158.

Hauschild, A.H.W., Aris, B.J. and Hilsheimer, R. (1975) *Clostridium botulinum* in marinated products. *Canadian Institute of Food Science and Technology Journal.* **8**, 84–87.

Hayashi, K., Sakaguchi, S. and Sakaguchi, G. (1986) Primary multiplication of *Clostridium botulinum* type A in mustard-miso stuffing of 'karashi-renkon' (deep-fried mustard-stuffed lotus root). *International Journal of Food Microbiology* **3**, 311–320.

Hobbs, G. (1976) *Clostridium botulinum* in fishery products. *Advances in Food Research* **22**, 135–185.

Huhtanen, C.N., Naghski, J., Custer, C.S. and Russell, R.W. (1976) Growth and toxin production by *Clostridium botulinum* in moldy tomato juice. *Applied and Environmental Microbiology* **32**, 711–715.

Huss, H.H. (1980) Distribution of *Clostridium botulinum. Applied and Environmental Microbiology* **39**, 764–769.

Huss, H.H., Pederson, A. and Cann, D.C. (1974a) The incidence of *Clostridium botulinum* in Danish trout farms. I. Distribution in fish and their environment. *Journal of Food Technology* **9**, 445–450.

Huss, H.H., Pederson, A. and Cann, D.C. (1974b) The incidence of *Clostridium botulinum* in Danish trout farms. II. Measures to reduce contamination of the fish. *Journal of Food Technology* **9**, 451–458.

Ikawa, J.Y. and Genigeorgis, C. (1987) Probability of growth and toxin production by non-proteolytic *Clostridium botulinum* in rockfish fillets stored under modified atmospheres. *International Journal of Food Microbiology* **4**, 167–181.

Ingram, D.M. and Robinson, R.H.M. (1951) A discussion of the literature on botulism in relation to acid food. *Proceedings of the Society for Applied Bacteriology* **14**, 73–84.

Insalata, N.E, Witzeman, S.E, Fredericks, G.J. and Sunga, F.C.A. (1969) Incidence study of spores of *Clostridium botulinum* in convenience foods. *Applied Microbiology* **17**, 542–544.

Jensen, M.J., Genigeorgis, C. and Lindroth, S. (1986) Probability of growth of *Clostridium botulinum* as affected by strain, cell and serologic type, inoculum size and temperature and time of incubation in a model broth system. *Journal of Food Safety* **8**, 109–126.

Johannsen, A. (1965) *Clostridium botulinum* type E in foods and the environment generally. *Journal of Bacteriology* **28**, 90–94.

Kautter, D.A., Lilly, T. and Lynt, R. (1978) Evaluation of the botulism hazard in fresh mushrooms wrapped in commercial polyvinylchloride film. *Journal of Food Protection* **41**, 120–121.

Kautter, D.A., Lilly, T., Solomon, H.M. and Lynt, R.K. (1982) *Clostridium botulinum* spores in infant foods: a survey. *Journal of Food Protection* **45**, 1028–1029.

Klarman, D. (1989) *Clostridium botulinum* in faecal samples of cattle and swine and samples of raw material and pulverized dehydrated meat of different rendering plants. *Berliner und Muenchener Tieraerztliche Wochenschrift* **102**, 84–87.

Lindroth, S.E. and Genigeorgis, C.A. (1986) Probability of growth and toxin production by nonproteolytic *Clostridium botulinum* in rockfish stored under modified atmospheres. *International Journal of Food Microbiology* **3**, 167–181.

Lund, B.M., George, S.M. and Franklin, J.G. (1987) Inhibition of type A and type B (Proteolytic) *Clostridium botulinum* by sorbic acid. *Applied and Environmental Microbiology* **53**, 935–941.

MacDonald, K.L., Spengler, R.F., Hatheway, C.L., Hargrett, N.T. and Cohen, M.L. (1985) Type A botulism from sautéed onions. *Journal of the American Medical Association* **253**, 1275–1278.

Marchal, A., De Mayeur, S. and Yde, M. (1985) Le botulisme en Belgique à propos de quelques foyers récents. *Revue Médicale de Bruxelles* **6**, 690–692.

Marshall, B.J., Ohye, D.F. and Christian, J.H.B. (1971) Tolerance of bacteria to high concentrations of NaCl and glycerol in the growth medium. *Applied Microbiology* **21**, 363–364.

Meyer, K.F. and Dubovsky, B.J. (1922a) The distribution of spores of *Bacillus botulinus* in California. II. *Journal of Infectious Diseases* **31**, 541–550.

Meyer, K.F. and Dubovsky, B.J. (1922b) The distribution of spores of *Bacillus botulinus* in the United States. IV. *Journal of Infectious Diseases* **31**, 559–568.

Midura, T.E, Snowden, S., Wood, R.M. and Arnon, S.S. (1979) Isolation of *Clostridium botulinum* from honey. *Journal of Clinical Microbiology* **9**, 282–283.

Montville, T.J. (1982) Metabiotic effects of *Bacillus licheniformis* on *Clostridium botulinum*: implications for home-canned tomatoes. *Applied and Environmental Microbiology* **44**, 334–338.

Notermans, S., Dufrenne, J. and Gerrits, J.P.G. (1989) Natural occurrence of *Clostridium botulinum* on fresh mushrooms (*Agaricus bisporus*). *Journal of Food Protection* **52**, 733–736.

Odlaug, T.E., and Pflug, I.J. (1979) *Clostridium botulinum* growth and toxin production in tomato juice containing *Aspergillus gracilis. Applied and Environmental Microbiology* **37**, 496–501.

Ohye, D.F. and Christian, J.H.B. (1966) Combined effects of temperature, pH and water activity on growth and toxin production by *Clostridium botulinum* types A, B, and E. *Proceedings of the 5th International Symposium on Food Microbiology*, pp. 136–143.

O'Mahony, M., Mitchell, E. Gilbert, R.J., Hutchinson, D.N., Begg, N.T., Rodhouse, J.C. and Morris, J.E. (1990) An outbreak of foodborne botulism associated with contaminated hazelnut yoghurt. *Epidemiology and Infection* **104**, 389–395.

Otofuji, T., Tokiwa, H. and Takahashi, K. (1987) A food poisoning incident caused by *Clostridium botulinum* toxin A in Japan. *Epidemiology and Infection* **99**, 167–172.

Raatjes, G.J.M. and Smelt, J.P.P.M. (1979) *Clostridium botulinum* can grow and form toxin at pH values lower than 4·6. *Nature* **281**, 398–399.

Riemann, H. (1967) The effect of the number of spores on growth and toxin formation of *Clostridium botulinum* type E in inhibitory environments. In: *Botulism 1966* ed. Ingram, M. and Roberts, T.A. pp. 150–157. London: Chapman and Hall.

Roberts, T.A. and Ingram, M. (1973) Inhibition of growth of *Clostridium botulinum* at different pH values by sodium chloride and sodium nitrite. *Journal of Food Technology* **8**, 467–475.

Roberts, T.A. and Smart, J.L. (1976) The occurrence and growth of *Clostridium* spp. in vacuum-packed bacon with particular reference to *Clostridium perfringens* (*welchii*) and *Clostridium botulinum. Journal of Food Technology* **11**, 229–244.

Roberts, T.A. and Smart, J.L. (1977). The occurrence of clostridia, particularly *Clostridium botulinum*, in bacon and pork. In *Spore Research 1976* ed. Barker, A.N., Wolf, J., Ellar, D.J.,

Dring, G.J. and Gould, G.W. pp. 911–915. New York: Academic Press.

Roberts, T.A., Gibson, A.M. and Robinson, A. (1981a) Factors controlling the growth of *Clostridium botulinum* types A and B in pasteurized, cured meats. I. Growth in pork slurries prepared from 'low' pH meat (pH 5·5–6·3). *Journal of Food Technology* **15**, 239–266.

Roberts, T.A., Gibson, A.M. and Robinson, A. (1981b) Factors controlling the growth of *Clostridium botulinum* types A and B in pasteurized, cured meats. II. Growth in pork slurries prepared from 'high' pH meat (range 6·3–6·8). *Journal of Food Technology* **16**, 267–281.

Roberts, T.A., Gibson, A.M. and Robinson, A. (1982) Factors controlling the growth of *Clostridium botulinum* types A and B in pasteurized, cured meats. III. The effect of potassium sorbate. *Journal of Food Technology* **17**, 307–326.

Rouhbakhsh-Khaleghdoust, A. (1975) The incidence of *Clostridium botulinum* type E in fish and bottom deposits in the Caspian Sea coastal waters. *Pahlavi Medical Journal* **6**, 550–556.

Rozanova, L.I., Zemlyakov, V.L. and Mazokhina, N.N. 1972 Factors affecting *Clostridium botulinum* contamination of vegetables intended for preservation and of materials used. *Gigiena i Sanitariya* **37**, 102–109.

Scott, W.J. (1957) Water relations of food spoilage microorganisms. *Advances in Food Research* **7**, 84–127.

Scott, V.N. and Taylor, S.L. (1981) Temperature, pH, and spore load effects on the ability of nisin to prevent the outgrowth of *Clostridium botulinum* spores. *Journal of Food Science* **46**, 121–126.

Segner, W.P., Schmidt, C.E. and Boltz, J.K. (1966) Effect of sodium chloride and pH on the outgrowth of spores of type E *Clostridium botulinum* at optimal and sub-optimal temperatures. *Applied Microbiology* **14**, 49–54.

Slater, P.E., Addiss, D.G., Cohen, A., Levanthal, A., Chassis, G., Zehavi, H., Bashari, A. and Costin, C. (1989) Foodborne botulism: an international outbreak. *International Journal of Epidemiology* **18**, 693–696.

Smelt, J.P.P.M., Raatjes, G.J.M., Crowther, J.S. and Verrips, C.T. (1982) Growth and toxin formation by *Clostridium botulinum* at low pH values. *Journal of Applied Bacteriology* **52**, 75–82.

Smoot, L.A. and Pierson, M.D. (1979) Effect of oxidation-reduction potential on the outgrowth and chemical inhibition of *Clostridium botulinum* 10755A spores. *Journal of Food Science* **44**, 700–704.

Spencer, R. (1967) Factors in cured meat and fish products affecting spore germination, growth and toxin production. In *Botulism 1966* ed. Ingram, M. and Roberts, T.A. pp. 123–135. London: Chapman and Hall.

Sperber, W.H. (1982) Requirements of *Clostridium botulinum* for growth and toxin production. *Food Technology* **36**(12), 89–94.

Solomon, H.G. and Kautter, D.A. (1986) Growth and toxin production by *Clostridium botulinum* in sautéed onions. *Journal of Food Protection* **49**, 618–620.

Solomon, H.G. and Kautter, D.A. (1988) Outgrowth and toxin production by *Clostridium botulinum* in bottled chopped garlic. *Journal of Food Protection* **51**, 862–865.

Solomon, H.G., Kautter, D.A., Lilly, T., Jr and Rhodehamel, E.J. (1990) Outgrowth of *Clostridium botulinum* types A and B in shredded cabbage at room temperature under a modified atmosphere. *Journal of Food Protection* **53**, 831–833.

St Louis, M.E., Peck, S.H.S., Bowering, D., Morgan, G.B., Blatherwick, J., Banerjee, S. *et al.* (1988) Botulism from chopped garlic; delayed recognition of a major outbreak. *Annals of International Medicine* **108**, 363–368.

Sugiyama, H., Mills, D.C. and Kuo, L.-J.C. (1978) Number of *Clostridium botulinum* spores in honey. *Journal of Food Protection* **41**, 848–850.

Taclindo, C., Midura, T., Nygaard, G.S. and Bodily, H.L. (1967) Examination of prepared foods in plastic packages for *Clostridium botulinum*. *Applied Microbiology* **15**, 426–430.

Tanaka, N. (1982) Toxin production by *Clostridium botulinum* in media at pH lower than 4·6. *Journal of Food Protection* **45**, 234–237.

Tanaka, N., Traisman, E., Plantinga, P., Finn, L., Flom, W., Meske, L. and Guggisberg (1986) Evaluation of factors involved in antibotulinal properties of pasteurized process cheese spreads. *Journal of Food Protection* **49**, 526–531.

Telzak, E.E., Bell, E.P., Kautter, D.A., Crowell, L., Budnick, L.D., Morse, L.D. and Shultz, S. (1990) An international outbreak of type E botulism due to uneviscerated fish. *Journal of Infectious Diseases* **161**, 340–342.

Terranova, W., Bremen, J.G., Locey, R.P. and Speck, S. (1978) Botulism type B: epidemiological aspects of an extensive outbreak. *American Journal of Epidemiology* **108**, 150–156.

Todd, E., Chang, P.C., Hauschild, A.H.W., Sharpe, A., Park, C. and Pivnick, H. (1974) Botulism from marinated mushrooms. *Proceedings of the IV International Congress of Food Science and Technology* **III**, 182–188.

Townsend, C.T., Yee, L. and Mercer, W.A. (1954) Inhibition of the growth of *Clostridium botulinum* by acidification. *Food Research* **19**, 536–542.

Tsai, S.J., Chang, Y.C. Wang, J.D. and Chou, J.H. (1990) Outbreak of type A botulism caused by a commercial food product in Taiwan: clinical and epidemiological investigations. *Clinical Medical Journal* **46**, 43–48.

Tsang, N., Post, L.S. and Solberg, M. (1985) Growth and toxin production by *Clostridium botulinum* in model acidified systems. *Journal of Food Science* **50**, 961–965.

Vergieva, V. and Incze, K. (1979) The ecology of *Clostridium botulinum*. *Husipar* **28**, 79–87.

Vicens, R., Rasolofonirina, N. and Coulanges, P. (1985) Premiers cas humains de botulisme alimentaire à Madagascar. *Archives Institut Pasteur de Madagascar* **52**, 11–22.

Weber, J.T., Hibbs, R.G., Darwish, A., Mishu, B., Corwin, A.L., Rakha, M. *et al.* (1993) A massive outbreak of type E botulism associated with traditional salted fish in Cairo. *Journal of Infectious Diseases* **167**, 451–454.

Wong, D.M., Young-Perkins, K.E. and Merson, R.L. (1988) Factors influencing *Clostridium botulinum* spore germination, outgrowth, and toxin formation in acidified media. *Applied and Environmental Microbiology* **54**, 1446–1450.

Young-Perkins, K.E. and Merson, R.L. (1987) *Clostridium botulinum* spore germination, outgrowth and toxin production below 4.6; interactions between pH, total acidity and buffering capacity. *Journal of Food Science* **52**, 1084–1088, 1096.

Journal of Applied Bacteriology Symposium Supplement 1994, **76**, 115S–128S

Heat resistance and recovery of spores of non-proteolytic *Clostridium botulinum* in relation to refrigerated, processed foods with an extended shelf-life

Barbara M. Lund and M.W. Peck[1]
8 The Walnuts, Branksome Road, Norwich, and [1]Institute of Food Research, Norwich Laboratory, Norwich Research Park, Colney, Norwich, UK

1. Introduction, 115S
2. The effect of heat treatment in phosphate buffer on survival of spores of non-proteolytic *Clostridium botulinum*, 116S
3. The ability of non-proteolytic *Clostridium botulinum* to multiply at low temperatures, 117S
4. The effect of heat treatment and incubation temperature on survival of, and growth from, spores of non-proteolytic *Clostridium botulinum* in a model food medium, 119S
5. Properties of lysozyme, and the nature of other enzymes that degrade peptidoglycan, 121S
6. Sublethal injury of spores by heat
   6.1 General aspects of sublethal injury, 122S
   6.2 Altered requirements of injured spores for non-nutrient germination stimulants, 122S
   6.3 Increased sensitivity of heat-injured spores to inhibitors, 123S
   6.4 Other effects, 124S
7. Conclusions, 124S
8. Acknowledgements, 125S
9. References, 125S

1. INTRODUCTION

The production of refrigerated processed (or pasteurized) foods with extended durability (REPFEDs), for use in catering or for retail sale, is a subject of considerable current interest (Lund and Notermans 1992). These foods are usually packaged under vacuum or modified atmosphere and/or in sealed containers. During production REPFEDs may be: (1) packaged then cooked; (2) cooked then packaged; (3) cooked, then packaged, then heated again. The term '*sous vide*' strictly refers to vacuum packing, without any indication of thermal processing, but it has become an accepted name for pasteurized ingredients or foods that are vacuum-packed prior to heat processing (Brown and Gould 1992). The requirements to ensure the safety of REPFEDs were reviewed by Mossel and Struijk (1991), who stressed that after production the temperature of these foods should not exceed 3°C for any significant period of time. The report on Vacuum Packaging and Associated Processes by the Advisory Committee on the Microbiological Safety of Food UK (1992) was concerned to a large extent with these foods. One of the main requirements in relation to REPFEDs is to ensure that their processing, composition and storage combine to give an adequate degree of protection against survival and growth of *Clostridium botulinum*. While storage below 3°C would prevent the growth of *Cl. botulinum*, it is difficult to ensure that this temperature is maintained, particularly in the case of foods for retail sale. For this reason it has been recommended that, in addition to refrigeration, further inhibitory factors should be incorporated into these foods so that their safety in relation to *Cl. botulinum* is not dependent solely on refrigeration.

Correspondence to: Dr B.M. Lund, 8 The Walnuts, Branksome Road, Norwich NR4 6SR, UK.

Sous vide foods have been produced in France for many years, and regulations for their hygienic production, distribution and sale and for their shelf-life have been published by the French Ministry of Agriculture (MAFR 1974; 1988). The storage temperature specified for these foods is between 0 and 3°C. The pasteurization specified in these regulations is expected to destroy vegetative bacteria, but not to kill spores (Rosset and Poumeyrol 1986); the core temperatures reached are usually 65°C or higher and the shelf life depends on the heat treatment (Lund and Notermans 1992). The Advisory Committee on the Microbiological Safety of Food UK (1992) recommended that for prepared chilled foods with an assigned shelf-life of more than 10 d further controlling factors, in addition to chill temperature, should be used singly or in combination to prevent the growth of psychrotrophic *Cl. botulinum*. These factors included:

(1) A heat treatment of 90°C for 10 min or equivalent lethality;
(2) A pH of 5 or less throughout the food and throughout all components of complex foods;

(3) A minimum salt level of 3·5% in the aqueous phase throughout the food and all components of complex foods;
(4) A water activity of 0·97 or less throughout the food and throughout all components of complex foods;
(5) A combination of heat and preservative factors that can be shown consistently to prevent growth and toxin production by psychrotrophic *Cl. botulinum*.

The highest temperatures used for pasteurization of these foods would have little effect on the heat-resistant spores of proteolytic, Group 1, strains of *Cl. botulinum*, which form toxins of type A, B or F (Hatheway 1992). These strains do not multiply at temperatures below 10°C, however, and as regulations specify that REPFEDs should be stored at temperatures well below 10°C (<3°C in France, <5–8°C in the UK), it may be considered that controls can be built in so that these foods are not consumed after storage at a temperature of 10°C or higher, and it may be possible to use time–temperature indicators to help ensure this. Group 2 non-proteolytic strains of *Cl. botulinum*, which form toxins of type B, E or F, can multiply and form toxin at temperatures as low as 3·3°C (Hatheway 1992). Spores of these psychrotrophic strains are less heat-resistant than those of Group 1 strains but are able to survive some heat treatments at temperatures between 65 and 95°C. As is described in the following sections of this paper, the presence of lysozyme in recovery media increases the number of survivors that result in growth after heat treatment. Lysozyme occurs at a high concentration in eggs and may also be present in other foods. It is relatively heat-resistant and may survive heat-processing.

The majority of studies of the heat-resistance of spores of non-proteolytic *Cl. botulinum* reported in the literature have not taken into account the effect of lysozyme in the recovery medium (Lund and Notermans 1992), and the minimum heat treatment required to give a substantial reduction in the survival of these spores has been the subject of debate (Brown and Gould 1992). The requirement to review conditions to ensure the safety of REPFEDs has led to the need for an improved evaluation of the heat-resistance of spores of non-proteolytic *Cl. botulinum*, and of the ability of surviving spores to result in growth of the organism. The aim of this paper is to contribute to this evaluation.

In studying the effect of suboptimal conditions, such as low temperature and low pH, on growth of *Cl. botulinum* and in determining the numbers of survivors after heat treatment it is important to ensure that traces of oxygen, which may have a determining influence on the results (Lund 1993), are eliminated. In experiments at this institute strictly anaerobic conditions under a head space of oxygen-free $N_2/H_2/CO_2$, 85 : 10 : 5 or H_2/CO_2, 90 : 10 have been used to produce media with a redox potential equivalent to < −200 mV at pH 7.

2. THE EFFECT OF HEAT TREATMENT IN PHOSPHATE BUFFER ON SURVIVAL OF SPORES OF NON-PROTEOLYTIC *CLOSTRIDIUM BOTULINUM*

In these experiments phosphate buffer was used as a standard medium in which to heat the spores. The technique of Kooiman and Geers (1975) was used because this enables almost instantaneous heating of spores and very rapid cooling. Heat treatment of spores of non-proteolytic type B strain 17B at 85°C for 1 min followed by plating on a peptone–yeast extract–glucose–starch (PYGS) medium, which gave the maximum count of unheated spores, and incubation at 30°C under H_2/CO_2 (90 : 10 v/v) resulted in more than a 5-log inactivation (Peck *et al.* 1992a). After heating spores of six non-proteolytic strains at 85°C for 10 min, plating the heat-treated spores on medium that contained lysozyme (10 μg ml^{-1}; 625 units ml^{-1}) increased the number of colonies formed by a factor of between 10^3 and 10^6 compared with the number on medium without added lysozyme (Peck *et al.* 1992a; Fig 1.). As little as 0·1 μg lysozyme ml^{-1} had some effect; 10–100 μg ml^{-1} gave the maximum effect for most strains and the highest concentration was not inhibitory. Medium containing egg yolk emulsion gave similar counts from heated spores to those on PYGS + 10 μg lysozyme ml^{-1}, this was probably due to the presence of lysozyme in the egg yolk emulsion. On horse blood agar the counts from heated spores were higher than those on PYGS medium without added lysozyme; this can be explained by the presence of a low concentration of lysozyme in the blood, resulting in about 0·1 μg lysozyme ml^{-1} in the blood agar.

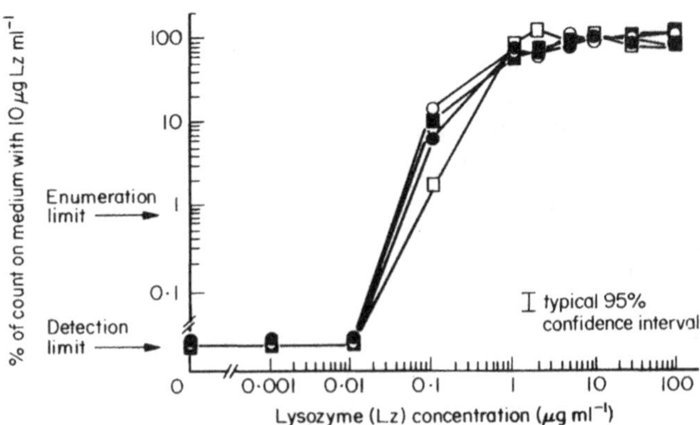

Fig. 1 The effect of lysozyme concentration in the plating medium on the recovery of spores of non-proteolytic *Clostridium botulinum* after heating at 85°C for 10 min (modified from Peck *et al.* 1992a). ■, FT50; ○, 2B; ●, 17B; □, Beluga

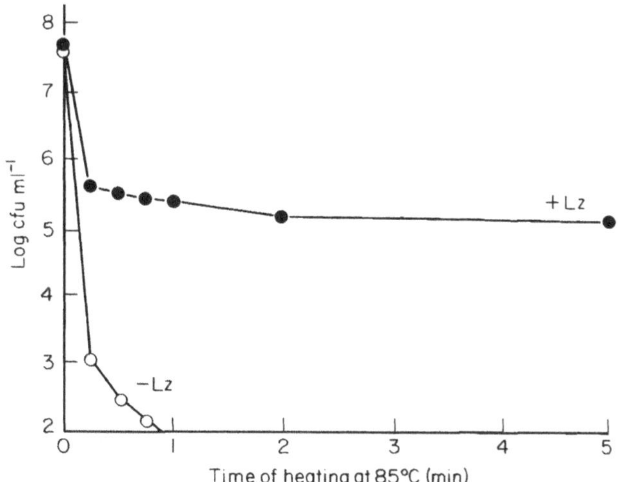

Fig. 2 The effect of lysozyme (Lz; 10 μg ml^{-1}) in the plating medium on the estimated heat-resistance of spores of *Clostridium botulinum* 17B (Peck *et al.* 1992a)

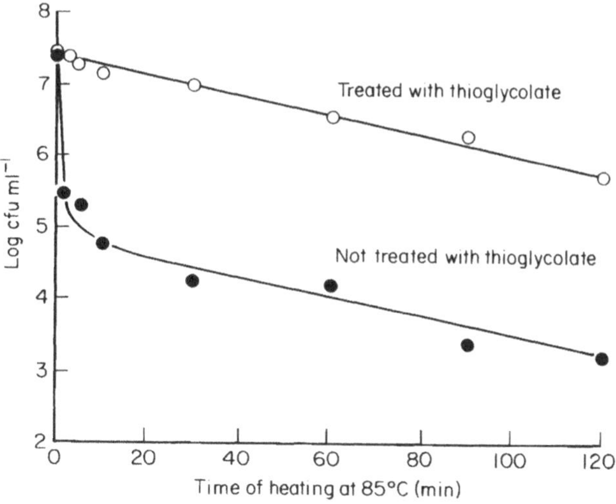

Fig. 3 The effect of treatment of heated spores with alkaline thioglycolate, prior to plating on medium + lysozyme (10 μg ml^{-1}), on the estimated heat-resistance of spores of *Clostridium botulinum* 17B (Peck *et al.* 1992b)

Table 1 *D*-values of lysozyme-permeable spores of non-proteolytic *Cl. botulinum* determined on medium + lysozyme (modified from Peck *et al.* 1993)

Strain	Temperature (°C)	*D* (min)
17B	85	100
	90	18·7
	95	4·4
Beluga (E)	85	45·6
	90	11·8
	95	2·8

D-value at 85°C for lysozyme-impermeable spores, <1 min.

The contrast between survival curves for strain 17B determined on medium with and without lysozyme is shown in Fig. 2. On the medium containing lysozyme, a biphasic survival curve was obtained. Tests were done to verify that this was not due to a failure to expose all the spores to the full heat treatment. The proportion of heated spores that formed colonies on medium containing lysozyme (10 μg ml^{-1}) after heating at 85°C for 1 min was usually between 0·1% and 1%, but in the case of one strain of type E (Foster B96) it was about 20% (Peck *et al.* 1993). The lysozyme in the recovery medium is probably able to diffuse through the coat of a proportion of the heated spores and degrade the cortex, replacing a germination system that has been inactivated by heat. Treatment of spores, after heating, with alkaline thioglycolate (22% w/v at pH 10 for 30 min at 45°C) before plating on medium containing lysozyme resulted in a marked increase in the number of colonies formed, and gave a logarithmic heat-inactivation curve (Fig. 3). This treatment is known to break disulphide bonds in spore coats, making the coats permeable to lysozyme. Electron microscopy showed that after heating and treatment with alkaline thioglycolate the spore coats were disrupted (Figs 4 and 5).

In estimating decimal reduction times (*D*-values) for the effect of heat treatment on spores of non-proteolytic *Cl. botulinum*, the results will depend on whether the *D*-values are based on survivor curves or on an end-point method, and the number of log cycles over which inactivation is measured. *D*-values for lysozyme-permeable and lysozyme-impermeable spores determined by plating on medium containing lysozyme are shown in Table 1. In the case of spores plated on medium without lysozyme, the *D*-value is based on a 10^5-fold reduction in numbers, whereas for the spores plated on medium containing lysozyme, the *D*-values are based on a reduction of between 10-fold and 10^5-fold, depending on the temperature.

3. THE ABILITY OF NON-PROTEOLYTIC *CLOSTRIDIUM BOTULINUM* TO MULTIPLY AT LOW TEMPERATURES

Incubation at low temperature may limit the proportion of spores that are able to undergo outgrowth and the proportion of vegetative bacteria that are able to grow. These effects will both influence the probability (*P*) that detectable growth and toxin will be produced from a single spore or vegetative bacterium in a given time. Values of *P* for spores of non-proteolytic types B, E and F incubated at 4°C for 60 d were approximately 1 in 10^6, 1 in 10^7 and 1 in 10^2, respectively (Ikawa *et al.* 1986). For two strains of non-proteolytic type B, grown at 30°C, *P* for growth from a single vegetative bacterium at 4°C in 28 d was between 1 in 10^4 and 1 in 10^5 (Jensen *et al.* 1987), and for a mixture of

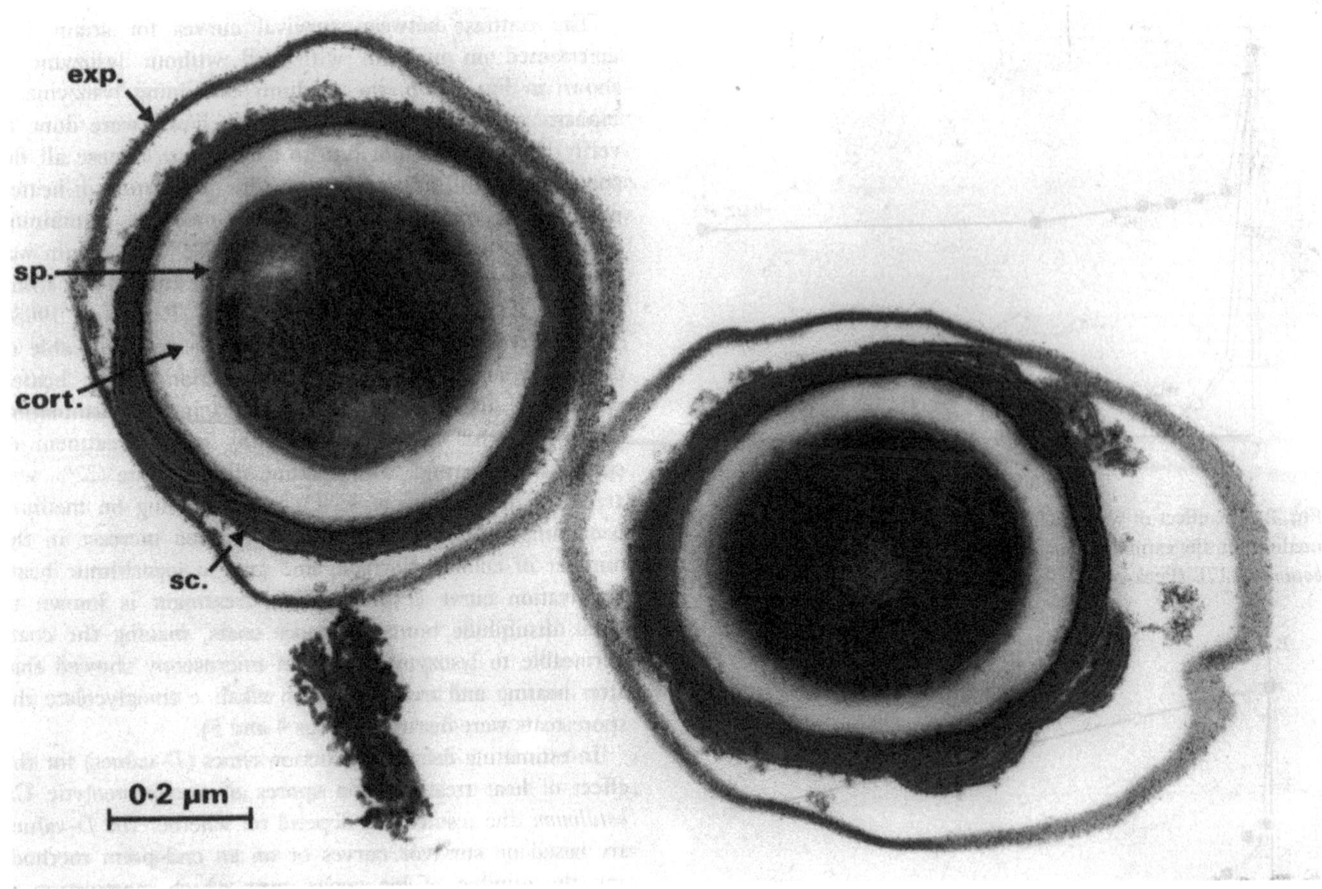

Fig. 4 Electron micrograph showing section through spores of *Clostridium botulinum* 17B. Spores were fixed in cacodylate buffer, 0·05 mol l^{-1}, pH 7·2, containing glutaraldehyde, 4%, and ruthenium red, 0·15%, and postfixed in solution of osmium tetroxide, 1%, and ruthenium red, 0·15%. Sections were stained sequentially with uranyl acetate and lead citrate: exp = exosporium, sc = spore coat, cort = cortex, sp = spore core

four strains of non-proteolytic type B, *P* for growth from a single vegetative bacterium at 6°C in 60 d was between 1 in 10^2 and 1 in 10^3 (Lund *et al.* 1990). In the report by Graham and Lund (1993) there was an indication that *P* for growth from a single vegetative bacterium of non-proteolytic type B at 4°C was about 1 in 10^4. In the reports of growth of non-proteolytic *Cl. botulinum* at temperatures below 4°C (Schmidt *et al.* 1961; Eklund *et al.* 1967a,b) the inoculum size was always greater than 10^5 bacteria/container. Nevertheless, an inoculum of 10^3 spores of type E per trout fillet survived hot smoking (maximum core temperature 60°C) and the vacuum-packed, smoked fillets were positive for toxin after storage for 35 d at 5–8°C (Dehof *et al.* 1989).

Spores of *Cl. botulinum* type E strain VH germinated within 24 h at temperatures between 2° and 50°C (Grecz and Arvay 1982). There was extensive and relatively rapid germination at 2°C, the lowest temperature tested, yielding about 60% phase-variable (partially germinated) spores by 18 h, which became phase-dark (fully germinated) after incubation for 26 h. The maximum number of fully germinated spores was obtained at 9°C, which appeared to be the optimum temperature for the final step in spore germination. This work demonstrated that the temperature range for germination of spores of non-proteolytic *Cl. botulinum* is wider than that allowing vegetative growth, and indicated that germination can occur relatively rapidly at temperatures in the range 2–10°C.

Strains of non-proteolytic *Cl. botulinum* of types B, E and F have been reported to multiply at a temperature as low as 3·3°C, but there is little quantitative information on the rate of growth. At 5, 10 and 20°C in optimum conditions, doubling times of 42·6, 7·2 and 1·9 h, respectively, have been reported for type E strains (Ohye and Scott 1957) and of 28·6, 5·7 and 1·3 h, respectively, for a type B strain (Graham and Lund 1993). Studies of the effect of temperature, pH and NaCl on the lag period and on the growth rate of non-proteolytic *Cl. botulinum* have been

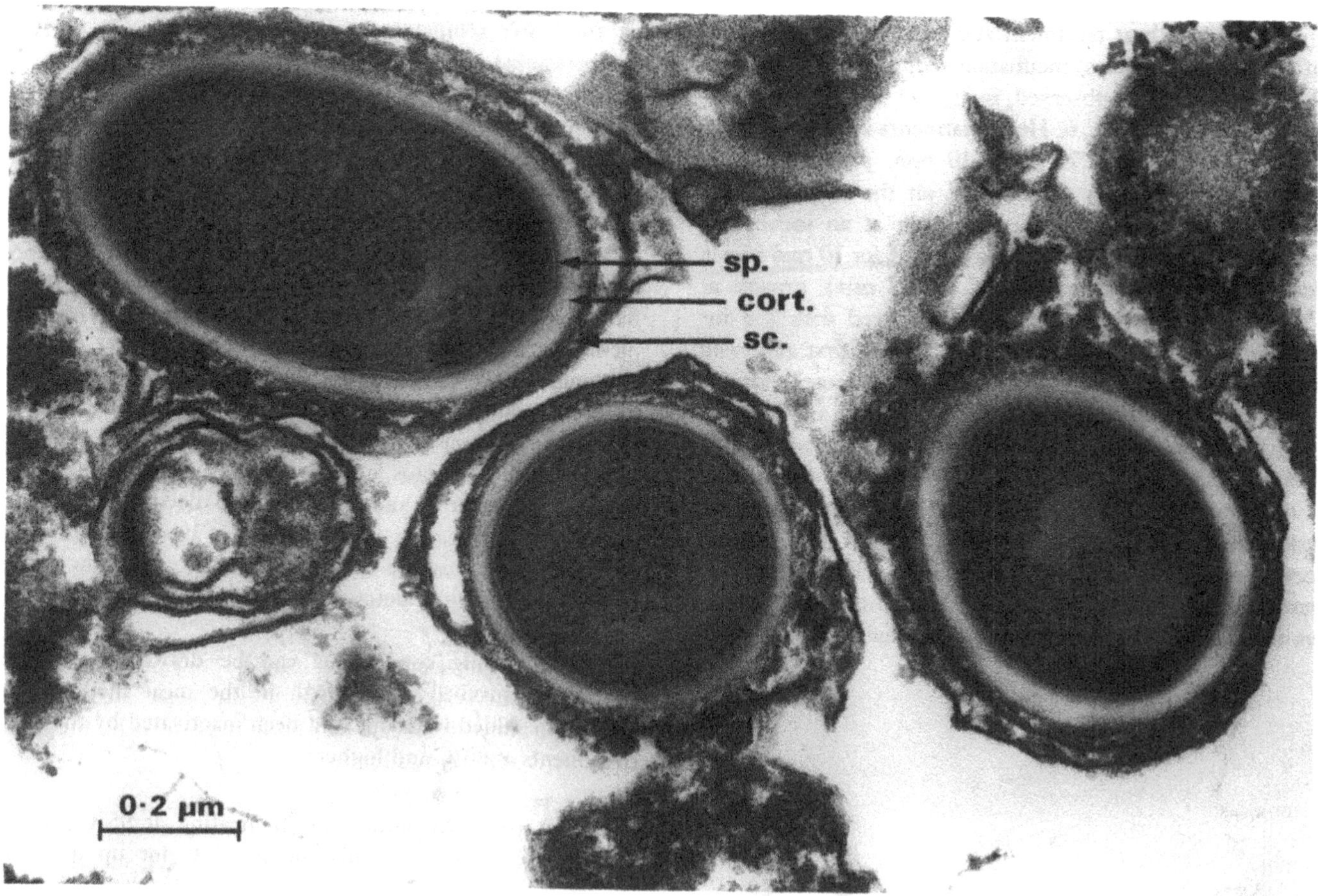

Fig. 5 Electron micrograph showing section through spores of *Clostridium botulinum* 17B after heating at 85°C for 10 min and treatment with alkaline thioglycolate. The method of preparation was as in Fig. 4: sc = spore coat (damaged), cort = cortex, sp = spore core

made at this institute under the MAFF Predictive Microbiology Programme.

4. THE EFFECT OF HEAT TREATMENT AND INCUBATION TEMPERATURE ON SURVIVAL OF, AND GROWTH FROM, SPORES OF NON-PROTEOLYTIC *CLOSTRIDIUM BOTULINUM* IN A MODEL FOOD MEDIUM

Experiments have been done to assess the effect of heat treatment and subsequent incubation temperature on the ability of spores of non-proteolytic *Cl. botulinum* to survive and give growth in an anaerobic meat medium intended as a model food (Peck *et al.* 1994). The medium consisted of: minced beef, 500 g; glucose, 10 g; NaCl, 10 g; soluble starch, 10 g; distilled water to 1000 g; and had a fat content of approximately 9·3% (w/w). The medium was prepared and distributed into 20 ml volumes using strictly anaerobic conditions under oxygen-free $H_2/CO_2/N_2$ (10 : 5 : 85 v/v). Immediately before an experiment a sterile solution of lysozyme, prepared under strictly anaerobic conditions, was added to give a concentration of 10 μg ml^{-1} (625 units ml^{-1}).

A spore suspension was used that contained an equal number of spores of each of 12 strains, five of type B, five of type E and two of type F. At the start of an experiment 0·2 ml of the spore suspension was added to each tube containing 20 ml of meat medium to give approximately 10^6 spores per tube. Inoculated tubes were then subjected to each of six heat treatments, cooled rapidly, and 10 tubes from each heat treatment were incubated at 6, 8, 10, 12 and 25°C for up to 60 d. The tubes were observed for growth, as shown by formation of gas, at regular intervals. The fact that growth was that of *Cl. botulinum* was confirmed by tests for toxin. Some tubes that did not show growth were removed before the full 60 d to check that toxin had not been formed. Growth was accompanied by formation of toxin and no toxin was found in tubes that did not show evidence of growth.

The effect of heat treatment and incubation temperature on the number of days' incubation required before growth of *Cl. botulinum* was observed, in at least one of each set of tubes, is shown in Fig. 6. Heat treatments at 65°C for 6 h, 70°C for 3 min, and 75°C for 10 min, plus heating-up times, had a relatively small effect on the time required before growth was first observed, even at an incubation temperature of 6°C. After heating at 80°C for 10 min plus a heat-up time (equivalent to 80°C for 23 min), growth at 6°C was first observed after 28 d, compared with 7 d for unheated spores, and growth at 25°C was first observed after 3 d, compared with 1 d for unheated spores. After a heat treatment equivalent to holding at 85°C for 18 min, growth was detected in tubes incubated at 6 and at 8°C after 53 d, and at 10, 12 and 25°C growth was detected in at least one tube after 42, 42 and 6 d, respectively. After a heat treatment equivalent to holding at 95°C for 15 min, growth was detected after 32 d in one of four tubes that were incubated at 25°C for the full period of 60 d but was not detected in 28 tubes that were incubated for the full 60 d at the lower temperatures (seven tubes at each of four temperatures). After tubes heated at 95°C had been incubated at 6, 8, 10° or 12°C for 60 d, four tubes incubated at each temperature were transferred to 25°C. Growth of *Cl. botulinum* occurred in two of these tubes in 14 d; thus survivors were detected in 3 out of 20 tubes heated at 95°C and incubated at 25°C. Types B, E, and F toxin were detected in tubes that had been heated at up to 75°C and were then incubated at 10, 12 or 25°C, but in tubes incubated at 6 or 8°C and those heated at the highest temperatures, the toxin detected most frequently was type B. Although lysozyme (10 μg ml^{-1}) had been added to all the tubes of medium in these experiments, later work indicated that when the spores were heated in meat medium at 80 or 85°C for 10 min plus heat-up time, similar results were obtained whether or not lysozyme had been added to the medium *before* heating. The addition of lysozyme *after* the heat treatment resulted in a marked decrease in the incubation time required before growth was observed from surviving spores.

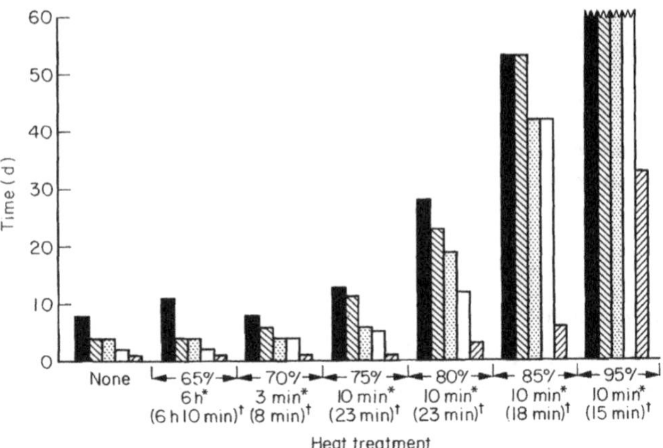

Fig. 6 The effect of heat treatment and incubation temperature on the survival and the number of days required for growth from spores of non-proteolytic *Clostridium botulinum* in meat medium at 6° (■), 8° (▧), 10° (▦), 12° (□) and 25° (▨) (modified from Peck *et al.* 1994). Tubes of medium prepared in strictly anaerobic conditions under $H_2/CO_2/N_2$ 10 : 5 : 85 (v/v) were inoculated with 10^6 spores/tube before the heat treatment. Fifty tubes of inoculated medium were subjected to each heat treatment; after cooling, 10 tubes were incubated at each temperature. Some tubes were removed for testing before the end of the incubation period. The histogram shows the incubation time required before the first tube in the set showed growth: * = time at target temperature; † = equivalent time at target temperature, calculated from the temperature – time data during heating from 20°C to the target temperature, and cooling to < 10°C. A *z*-value of 8°C was assumed (Lund and Notermans 1992)

The following conclusions can be drawn from these studies of survival and growth in the meat medium, in which the added lysozyme had been inactivated by the heat treatments at 80°C and higher:

(1) Heat treatment equivalent to heating at 95°C for 15 min followed by incubation at 25°C for up to 60 d failed to inactivate 10^6 spores of non-proteolytic *Cl. botulinum*;
(2) Heat treatment equivalent to heating at 95°C for 15 min followed by incubation at 12°C or lower for 60 d resulted in inactivation/inhibition of 10^6 spores of non-proteolytic *Cl. botulinum*;
(3) Heat treatment equivalent to heating at 85°C for 18 min followed by incubation at 6–12°C for 60 d failed to give inactivation/inhibition of 10^6 spores of non-proteolytic *Cl. botulinum*.

Although lysozyme was inactivated by the heat-treatment at 80 or 85°C for 10 min plus heat-up time in these experiments, the concentration added was relatively low and the heat-resistance of lysozyme is very dependent on the nature of the medium in which it is heated. In foods of a different composition, lysozyme may remain active after this heat treatment.

On the basis of the studies of the heat-resistance of spores in phosphate buffer (Section 2) and recovery of spores on medium without added lysozyme, an estimated 6-log reduction in numbers of viable spores would be obtained by heating at 85°C for 2 min. This contrasts with the finding that heating at 95°C for 15 min failed to inactivate 10^6 spores in the meat medium. The apparently higher heat resistance in the meat medium may have been due mainly to a protective effect of some components of the medium.

5. PROPERTIES OF LYSOZYME, AND THE NATURE OF OTHER ENZYMES THAT DEGRADE PEPTIDOGLYCAN

Lysozyme is classified as a muramidase (E.C. 3.2.1.17) and splits the β-1,4 linkage between *N*-acetylmuramic acid and N-acetylglucosamine in the peptidoglycan of bacterial cell walls (Strominger and Tipper 1974) and of the spore cortex (Warth and Strominger 1972). Its activity is usually measured by its ability to lyse cells of *Micrococcus luteus*. Its effect on the recovery of heated spores probably results from its ability to replace a spore germination system that has been inactivated by the heat treatment (Gould 1984). Enzymes with lysozyme activity are distributed widely in nature and have been isolated from many sources of food including the eggs of birds, mammalian tissue and milk, fish, molluscs, crustacea and plants (Table 2), and also from fungi, bacteria and bacteriophages. In many cases the concentrations present in raw foods are much higher than those required to initiate the growth of heat-damaged spores of non-proteolytic *Cl. botulinum*. A particularly high concentration is present in hen egg-white, which is the major source of commercial preparations of the enzyme. Hen egg-white lysozyme has a molecular weight of approximately 14 300–14 600 and contains four disulphide bonds (Proctor and Cunningham 1988); its tertiary structure has been elucidated. It is relatively heat stable at pH values below 7 (Fig. 7). Even relatively low concentrations in phosphate buffer at pH 6·2 lost only 5% activity after heating at 80°C for 30 min and approximately 25% activity after heating at 100°C for 20 min (Smolelis and Hartsell 1952). The heat-stability of lysozyme can be decreased in the presence of other proteins such as ovalbumin and ovotransferrin, or increased in the presence of glucose, sucrose, sorbitol, low concentrations of sodium chloride, and some polysaccharides (Proctor and Cunningham 1988).

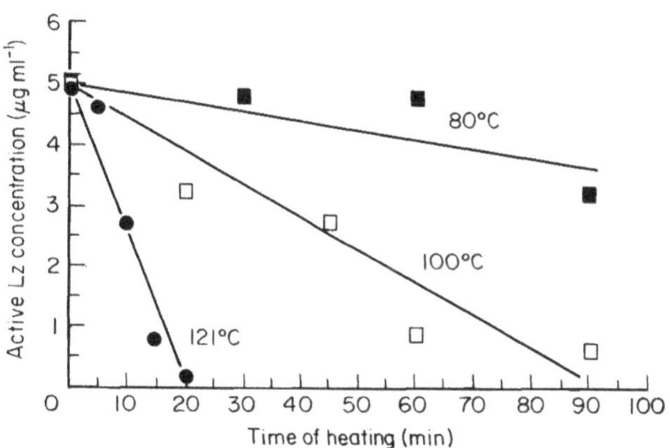

Fig. 7 The effect of heating in phosphate buffer, pH 6·2, on the activity of hen egg-white lysozyme (Lz) (calculated from Smolelis and Hartsell 1952)

Other enzymes are also able to degrade bacterial peptidoglycan (Strominger and Tipper 1974; Rogers *et al.* 1980; Table 3). Some lysozymes also show chitinase activity, and some of the enzymes described as lysozymes from plants appear to be primarily chitinases (Nord and Wadström 1972; Jollès *et al.* 1974; Boller 1986; Jekel *et al.* 1991). In addition to lysozymes, *endo-β-N*-acetylglucosaminidases, transglycosylases, *N*-acetylmuramyl-L-

Table 2 The occurrence in raw foods of enzymes with lysozyme activity

Source of lysozyme	Examples (with concentration (µg/g) where known)	References
Eggs of birds and reptiles	Hen egg (3000), eggs of duck (1000), goose (1200), turkey, quail, pheasant, guinea fowl, swan	Jollès 1967; Jollès and Jollès 1984; Proctor and Cunningham 1988
Mammalian tissue and milk	Cow milk (0·3), human milk (60), rabbit serum (0·8), rabbit spleen (400), dog kidney (15), cattle organs	Jollès 1967; Carroll and Martinez 1979; Jollès and Jollès 1984; Proctor and Cunningham 1988
Fish	Rainbow trout (10), salmon, plaice, cod, haddock, carp, flounder, halibut, sea trout, catfish, turbot	Murray and Fletcher 1976; Takahashi *et al.* 1986; Grinde *et al.* 1988; Lie *et al.* 1989
Plant	Cauliflower (28), broccoli (8), cabbage (2), turnip (2), bean, papaya (8), radish (5), parsnip (2), rutabaga (9), fig, wheat, maize, barley, kiwi fruit	Chandan and Ereifij 1981; Boller 1986; Roberts and Selitrennikoff 1988; Lynn 1989
Mollusc	Mussel (48), pearl oyster (58), black abalone, American oyster, snail (2), Japanese manila clam, venus clam, Asiatic clam	Rodrick and Cheng 1974; McHenery and Birkbeck 1979, 1982; Mochizuki and Matsumiya 1983; Ottaviani 1991
Crustacea	Crayfish (200), lobster, crab	Fenouil and Roch 1991

Table 3 Enzymes that degrade peptidoglycan

1.	*endo-β-N*-Acetylmuramidases (lysozymes, E.C. 3.2.1.17)
2.	*endo-β-N*-Acetylglucosaminidases
3.	Chitinases (E.C. 3.2.1.14)
4.	Transglycosylases
5.	*N*-Acetylmuramyl-L-alanine amidases (E.C. 3.5.1.28)
6.	*endo*-Peptidases

alanine amidases and *endo*-peptidases are lysins formed by vegetative bacteria (Rogers *et al.* 1980; Höltje and Tuomanen 1991). Vegetative cells of proteolytic *Cl. botulinum* type A produced two autolytic enzymes, *N*-acetylmuramyl-L-alanine amidase and *endo-β-N*-acetylglucosaminidase (Takumi *et al.* 1971). An extracellular protein produced by some strains of *Clostridium perfringens*, which initiated germination of heat-damaged spores of this organism, was identified as an *N*-acetylmuramyl-L-alanine amidase (Tang and Labbe 1987). Cortex-lytic enzymes are normally present in the spore cortex in an inactive state, and are activated when germination is triggered (Foster and Johnstone 1989). Enzymes capable of lysing cortex peptidoglycan have been isolated from both dormant and germinating spores (Gombas and Labbe 1985; Foster and Johnstone 1988) and were released as exudate from spores of *Cl. perfringens* during germination (Ando 1979). The main cortex-lytic enzymes detected in spores of *Bacillus subtilis* and *Bacillus cereus* were *endo-β-N*-acetylglucosaminidase and *N*-acetylmuramyl-L-alanine amidase (Warth 1972). In view of these reports it is likely that other enzymes, in addition to lysozyme, may permeate heated spores of non-proteolytic *Cl. botulinum* and induce germination.

It is clear that both lysozyme and chitinase may be present naturally in foods, and that other enzymes capable of degrading peptidoglycan may be released by any microorganisms that have survived processing and are able to multiply in a food. Proposals have been made to add lysozyme and other bacteriolytic enzymes to foods as preservative factors and Japanese workers have patented several processes for the use of lysozyme (Proctor and Cunningham 1988; Cunningham *et al.* 1991; Nielsen 1991). The use of low concentrations of this enzyme to prevent the 'late-blowing' of cheese due to butyric acid bacteria, has been permitted in several countries (Lund and Notermans 1992). A modified bacteriophage T4 lysozyme, in which a new disulphide bond has been incorporated to increase its heat stability, has been developed for use in cheese-making and for other applications (Proctor and Cunningham 1988). The use of plant chitinase as a food preservative, mainly as an antifungal agent, has also been proposed (Teichgräber *et al.* 1993).

6. SUBLETHAL INJURY OF SPORES BY HEAT

6.1 General aspects of sublethal injury

It has been known for many years that heat treatment can cause sublethal as well as lethal damage to bacterial spores. The primary effect of sublethal injury can be on (1) the germination system, or (2) outgrowth and growth of vegetative bacteria. In the case of spores of non-proteolytic *Cl. botulinum*, most of the studies of the mechanism of germination have concerned type E. Freshly prepared spores of *Cl. botulinum* type E were reported to show slow and incomplete germination after a lag period, whereas storing the spore suspension at 4°C increased the rate of germination to a maximum after about 20 d (Ando and Iida 1970). Germination of normal spores in a complex medium occurred in aerobic conditions and was not affected by the initial redox potential. Post-germinative development was strongly affected by oxygen and redox potential. L-Alanine was important in the initiation of germination (Ando 1971). The germination pathway may, in general, resemble that suggested for *Bacillus megaterium* KM, comprising: (1) interaction of L-alanine with a receptor; (2) activation of proteolytic activity; (3) activation of a cortex lytic enzyme; (4) selective cortex hydrolysis and early germination events; (5) late germination events (Foster and Johnstone 1989). Factors that inhibit the germination of bacterial spores have been reviewed by Smoot and Pierson (1982).

Sublethal injury of spores, like that of vegetative bacteria, is characterized by the inability of the spores to give visible signs of growth in conditions that are suitable, or even optimum, for normal undamaged spores. Many of the types of injury that occur have been described by Adams (1978) and include: (1) altered requirements by survivors for non-nutrient germination stimulants; (2) increased sensitivity of survivors to inhibitors; and (3) other effects.

6.2 Altered requirements of injured spores for non-nutrient germination stimulants

The fact that the addition of lysozyme to the culture medium increased the recovery of heated spores of non-proteolytic *Cl. botulinum*, but not of unheated spores, was first reported by Sebald and Ionesco (1972) and confirmed by Alderton *et al.* (1974), Hauschild and Hilsheimer (1977), Smelt (1980), Scott and Bernard (1985), and Peck *et al.* (1992a). A similar effect has been described in the case of spores of several species of *Bacillus* and *Clostridium* but only a slight effect was found with spores of proteolytic *Cl. botulinum* type A (Alderton *et al.* 1974). In the case of heated spores that require lysozyme for germination, despite the presence of a medium that supplies the nutrients that allow germination of uninjured spores, spores that have been injured but not completely inactivated remain dormant and fail to germinate. The mechanism of this effect has been investigated most extensively in the case of spores of *Cl. perfringens*. Cassier and Sebald (1969) first reported that the addition of egg yolk or lysozyme to the recovery medium increased the recovery of heat-treated spores of *Cl. perfringens* type A , whereas unheated spores were not affected. Cultures of some strains of *Cl. perfringens* produced an 'initiation protein', later identified as an *N*-acetylmuramyl-L-alanine amidase (Tang and Labbe 1987), which also increased the recovery of heat-damaged spores. Duncan *et al.* (1972) found that treatment with alkali, which removed coat protein material, reduced the apparent viability of the spores by a factor of $>10^5$, but addition of lysozyme to the enumeration medium resulted in growth from 90–95% of the alkali-treated spores. This work suggested that a component of the natural germination system was associated with the alkali-soluble spore coat protein and that removal of this protein both inactivated the germination system and made the spores permeable to lysozyme. A spore-lytic enzyme was shown to be released from spores of *Cl. perfringens* during germination (Ando 1979) and activation of a precursor of this enzyme was energy-dependent (Ando and Tsuzuki 1984). At least two spore-lytic enzymes were later purified from spores of *Cl. perfringens* (Gombas and Labbe 1985).

Biphasic log survivor–time curves were obtained when heat-treated spores of *Cl.* perfringens were plated on medium that contained lysozyme, and the results indicated that 1–2% of the heat-treated spores responded to the presence of lysozyme in the medium (Adams 1973). Treatment of heated spores of *Cl. perfringens* with EDTA increased the number that formed colonies on medium containing lysozyme by between 10 and 100-fold and transformed biphasic log survivor-time curves into straight lines (Adams 1974). A similar treatment with EDTA failed to increase the number of heated spores of non-proteolytic *Cl. botulinum* that gave colonies on medium containing lysozyme (Peck *et al.* 1992b). Heat-injured spores of *Cl. perfringens* failed to grow on selective media used for isolation of this bacterium (Barach *et al.* 1974). The addition of lysozyme to these media was advocated and D-cycloserine was reported not to interfere with the action of lysozyme and thus to have an advantage over other selective agents.

In most cases the normal spore coat prevents diffusion of lysozyme into spores; the ability of lysozyme to permeate spores results from a deficiency of the spore coat. A small proportion of spores of *Cl. perfringens* and *Clostridium difficile* were reported to germinate in response to lysozyme without pre-treatment (Adams 1974; Ionesco 1978) while all undamaged spores of *B. megaterium* ATCC 9884 were permeable to lysozyme (Suzuki and Rode 1969). Spores of mutant strains of *Cl. perfringens* and *B. subtilis* have been described that were permeable to lysozyme and had defective spore coats (Cassier and Ryter 1971; Zheng *et al.* 1988). Chemical treatment with alkali or a reducing agent has also been found to increase the permeability of spores of many strains of *Bacillus* and *Clostridium* spp. (including *Cl. perfringens*, *Cl. difficile* and non-proteolytic *Cl. botulinum*) to lysozyme (Gould and Hitchins 1963; Cassier and Ryter 1971; Duncan *et al.* 1972; Kamiya *et al.* 1989; Peck *et al.* 1992b). Thus in many instances lysozyme can permeate spores that lack, or have a defective, spore coat. These spores all possess a normal cortex and are refractile and fully heat-resistant. During the germination of undamaged spores, lytic enzymes capable of degrading the cortex are activated by release from an inactive form (Foster and Johnstone 1989). The requirement for lysozyme for germination and colony formation by heat-treated spores presumably results either from damage to the spore lytic enzyme or to the mechanism responsible for its activation. In the case of spores with permeable spore coats lysozyme is able to diffuse into the spore, replace the spore lytic enzyme and hydrolyse cortex peptidoglycan.

Lysozyme and other lytic enzymes are not the only non-nutrient germination stimulants for which a requirement is induced by heat treatment of spores. Spores of *B. subtilis* A that were severely injured by UHT heat treatment failed to form colonies on fortified nutrient agar unless calcium chloride and sodium dipicolinate were added (Edwards *et al.* 1965). The mechanism of the effect of these compounds is still unknown.

6.3 Increased sensitivity of heat-injured spores to inhibitors

Growth and formation of colonies from spores that survive heat treatments are often more sensitive to inhibitors than growth from unheated spores (Adams 1978). The recovery of heated spores of several strains of *Bacillus* and *Clostridium* spp. is increased by the presence of starch in the recovery medium. This effect is thought to be due to the binding by starch of inhibitory compounds in the medium; it was suggested that these inhibitors were trace levels of long chain fatty acids, but later work cast doubt on this hypothesis (Roberts 1970).

Heat treatment of spores of *Clostridium sporogenes* PA3679 at 115°C for 5 min resulted in a 2·5-log inactivation as estimated by plating on a nutrient medium (Roberts and Ingram 1966). The presence of 3% (w/v) NaCl in the plating medium for the heated spores gave an overall 4·1-log inactivation/inhibition, but this concentration of NaCl had no effect on growth from heat-activated spores. In a further experiment, heating at 115°C for 5 min resulted in approximately a 2·0-log inactivation as estimated by plating on a nutrient medium at pH 6·8; the presence of 50 ppm sodium nitrite in the plating medium resulted in an overall 5·0-log inactivation/inhibition, although this concentration of sodium nitrite had less effect on growth from heat-activated spores. Similar results have been obtained with spores of other strains of *Clostridium* and *Bacillus* spp.

Heat treatment of spores of *Cl. perfringens* NCTC 8798 at 105°C for 8·5 min damaged both the germination and outgrowth systems of most spores (Barach *et al.* 1975). The damage to outgrowth was detected by the ability of heated spores to form colonies on a non-selective medium but not in the presence of antibiotics (polymyxin and neomycin) used in selective enumeration media. Repair of this damage occurred in a non-selective laboratory medium and in foods, and took place during outgrowth (Flowers and Adams 1976). A proportion of the heated spores treated with alkali and incubated in liquid medium with lysozyme were osmotically unstable and could be stabilized by dextran. These results, and the apparent failure of inhibitors of synthesis of cell walls, protein, ribonucleic acid and deoxyribonucleic acid to prevent repair, led the authors to suggest that the site of damage to the outgrowth system was the spore plasma membrane. After heating at 90°C for 6 h, growth from spores of this strain of *Cl. perfringens* was much more sensitive to inhibition by nitrite, NaCl, polymyxin and neomycin than was growth from heat-activated spores (Chumney and Adams 1980). Repair of this damage occurred in a nutrient medium within 2 h at 35°C in the absence of multiplication. The suggestion was made that the increased sensitivity of the heated spores to nitrite and to NaCl, as well as to the antibiotics resulted from damage affecting the spore plasma membrane.

6.4 Other effects

For spores that have survived severe heat treatments the optimum temperature for colony formation may be lower than that for unheated spores (Adams 1978). While growth of proteolytic type A and B strains of *Cl. botulinum* occurs more rapidly at 37°C than at lower temperatures, after heat-treatment of spores greater numbers were recovered at 24–27°C than at 31° or 37°C (Williams and Reed 1942; Olsen and Scott 1950; Sugiyama 1951). Several workers have reported that growth from spores that have survived severe heat treatments requires nutrients not required by unheated spores. Supplements that have been reported to improve the recovery of heated spores include glucose, blood, yeast and liver extracts, vitamins and amino acids (Adams 1978). The redox potential of the recovery medium may be more critical for heated spores of *Clostridium* spp. than for unheated spores. Heat-treated spores of proteolytic *Cl. botulinum* types A and B showed a greater requirement for thioglycolate (0·01%) in the recovery medium than unheated spores (Olsen and Scott 1950). The authors suggested that this was probably due to a specific effect of the thioglycolate, but it is also possible that it was due to the effect of the thioglycolate in lowering and poising the redox potential. The presence of bicarbonate has been shown to reduce the incubation time for formation of colonies from spores of eight strains (probably proteolytic) of *Cl. botulinum* and a strain of *Cl. sporogenes* (Andersen 1951). Several other workers have shown that bicarbonate or carbon dioxide was essential for rapid germination of spores of proteolytic and non-proteolytic strains of *Cl. botulinum* (Treadwell *et al.* 1958; Ando and Iida 1970; Rowley and Feehery 1970). According to Roberts (1970) bicarbonate had no effect on colony counts from unheated spores of proteolytic *Cl. botulinum* type A and other *Clostridium* spp., but increased colony counts from heated spores by between 4 and 130-fold. There are also reports that growth from heated spores is more sensitive to pH than growth from unheated spores (Adams 1978).

7. CONCLUSIONS

The failure to use optimum conditions for recovery of heated spores of non-proteolytic *Cl. botulinum* has probably contributed to the delays reported before growth and to low estimates of heat-resistance (Lynt *et al.* 1983). In many studies of the heat-resistance of these spores the effect of lysozyme has not been considered. Pasteurization of seafood, in particular of meat from the blue crab (*Callinectes sapidus*), has been practised in the USA for many years. The current standard practice of heating cans of crab meat in a water bath at 88°C until cold-point temperatures reach 85°C for 1 min was published in 1970; this extends the shelf-life from 6–10 d for fresh crab meat to 6–18 months for properly pasteurized crab meat (Rippon and Hackney 1992). Storage temperatures of 2·2°C or below are said to be necessary for both shelf-life and safety. This process was based on historical data that gave the desired shelf-life, and not on any target organism. Work by Cockey and Tatro (1974), using inoculated crab meat, was reported to show that this process achieved an 8-log reduction in the number of surviving spores of non-proteolytic *Cl. botulinum* type E. The number of spores that survived pasteurization was determined in medium without lysozyme, and cans of pasteurized crab were incubated at 4·4°C for 6 months to test for formation of toxin. Our results indicate that in such a process many spores of non-proteolytic *Cl. botulinum* may be damaged, and remain dormant but not completely inactivated. In practice the traditional pasteurization process is stated to provide a heating/cooling curve that gives an average $F_{85°C}$ of 31 min (assuming a *z*-value of 8·9°C), and many producers are using processes equivalent to $F_{85°C} = 60–120$ min (Rippon and Hackney 1992).

Eklund *et al.* (1988) investigated the heat treatment required to inactivate spores of non-proteolytic *Cl. botulinum* type B and E in vacuum-packed, hot-process, smoked salmon. The aim was to investigate the feasibility of using a pasteurization process that would inactivate the spores without altering the sensory characteristics of the product. When samples of fish inoculated with spores were pasteurized by heating in a bath at 85°C for 85 min and then stored at 25°C for 21 d, 0/21 of those inoculated with 10^6 type E spores were toxic but 37/41 of those inoculated with 10^6 non-proteolytic type B spores were toxic. After heating samples inoculated with the type B spores in a bath at 85°C for 165 min and 175 min, and incubation at 25°C for 21 d, 4/42 and 0/41 samples respectively were toxic. The time required for the samples to reach the temperature of the bath was reported as 29 min. When samples inoculated with 10^6 type B spores were heated at 92·2°C for 40, 45, 55 and 65 min and stored at 25°C for 21 d (14 d in the case of samples heated for 40 min) the numbers of samples that became toxic were 1/4, 1/74, 1/40 and 0/40 respectively; the time required for samples to reach 92·2°C was reported as 28 min. In the light of these results and sensory studies, Eklund *et al.* (1988) concluded that it was feasible to pasteurize products made from fillets, steaks or other sections of fish at 92·2°C for times up to 100 min, but because this would not inactivate spores of proteolytic strains of *Cl. botulinum* they advised that the packs should carry directions to store below 3·3°C.

The fact that heat-damaged spores are more susceptible to inhibitors than unheated spores offers the opportunity to use the combination of heat treatment, refrigeration and control of shelf-life, together with preservative factors where possible, to prevent outgrowth and formation of toxin and to build in acceptable safety factors against non-proteolytic *Cl. botulinum* in REPFEDs. The satisfactory use of these systems requires an understanding of the heat-resistance of spores of these bacteria and of the effect of inhibitory factors on damaged spores.

In this paper the authors have sought to provide an increased understanding of the heat-resistance of, and subsequent growth from, spores of non-proteolytic *Cl. botulinum*, in order to provide an improved basis for decisions about the heat processing of REPFEDs. Further work is required to determine the effect of sporulation conditions on the heat-resistance of these spores, to determine the heat-resistance of spores in foods, to investigate the occurrence and heat-resistance of lysozyme in foods, and to investigate the relevance of other enzymes capable of degrading peptidoglycan.

8. ACKNOWLEDGEMENTS

We wish to thank David A. Fairbairn for his major contribution to the experiments reported at IFRN, Mary L. Parker for electron microscopy, and Nestlé Research and Development and the Department of Health for funding this work.

9. REFERENCES

Adams, D.M. (1973) Inactivation of *Clostridium perfringens* type A spores at ultrahigh temperatures. *Applied Microbiology* **26**, 282–287.

Adams, D.M. (1974) Requirement for and sensitivity to lysozyme by *Clostridium perfringens* spores heated at ultrahigh temperatures. *Applied Microbiology* **27**, 797–801.

Adams, D.M. (1978) Heat injury of bacterial spores. *Advances in Applied Microbiology* **23**, 245–261.

Advisory Committee on the Microbiological Safety of Food UK (1992) *Report on Vacuum Packaging and Associated Processes*. London: HMSO.

Alderton, G., Chen, J.K. and Ito, K.A. (1974) Effect of lysozyme on the recovery of heated *Clostridium botulinum* spores. *Applied Microbiology* **27**, 613–615.

Andersen, A.A. (1951) A rapid plate method of counting spores of *Clostridium botulinum*. *Journal of Bacteriology* **62**, 425–432.

Ando, Y. (1971) The germination requirements of spores of *Clostridium botulinum* type E. *Japanese Journal of Microbiology* **15**, 515–525.

Ando, Y. (1979) Spore lytic enzyme released from *Clostridium perfringens* spores during germination. *Journal of Bacteriology* **140**, 59–64.

Ando, Y. and Iida, H. (1970) Factors affecting the germination of spores of *Clostridium botulinum* type E. *Japanese Journal of Microbiology* **14**, 361–370.

Ando, Y. and Tsuzuki, T. (1984) Energy-dependent activation of spore-lytic enzyme precursor by germinated spores. *Biochemical and Biophysical Research Communications* **123**, 463–467.

Barach, J.T., Adams, D.M. and Speck, M.L. (1974) Recovery of heated *Clostridium perfringens* Type A spores on selective media. *Applied Microbiology* **28**, 793–797.

Barach, J.T., Flowers, R.S. and Adams, D.M. (1975) Repair of heat-injured *Clostridium perfringens* spores during outgrowth. *Applied Microbiology* **30**, 873–875.

Boller, T. (1986) Chitinase: A defense of higher plants against pathogens. In *Chitin in Nature and Technology* ed. Muzzarelli, R., Jeuniaux, C. and Gooday, G.W. pp. 223–233. New York: Plenum Press.

Brown, M.H. and Gould, G.W. (1992) Processing. In *Chilled Foods. A Comprehensive Guide* ed. Dennis, C. and Stringer, M. pp. 111–146. New York: Ellis Horwood.

Carroll, S.F. and Martinez, R.J. (1979) Role of rabbit lysozyme in *in vitro* serum and plasma serum bactericidal reactions against *Bacillus subtilis*. *Infection and Immunity* **25**, 810–819.

Cassier, M. and Ryter, A. (1971) Sur un mutant de *Clostridium perfringens* donnant des spores sans tuniques a germination lysozyme-dépendante. *Annales de l'Institut Pasteur, Paris* **121**, 717–731.

Cassier, M. and Sebald, M. (1969) Germination lysozyme-dépendante des spores de *Clostridium perfringens* ATCC 3624 apres traitment thermique. *Annales de l'Institut Pasteur, Paris* **117**, 312–324.

Chandan, R.C. and Ereifej, K.I. (1981) Determination of lysozyme in raw fruits and vegetables. *Journal of Food Science* **46**, 1278–1279.

Chumney, R.K. and Adams, D.M. (1980) Relationship between the increased sensitivity of heat injured *Clostridium perfringens* spores to surface active antibiotics and to sodium chloride and sodium nitrite. *Journal of Applied Bacteriology* **49**, 55–63.

Cockey, R.R. and Tatro, M.C. (1974) Survival studies of *Clostridium botulinum* type E in pasteurized meat of the blue crab *Callinectes sapidus*. *Applied Microbiology* **27**, 629–633.

Cunningham, F.E., Proctor, V.A. and Goetsch, S.J. (1991) Egg-white lysozyme as a food preservative: an overview. *World's Poultry Science Journal* **47**, 141–163.

Dehof, E., Greuel, E. and Kramer, J. (1989) [Tenacity of *Clostridium botulinum* type E in hot-smoked vacuum-packed trout fillets] *Archiv fur Lebensmittelhygiene* **40**(2) 27–29.

Duncan, C.L., Labbe, R.G. and Reich, R.R. (1972) Germination of heat-and alkali-altered spores of *Clostridium perfringens* type A by lysozyme and an initiation protein. *Journal of Bacteriology* **109**, 550–559.

Edwards, J.L., Jr, Busta, F.F. and Speck, M.L. (1965) Heat injury of *Bacillus subtilis* spores at ultra high temperatures. *Applied Microbiology* **13**, 858–864.

Eklund, M.W., Poysky, F.T. and Wieler, D.I. (1967a) Characteristics of *Clostridium botulinum* type F isolated from the Pacific coast of the United States. *Applied Microbiology* **15**, 1316–1323.

Eklund, M.W., Wieler, D.I. and Poysky, F.T. (1967b) Outgrowth and toxin production of non-proteolytic type B *Clostridium botulinum* at 3·3°C to 5·6°C. *Journal of Bacteriology* **93**, 1461–1462.

Eklund, M.W., Peterson, M.E., Paranjpye, R. and Pelroy, G.A. (1988) Feasibility of a heat-pasteurization process for the inactivation of non-proteolytic *Clostridium botulinum* types B and E in vacuum-packaged, hot-process (smoked) fish. *Journal of Food Protection* **51**, 720–726.

Fenouil, E. and Roch, Ph. (1991) Evidence and characterization of lysozyme in six species of freshwater crayfishes from *Astacidae* and *Cambaridae* families. *Comparative Biochemistry and Physiology* **99B**, 43–49.

Flowers, R.S. and Adams, D.M. (1976) Spore membrane as the site of damage within heated *Clostridium perfringens* spores. *Journal of Bacteriology* **125**, 429–434.

Foster, S.J. and Johnstone, K. (1988) Germination-specific cortex-lytic enzyme is activated during triggering of *Bacillus megaterium* KM spore germination. *Molecular Microbiology* **2**, 727–733.

Foster, S.J. and Johnstone, K. (1989) The trigger mechanism of bacterial spore germination. In *Regulation of Procaryotic Development* ed. Smith, I., Slepecky, R. and Setlow, P. pp. 89–108. American Society for Microbiology: Washington, DC.

Gombas, D.E. and Labbe, R.G. (1985) Purification and properties of spore-lytic enzymes from *Clostridium perfringens* type A spores. *Journal of General Microbiology* **131**, 1487–1496.

Gould, G.W. (1984) Injury and repair mechanisms in bacterial spores. In *The Revival of Injured Microbes*. ed. Andrew, M.H.E. and Russell, A.D. pp. 199–220. London: Academic Press.

Gould, G.W. and Hitchins, A.D. (1963) Sensitization of bacterial spores to lysozyme and hydrogen peroxide with reagents which rupture disulphide bonds. *Journal of General Microbiology* **33**, 413–423.

Graham, A.F. and Lund, B.M. (1993) The effect of temperature on the growth of non-proteolytic type B *Clostridium botulinum*. *Letters in Applied Microbiology* **16**, 158–160.

Grecz, N. and Arvay, L.H. (1982) Effect of temperature on spore germination and vegetative cell growth of *Clostridium botulinum*. *Applied and Environmental Microbiology* **43**, 331–337.

Grinde, B., Lie, O., Poppe, T. and Salte, R. (1988) Species and individual variation in lysozyme activity in fish of interest in aquaculture. *Aquaculture* **68**, 299–304.

Hatheway, C.L. (1992) *Clostridium botulinum* and other *Clostridia* that produce botulinum toxin. In *Clostridium botulinum: Ecology and Control in Foods* ed. Hauschild, A.H.W. and Dodds, K.L. pp. 3–20. New York: Marcel Dekker.

Hauschild, A.H.W. and Hilsheimer, R. (1977) Enumeration of *Clostridium botulinum* spores in meats by a pour-plate procedure.*Canadian Journal of Microbiology* **23**, 829–832.

Höltje, J.-V. and Tuomanen, E.I. (1991) The murein hydrolases of *Escherichia coli*: properties, functions and impact on the course of infections *in vivo*. *Journal of General Microbiology* **137**, 441–454.

Ikawa, Y., Genigeorgis, C. and Lindroth, S. (1986) Temperature and time effect on the probability of *Clostridium botulinum* growth in a model broth. In *Proceedings of the 2nd World Congress, Foodborne Infections and Intoxications*, Vol. 1, pp. 370–374. Berlin: Institute of Veterinary Medicine—Robert von Ostertag-Institute.

Ionesco, H. (1978) Initiation de germination des spores de *Clostridium difficile* par le lysozyme. *Compte rendu de l'Academie des sciences, Paris (Serie D)* **287**, 659–661.

Jekel, P.A., Hartmann, B.H. and Beintema, J.J. (1991) The primary structure of hevamine, an enzyme with lysozyme/chitinase activity from *Hevea brasiliensis* latex. *European Journal of Biochemistry* **200**, 123–130.

Jensen, M.J., Genigeorgis, C. and Lindroth, S. (1987) Probability of growth of *Clostridium botulinum* as affected by strain, cell and serological type, inoculum size and temperature and time of incubation in a model broth system. *Journal of Food Safety* **8**, 109–126.

Jollès, P. (1967) Relationship between chemical structure and biological activity of hen egg-white lysozyme and lysozymes of different species. *Proceedings of the Royal Society, London. Series B* **167**, 350–364.

Jollès, P. and Jollès, J. (1984) What's new in lysozyme research? *Molecular and Cellular Biochemistry* **63**, 165–189.

Jollès P., Bernier, I., Berthou, J., Charlemagne, D., Faure, A., Hermann, J., Périn, J.-P. and Saint-Blancard, J. (1974) From lysozymes to chitinases: structural, kinetic and crystallographic studies. In *Lysozyme* ed. Osserman, E.F., Canfield, R.E. and Beychok, S. pp. 31–54. New York: Academic Press.

Kamiya, S., Yamakawa, K., Ogura, H. and Nakamura, S. (1989) Recovery of spores of *Clostridium difficile* altered by heat or alkali. *Journal of Medical Microbiology* **28**, 217–221.

Kooiman, W.J. and Geers, J.M. (1975) Simple and accurate technique for the determination of heat resistance of bacterial spores. *Journal of Applied Bacteriology* **38**, 185–189.

Lie, O., Evensen, O., Sorensen, A. and Froysadal, E. (1989) Study on lysozyme activity in some fish species. *Diseases of Aquatic Organisms* **6**, 1–5.

Lund, B.M. (1993) Quantification of factors affecting the probability of development of pathogenic bacteria, in particular *Clostridium botulinum*, in foods. In *Application of Predictive Microbiology and Computer Modelling Techniques to the Food Industry. Journal of Industrial Microbiology* **12**, 144–155.

Lund, B.M. and Notermans, S.H.W. (1992) Potential Hazards associated with REPFEDs (refrigerated, processed foods of extended durability). In *Clostridium botulinum: Ecology and Control in Foods* ed. Hauschild, A.H.W. and Dodds, K.L. pp. 279–303. New York: Marcel Dekker.

Lund, B.M., Graham, A.F., George, S.M. and Brown, D. (1990) The combined effect of incubation temperature, pH and sorbic acid on the probability of growth of non-proteolytic, type B *Clostridium botulinum*. *Journal of Applied Bacteriology* **69**, 481–492.

Lynn, K.R. (1989) A lysozyme from the fruit of *Actinida chinensis*. *Phytochemistry* **28**, 2267–2268.

Lynt, R.K., Kautter, D.A. and Solomon, H.M. (1983) Effect of delayed germination by heat-damaged spores on estimates of heat resistance of *Clostridium botulinum* types E and F. *Journal of Food Science* **48**, 226–229.

MAFR (Ministry of Agriculture, French Republic) (1974) Regulations for the hygienic conditions concerning preparation, preservation, distribution and sale of ready-to-eat meals. *Journal Officiel de la Republique Française*. Order of the 26th June, 1974. Paris.

MAFR (Ministry of Agriculture, French Republic) (1988) *Extension of the Shelf-Life of Ready-to-Eat Meals, Alteration to the Protocol for Obtaining Authorization*. Memorandum DGAL/SVHA/N88/No. 8106, 31st May, 1988. Paris.

McHenery, J.G. and Birkbeck, T.H. (1979) Lysozyme of the mussel, *Mytilus edulis* (L.). *Marine Biology Letters* **1**, 111–119.

McHenery, J.G. and Birkbeck, T.H. (1982) Characterization of the lysozyme of *Mytilus edulis* (L.). *Comparative Biochemistry and Physiology* **71B**, 583–589.

Mochizuki, A. and Matsumiya, M. (1983) Lysozyme activity in shellfishes. *Bulletin of the Japanese Society of Scientific Fisheries* **49**, 131–135.

Mossel, D.A.A. and Struijk, C.B. (1991) Public health implications of refrigerated, pasteurized ('sous vide') foods. *International Journal of Food Microbiology* **13**, 187–206.

Murray, C.K. and Fletcher, T.C. (1976) The immunohistochemical localization of lysozyme in plaice (*Pleuronectes platessa* L.) tissues. *Journal of Fish Biology* **9**, 329–334.

Nielsen, H.K. (1991) Novel bacteriolytic enzymes and cyclodextrin glycosyl transferase for the food industry. *Food Technology* **45**(1), 102–104.

Nord, C.E. and Wadström T. (1972) Chitinase activity and substrate specificity of three bacteriolytic *endo-β-N*-acetylmuramidases and one *endo-β-N*-acetylglucosaminidase. *Acta Chemica Scandinavica* **26**, 653–660.

Ohye, D.F. and Scott, W.J. (1957) Studies in the physiology of *Clostridium botulinum* type E. *Australian Journal of Biological Science* **10**, 85–94.

Olsen, A.M. and Scott, W.J. (1950) The enumeration of heated bacterial spores 1. Experiments with *Clostridium botulinum* and other species of *Clostridium*. *Australian Journal of Scientific Research Series B* **3**, 219–233.

Ottaviani, E. (1991) Tissue distribution and levels of natural and induced serum lysozyme immunoreactive molecules in a freshwater snail. *Tissue and Cell* **23**, 317–324.

Peck, M.W., Fairbairn, D.A. and Lund, B.M. (1992a) The effect of recovery medium on the estimated heat-inactivation of spores of non-proteolytic *Clostridium botulinum*. *Letters in Applied Microbiology* **15**, 146–151.

Peck, M.W., Fairbairn, D.A. and Lund, B.M. (1992b) Factors affecting growth from heat-treated spores of non-proteolytic *Clostridium botulinum*. *Letters in Applied Microbiology* **15**, 152–155.

Peck, M.W., Fairbairn, D.A. and Lund, B.M. (1993) Heat-resistance of spores of non-proteolytic *Clostridium botulinum* estimated on medium containing lysozyme. *Letters in Applied Microbiology* **16**, 126–131.

Peck, M.W., Fairbairn, D.A., Lund, B.M., Kaspersson, A.S. and Underland, P. (1994) Effect of heat treatments on survival of, and growth from, spores of non-proteolytic *Clostridium botulinum* at chilled temperatures. *Applied and Environmental Microbiology* (in preparation).

Proctor, V.A. and Cunningham, F.E. (1988) The chemistry of lysozyme and its use as a food preservative and a pharmaceutical. *CRC Critical Reviews in Food Science and Nutrition* **26**, 359–395.

Rippon, T.E. and Hackney, C.R. (1992) Pasteurization of seafood: potential for shelf-life extension. *Food Technology* **46**(12), 88–94.

Roberts, T.A. (1970) Recovering spores damaged by heat, ionizing radiations or ethylene oxide. *Journal of Applied Bacteriology* **33**, 74–94.

Roberts, T.A. and Ingram, M. (1966) The effect of sodium chloride, potassium nitrate and sodium nitrite on the recovery of heated bacterial spores. *Journal of Food Technology* **1**, 147–163.

Roberts, W.K. and Selitrennikoff, C.P. (1988) Plant and bacterial chitinases differ in antifungal activity. *Journal of General Microbiology* **134**, 169–176.

Rodrick, G.E. and Cheng, T.C. (1974) Kinetic properties of lysozyme from the hemolymph of *Crassostrea virginica*. *Journal of Invertebrate Pathology* **24**, 41–48.

Rogers, H.J., Perkins, H.R. and Ward, J.B. (1980) The bacterial autolysins. In *Microbial Cell Walls and Membranes*, pp. 437–460. London: Chapman and Hall.

Rosset, R. and Poumeyrol, G. (1986) Modern processes for the preparation of ready-to-eat meals by cooking before or after *sous-vide* packaging. *Sciences des Aliments* 6 H.S. VI, 161–167.

Rowley, D.B. and Feeherry, F. (1970) Conditions affecting germination of *Clostridium botulinum* 62A spores in a chemically defined medium. *Journal of Bacteriology* **104**, 1151–1157.

Schmidt, C.F., Lechowich, R.V. and Folinazzo, J.F. (1961) Growth and toxin production by type E *Clostridium botulinum* below 40°F. *Journal of Food Science* **26**, 626–630.

Scott, V.N. and Bernard, D.T. (1985) The effect of lysozyme on the heat resistance of non-proteolytic type B *Clostridium botulinum*. *Journal of Food Safety* 7, 145–154.

Sebald, M. and Ionesco, H. (1972) Germination lzp-dépendante des spores de *Clostridium botulinum* type E. *Comptes rendue de l'Academie des sciences, Paris (Serie D)* **275**, 2175–2177.

Smelt, J.P.P.M. (1980) Heat resistance of *Clostridium botulinum* in acid ingredients and its significance for the safety of chilled foods. Ph.D. Dissertation, University of Utrecht, The Netherlands.

Smolelis, A.N. and Hartsell, S.E. (1952) Factors affecting the lytic activity of lysozyme. *Journal of Bacteriology* **63**, 665–674.

Smoot, L.A. and Pierson, M.D. (1982) Inhibition and control of bacterial spore germination. *Journal of Food Protection* **45**, 84–92.

Strominger, J. and Tipper, D.J. (1974) Structure of bacterial cell walls: the lysozyme substrate. In *Lysozyme*. ed. Osserman, E.E., Canfield, R.E. and Beychok, S. pp. 169–184. New York: Academic Press.

Sugiyama, H. (1951) Studies on factors affecting the heat resistance of spores of *Clostridium botulinum*. *Journal of Bacteriology* **62**, 81–96.

Suzuki, Y. and Rode, L.J. (1969) Effect of lysozyme on resting spores of *Bacillus megaterium*. *Journal of Bacteriology* **98**, 238–245.

Takahashi, Y., Itami, T. and Konegawa, K. (1986) Enzymic properties of partially purified lysozyme from the skin mucus of carp. *Bulletin of the Japanese Society of Scientific Fisheries* **52**, 1209–1214.

Takumi, K., Kawata, T. and Hisatsune, K. (1971) Autolytic system of *Clostridium botulinum*. II. Mode of action of autolytic enzymes in *Clostridium botulinum* type A. *Japanese Journal of Microbiology* **15**, 131–141.

Tang, S.S. and Labbe, R.G. (1987) Mode of action of *Clostridium perfringens* initiation protein (spore-lytic enzyme). *Annales de l'Institute Pasteur/Microbiologie* **138**, 597–608.

Teichgräber, P., Zache, U. and Knorr, D. (1993) Enzymes from germinating seeds—potential applications in food processing. *Trends in Food Science and Technology* **4**, 145–149.

Treadwell, P.E., Jann, G.J. and Salle, A. (1958) Studies on factors affecting the rapid germination of spores of *Clostridium botulinum*. *Journal of Bacteriology* **76**, 549–556.

Warth, A.D. (1972) Action of spore lytic enzymes on the cortex. In *Spores V* eds Halvorson, H.O., Hanson, R. and Campbell, L.L. pp. 28–34. American Society for Microbiology: Washington, DC.

Warth, A.D. and Strominger, J.L. (1972) Structure of the peptidoglycan from spores of *Bacillus subtilis*. *Biochemistry* **11**, 1389–1396.

Williams, O.B. and Reed, J.M. (1942) The significance of the incubation temperature of recovery cultures in determining spore resistance to heat. *Journal of Infectious Diseases* **71**, 225–227.

Zheng, L., Donovan, W.P., Fitz-James, P.C. and Losick, R. (1988) Gene encoding a morphogenic protein required in the assembly of the outer coat of the *Bacillus subtilis* endospore. *Genes and Development* **2**, 1047–1054.

Journal of Applied Bacteriology Symposium Supplement 1994, **76**, 129S–134S

Bioluminescence and spores as biological indicators of inimical processes

P.J. Hill, L. Hall, D.A. Vinicombe[1], C.J. Soper[1], P. Setlow[3], W.M. Waites, S. Denyer[2] and G.S.A.B. Stewart

Department of Applied Biochemistry and Food Science, University of Nottingham, Sutton Bonington Campus, Sutton Bonington, Leicestershire, [1]School of Pharmacy and Pharmacology, University of Bath, [2]University of Brighton, Department of Pharmacy, Cockroft Building, Moulsecoomb, Brighton, UK and [3]Department of Biochemistry, University of Connecticut Health Center, Farmington, CT, USA

1. Introduction, 129S
2. Studies using bioluminescent *Bacillus megaterium* KM
   2.1 The bioluminescence of *Bacillus megaterium* [pOCMS394], 130S
   2.2 Comparison of heat resistance of bioluminescent and non-bioluminescent spores, 130S
   2.3 Bioluminescence as a measure of loss of viability during heating, 130S
   2.4 The effect of food preservatives as reflected by bioluminescence, 130S
3. Highly bioluminescent *Bacillus subtilis*
   3.1 Plasmid vectors for high level bioluminescence in Gram-positive bacteria, 131S
   3.2 Control of light emission from *Bacillus subtilis*, 132S
   3.3 *lux*$^+$ *Bacillus subtilis* spores as monitors of ethylene oxide sterilization, 133S
4. References, 134S

1. INTRODUCTION

The detection and enumeration of micro-organisms, coupled with the detection and assay of antimicrobial substances, represent key industrial aspects of microbiology. The development of rapid methods to facilitate these assays is of major significance both to provide new and simple protocols and to bring microbial assays into real-time, rather than their current retrospective position. *In vivo* bioluminescence is an emerging tool for the microbiologist which may address many of these needs. Genetic engineering allows the introduction of a bioluminescent phenotype into industrially important bacteria and, because the emission of light depends upon bacterial viability, such constructs represent novel reagents to monitor *in situ* industrial biocides, residual antibiotics, starter culture activity and the efficiency of processing parameters such as pasteurization and sterilization (Stewart and Williams 1992).

At present, important biological indicators of these inimical processes consist of *Bacillus* spore preparations. Following exposure, the spores are incubated in a growth medium to facilitate germination and outgrowth of survivors; viability is typically ascertained by dye reduction or enumeration by classical microbiological counting methods, a time consuming exercise.

Correspondence to: Prof. G.S.A.B. Stewart, Department of Applied Biochemistry and Food Science, University of Nottingham, Sutton Bonington Campus, Sutton Bonington, Leicestershire LE12 5RD, UK.

Carmi *et al.* (1987) were the first to introduce the genes for *in vivo* bioluminescence into *Bacillus*. When these bacilli were allowed to undergo the sporulation process it soon became evident that a new rapid method for monitoring inimical processes was available (Stewart *et al.* 1989). Spores of *Bacillus megaterium* KM and *Bacillus subtilis* obtained from phenotypically bioluminescent vegetative cells are dark. This is not surprising since dormant bacterial spores exhibit no detectable metabolism, do not have detectable ATP or electron transport and hence have no energy with which to drive the light reaction. The onset of electron transport and the initiation of metabolism are, however, very early events during germination (Scott and Ellar 1978), a process which ultimately converts the spore back into a vegetative cell. For *lux* containing spores, germination is accompanied by the emergence of bioluminescence, providing for a sensitive real-time monitor of the germination and outgrowth process.

This study initially describes the characterization of spore preparations from bioluminescent *B. megaterium* KM. Although the low light output of these spores coupled with their heat sensitivity makes them inappropriate for their use as a commercial reagent, they have provided a model system against which the development of highly bioluminescent bacterial spores as biological indicators may be gauged. These initial studies have led to the construction of stable, highly bioluminescent *B. subtilis* capable of expressing a *luxAB* fusion gene from vegetative or sporulation specific promoters. Spores prepared from these strains provide significantly different germination associated bioluminescence profiles and are evaluated as biological indicators of ethylene oxide sterilization efficacy.

2. STUDIES USING BIOLUMINESCENT *BACILLUS MEGATERIUM* KM

2.1 The bioluminescence of *Bacillus megaterium* [pOCMS394]

The organism used in this study, *B. megaterium* KM, was transformed to a bioluminescent phenotype by plasmid pOCMS394 as previously described (Carmi *et al.* 1987). During germination and outgrowth of spores of *B. megaterium* [pOCMS394], it was found that light was produced after a lag of only 4·5 min. The initial rapid rate of light production was near-exponential and lasted for 10–20 min. This was followed by a 50 min period of slow increase and a second exponential rise until 180 min, after which the rise in light production again slowed. Microscopic examination of the spores suggested that the first rapid increase was associated with germination, the slower rise with swelling and the second rapid increase with elongation and division.

Inhibitors of protein synthesis such as chloramphenicol and rifampicin did not inhibit the initial rapid increase in bioluminescence but prevented the second. This suggests that the initial bioluminescence observed is due to preformed luciferase which was packaged into the bacterial spore, a supposition supported by the rapid acquisition of bioluminescence by the germinating spores. The second rapid increase in bioluminescence is attributed to *de novo* synthesis of luciferase by the recombinant organism.

2.2 Comparison of heat resistance of bioluminescent and non-bioluminescent spores

A comparison of the heat resistance of the non-bioluminescent and bioluminescent spores was necessary to ensure that the insertion of the *lux* genes had no effect upon the thermal stability of spores of *B. megaterium*. At 90°C destruction of the spores was clear, with the viable count being rapidly reduced. D_{90} values of 11 min for non-bioluminescent and 10 min for bioluminescent spores were obtained. It appears that the presence of plasmid-borne *lux* genes in *B. megaterium* and the use of the plasmid selective antibiotic chloramphenicol during the sporulation process does not markedly affect spore heat resistance.

2.3 Bioluminescence as a measure of loss of viability during heating

The effects of heat upon spore preparations of bioluminescent *B. megaterium* were measured using both bioluminescence and viable counts after heating at 80°C and 90°C for various times (0, 15, 30, 60, 90 and 150 min). Bioluminescence readings were seen to closely reflect the viable count. On plotting light output after 3-h recovery against percentage survivors, a straight line correlation was observed with a correlation coefficient of $r = 0{\cdot}98$ (Fig. 1). This gives great confidence in the use of bioluminescence as a rapid biological indicator of sporicidal efficacy.

2.4 The effect of food preservatives as reflected by bioluminescence

When *B. megaterium* [pOCMS394] spores germinate and outgrow there is a distinctive pattern of light production which appears to be associated with the various stages of transition from spores to vegetative cells. In this study, the food preservatives sodium benzoate, sodium chloride and potassium sorbate were used to halt the development of bioluminescent *B. megaterium* KM at various stages in the development of vegetative cells. The effect of each compound on the bioluminescence produced by the germinating and outgrowing spores, and by vegetative cells, was examined. Table 1 gives a list of the effects of these food preservatives upon *Bacillus* species, as postulated by Gould (1964).

Bioluminescence during germination and outgrowth in the presence of sodium benzoate was examined. As can

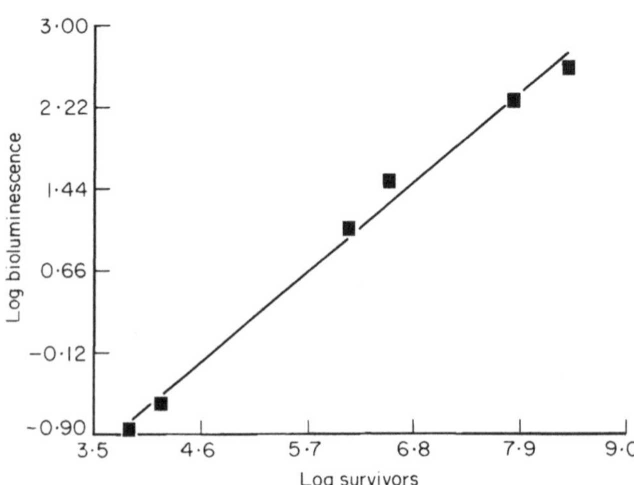

Fig. 1 Correlation between bioluminescence after 3-h recovery and viable count after heating *Bacillus megaterium* KM [pOCMS394] at 90°C for 0, 15, 30, 60, 90 and 150 min

Table 1 Effects of food preservatives upon the development of *Bacillus* spores

Compound	Developmental stage inhibited
Sodium benzoate	Outgrowth
Potassium sorbate	Division (germination at high concentration)
Sodium chloride	Division (germination at high concentration)

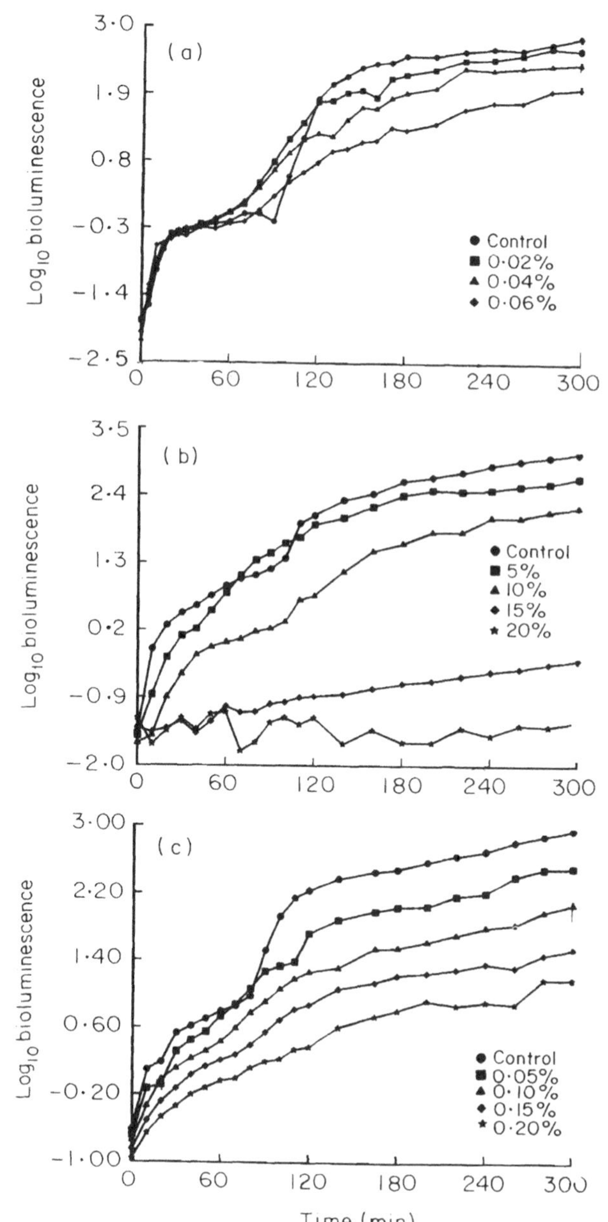

Fig. 2 Effect of food preservatives on the bioluminescence of germinating and outgrowing spores of *Bacillus megaterium* KM [pOCMS394]: (a) sodium benzoate, (b) sodium chloride, (c) potassium sorbate

clearly be seen in Fig. 2a, the initial rapid production of bioluminescence was unaffected by sodium benzoate at all chosen concentrations, indicating that the germination process is unaffected. The subsequent increase in bioluminescence associated with elongation was, however, inhibited with increasing concentrations of sodium benzoate.

Figure 2b shows the effect of sodium chloride upon light emission by germinating *B. megaterium* KM. In contrast with benzoate, the major effect of sodium chloride appears to be suppression of the initial light emission associated with germination. This could be correlated with the microscopic appearance of the germinating spores: the percentage of germinating spores is reduced with increasing sodium chloride concentration, suggesting that germination is the major developmental stage inhibited by sodium chloride.

The effect of increasing concentrations of potassium sorbate is shown in Fig. 2c. At low concentrations the bioluminescence profile suggests that there is little effect upon germination *per se*, but that elongation and division are slowed. At higher concentrations of sorbate there appears to be inhibition of the germination process in addition to inhibition of outgrowth, suggesting that sorbate inhibits both of these developmental stages.

These studies show, by their agreement with previous studies, that *in vivo* bioluminescence is an effective, non-invasive, real-time method for ascertaining the effect of chemicals upon the different developmental stages of the germinating bacterial endospore. In contrast with more 'traditional' approaches, the ease and 'low tech' aspects of the bioluminescence assay provide for rapid results with little requirement for expertise.

3. HIGHLY BIOLUMINESCENT *BACILLUS SUBTILIS*

3.1 Plasmid vectors for high level bioluminescence in Gram-positive bacteria

In order to achieve a high level of expression of heterologous genes in Gram-positive bacteria, a number of criteria must be met by the cloning vector: (1) the vector must be stably maintained; (2) a stringent Gram-positive ribosome binding site (RBS) is required; (3) a powerful Gram-positive promoter must be present. Initial attempts at attaining high expression of bacterial luciferase in *B. subtilis* even while fulfilling the above criteria, have failed to achieve the magnitude of bioluminescence possible from Gram-negative bacteria. Reasons suggested for the low light levels from Gram-positive bacteria include the incapacity to regenerate sufficient $FMNH_2$, or to supply oxygen, both vital substrates for the bioluminescent reaction (Karp 1989).

In 1991, Jacobs *et al.* showed that *B. subtilis* cells are indeed able to supply the substrates for bioluminescence and that high level output of light is possible from Gram-positive bacteria, provided that efficient expression of the luciferase genes is achieved. This is made possible by transcription from a powerful Gram-positive promoter, and translational coupling of *luxA* to another truncated gene product (Jacobs *et al.* 1991). In addition, the use of a *luxAB* fusion gene obviates the need for an additional Gram-positive RBS upstream of *luxB*.

A further desirable property of a cloning vector is broad host range, allowing constructions to be selected for and stably maintained in *Escherichia coli* before insertion of the final construct into the Gram-positive host. This property is required as competent *B. subtilis* cannot easily be transformed by plasmid monomers such as are created by ligation of DNA into cloning vectors (Canosi *et al.* 1978). The plasmid used by Jacobs *et al.* (1991), whilst providing for translationally coupled expression of a *luxAB* fusion gene from a powerful promoter followed by a transcription terminator cannot be maintained in Gram-negative bacteria. This, coupled with a lack of suitable cloning sites for insertion of other promoters makes this plasmid an inappropriate choice for further genetic manipulation.

Three new *lux* expression vectors, pSB322, pSB327 and pSB330, based upon the pMK4 replicon have been reported (Hill *et al.* 1993). These plasmids, with their modified sequences upstream of a strong Gram-positive RBS which lies upstream of a *Vibrio harveyi luxAB* fusion, provide an *Eco*RI site for the cloning of promoters in any reading frame in order to permit translational coupling to the *lux* genes. The above promoterless constructs are sufficiently dark in *E. coli* and *B. subtilis* to allow rapid detection of promoter elements cloned into this background. Although these vectors have proved effective in attaining high light emission from a number of Gram-positive organisms, their utility in spore construction is limited by the segregation to the mother cell of pMK4-based vectors during sporulation of *B. subtilis* (Sullivan *et al.* 1984).

In order to circumvent such problems, a new family of *lux* expression vectors based upon shuttle vector pHPS9 (Haima *et al.* 1990) has been constructed: pSB354, pSB355 and pSB356 (Fig. 5). In common with previous constructs these plasmids allow the insertion of promoter elements in any reading frame to give translational coupling (a prerequisite for high light expression). These new plasmids can be used in *E. coli*, which facilitates cloning steps, while in *B. subtilis* they are maintained using the replicon of pTA1060 (Bron *et al.* 1987). Constructs based upon this vector show very high stability in *B. subtilis*, even during sporulation and germination.

3.2 Control of light emission from *Bacillus subtilis*

Temporal control of bioluminescence from recombinant *B. subtilis* has been achieved by insertion of specific promoter

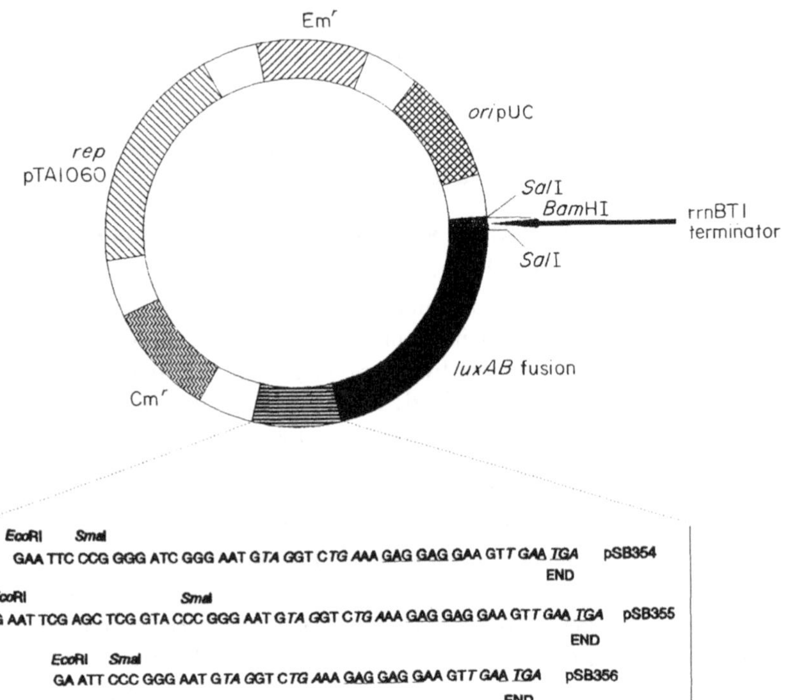

Fig. 3 The *Bacillus subtilis luxAB* promoter probe vectors

elements upstream of *luxAB*. A 460 base pair DNA fragment containing two overlapping vegetative promoter elements termed P43 (Wang and Doi 1984), was amplified with *Eco*RI flanking sequences using the polymerase chain reaction (PCR). The *Eco*RI fragment was inserted into pSB354 in order to provide for translational coupling to the *luxAB* gene. The promoter of the *B. subtilis sspB* gene, which codes for a spore associated small acid soluble protein, SASP-2 (Connors *et al.* 1986), was also inserted into pSB354 as an *Eco*RI fragment following its excision from plasmid p1273 (Setlow, unpublished).

As can be seen in Fig. 4a, *P43* does not appear to direct transcription of the *lux* genes until germination and outgrowth are complete, as would be expected from a vegetative promoter element. In contrast, Fig. 4b shows that light emission from germinating spores containing *luxAB* under the control of P_{sspB} occurs almost instantaneously upon the start of germination, indicating that luciferase has been packaged into the spore during sporulation. It is anticipated that the use of a germination specific promoter element will

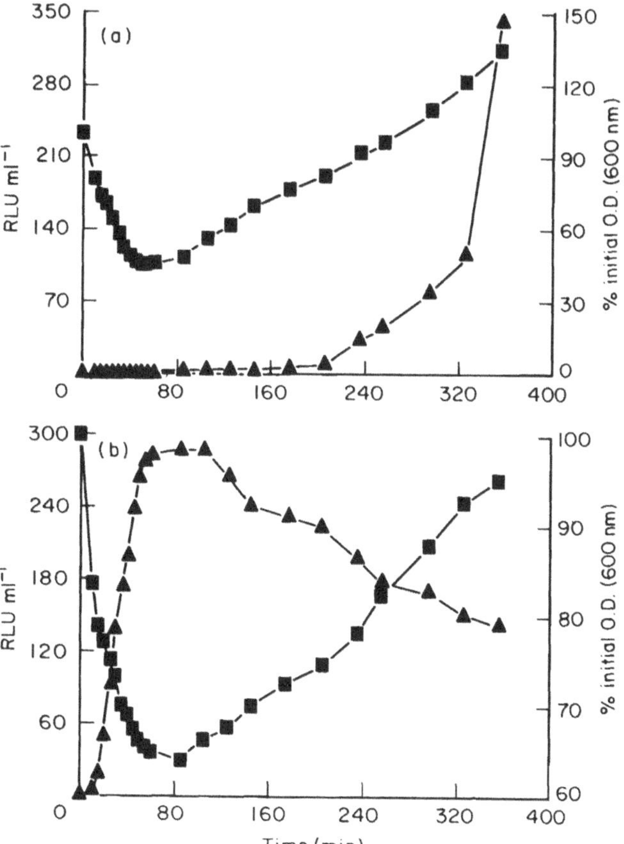

Fig. 4 Light emission by germinating *Bacillus subtilis* spores containing *luxAB* under the control of different promoters: (a) *P43*, (b) P_{sspB}. Squares denote optical density, triangles denote bioluminescence data, RLU, relative light units

enable the presence of sublethally injured spores, which may not be able to outgrow under the recovery conditions, to be detected. Such injured cells are non-recoverable and at present are undetectable by any other non-destructive method, but may represent a significant proportion of the spore population following a sub-optimal sterilization regime.

3.3 *lux*$^+$ *Bacillus subtilis* spores as monitors of ethylene oxide sterilization

Recombinant *B. subtilis* spores containing *luxAB* under control of the *P43* promoter have been assessed as biological indicators of ethylene oxide sterilization efficacy. Spores deposited on aluminium carriers were tested under conditions defined in the current UK Department of Health specification, DHSS Specification No. TSS/5/330.012—Specification for biological monitors for the control of ethylene oxide sterilization (475 mg l^{-1} ethylene oxide, 29°C). The apparatus used for this protocol was as described by Dadd and Daley (1980).

The inoculated carriers were placed in the exposure apparatus which had been prewarmed to 29°C. An initial vacuum of 500 mmHg was drawn over a period of 60 s and sufficient cold liquid ethylene oxide was admitted to attain a concentration of 475 mg l^{-1}. The apparatus was maintained at 29°C and at the end of exposure period the ethylene oxide was removed by evacuation and air admitted through a sterile filter.

Figure 5 shows the recovery of bioluminescence by *B. subtilis* spores following exposure to ethylene oxide for periods of 25, 50, 75 and 100 min. The data indicate that

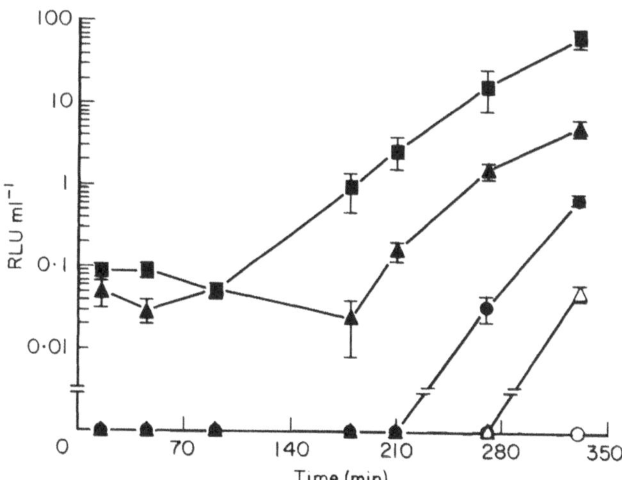

Fig. 5 Recovery of bioluminescence after challenge of *lux*$^+$ *Bacillus subtilis* spores with ethylene oxide. RLU, relative light units. ■, No challenge; ▲, 25 min challenge; ●, 50 min challenge; △, 75 min challenge; ○, 100 min challenge

light emission during the germination and outgrowth of spores after exposure to ethylene oxide is dramatically reduced, commensurate with injury and destruction of spores. The *D*-value calculated from direct enumeration via 2-d growth for plate count was approximately 31 min; the *D*-value calculated from light emission after 7 h was approximately 32 min, giving great confidence in the future research and commercial development of *lux*$^+$ recombinant bacilli as biological indicators of inimical processes.

4. REFERENCES

Bron, S., Bosma, P. Vanbelkum, M. and Luxen, E. (1987) Stability function in the *Bacillus subtilis* plasmid pTA1060. *Plasmid* **18**, 8–15.

Carmi, O.A., Stewart, G.S.A.B., Ulitzur, S. and Kuhn, J. (1987) Use of bacterial luciferase to establish a promoter probe vehicle capable of nondestructive real-time analysis of gene expression in *Bacillus* spp. *Journal of Bacteriology* **169**, 2165–2170.

Canosi, U., Morelli, G. and Trautner, T.A. (1978) The relationship between molecular structure and transformation efficiency of some *S. aureus* plasmids isolated from *B. subtilis*. *Molecular and General Genetics* **166**, 259–267.

Connors, M.J., Mason, J.M. and Setlow, P. (1986) Cloning and nucleotide sequencing of genes for three small, acid soluble proteins from *Bacillus subtilis* spores. *Journal of Bacteriology* **166**, 417–425.

Dadd, A.H. and Daley, C.M. (1980) Resistance of microorganisms to inactivation by gaseous ethylene oxide *Journal of Applied Bacteriology* **49**, 80–101.

Gould, G.W. (1964) Effect of food preservatives on the growth of bacteria from spores. In *Microbial Inhibitors in Food 4th International Symposium on Food Microbiology* ed. Molin, N. pp. 17–24. Sweden: Almvist and Wiksell.

Haima, P., Bron, S. and Venema, G. (1990) Novel plasmid marker rescue transformation system for molecular cloning in *Bacillus subtilis* enabling direct selection of recombinants. *Molecular and General Genetics* **223**, 185–191.

Hill, P.J., Rees, C.E.D., Winson, M.K. and Stewart, G.S.A.B. (1993) The application of *lux* genes. *Biotechnology and Applied Biochemistry* **17**, 3–14.

Jacobs, M., Hill, P.J. and Stewart, G.S.A.B. (1991) Highly bioluminescent *Bacillus subtilis* obtained through high-level expression of a *luxAB* fusion gene. *Molecular and General Genetics* **230**, 251–256.

Karp, M. (1989) Expression of bacterial luciferase genes from *Vibrio harveyi* in *Bacillus subtilis* and *Escherichia coli*. *Biochimica et Biophysica Acta* **1007**, 84–90.

Scott, I.R. and Ellar, D.J. (1978) Metabolism and the triggering of germination of *Bacillus megaterium*. *Journal of Biochemistry* **174**, 627–640.

Stewart, G.S.A.B. and Williams, P. (1992) *lux* Genes and the applications of bacterial bioluminescence. *Journal of General Microbiology* **138**, 1289–1303.

Stewart, G.S.A.B., Smith, A.T. and Denyer, S.P. (1989) Genetic engineering of bioluminescent bacteria. *Food Science and Technology Today* **3**, 19–22.

Sullivan, M.A., Yasabin, R.E. and Young, F.E. (1984) New shuttle vectors for *Bacillus subtilis* and *Escherichia coli* which allow rapid detection of inserted fragments. *Gene* **29**, 21–26.

Wang, P.-Z. and Doi, R.H. (1984) Overlapping promoters transcribed by *Bacillus subtilis* sigma–55 and sigma–37 RNA polymerase holoenzymes during growth and stationary phases. *Journal of Biological Chemistry* **259**, 8619–8625.

Journal of Applied Bacteriology Symposium Supplement 1994, **76**, 135S–138S

Subject index

Acidification, 108, 109
Advisory Committee on the Microbiological Safety of Food, 115
AGFK, 10, 11
Alanine, 10, 11
Alicyclobacillus, 2, 5
Alkaline thioglycolate, treatment with, 117
Amphibacillus, 2, 5
Antibiotics, 28–31, 43, 124
Aquatic environments, *Cl. botulinum* spores in, 106
Aseptic filling systems, 67, 69
Aseptic food processes, requirements for, 77
Aspergillus, 108
Autolysins, bacterial, 30, 31, 33, 34
Autolysis, 30

β-lactam antibiotics, 28
Bacillus, 2–5, 17, 25–36, 49–58, 81, 100
Bacillus
  acidocaldarius, 2
  acidoterrestris, 2
  anthracis, 6
  cereus, 6, 29, 32, 35, 50, 53, 56, 61–66, 83, 122
  circulans, 2, 84
  coagulans, 83
  cycloheptanicus, 2
  licheniformis, 108
  megaterium, 2, 18–21, 28, 29, 32, 35, 36, 50, 51, 53, 56
  pasteurii, 2, 4
  polymyxa, 77
  pumilis, 92
  sphaericus, 2, 27, 29, 93
  stearothermophilus, 45, 70, 72, 75, 77
  subtilis, 2, 9, 10, 14, 15, 18, 19, 25–36, 40, 43, 45, 46, 50, 53, 54, 57, 86, 87, 92, 96, 99, 100, 108, 122
  subtilis var. *niger*, 76, 77
Bacillus 'F', 84, 85
Bacillus megaterium
  bioluminescent, 129–134
  spores of, 40, 43–46, 122
Bacillus subtilis, bioluminescent, 129, 131–134
Bergey's Manual, 5
Biocides, chemical, resistance of spores to, 91–102
Bioluminescence, 129–134
Botulism, 105–112, 74

Canned foods, toxin production in, 108
Canning, 67, 69, 77
Capillary tube technique, 70–72
Carboxypeptidases, 28
Caryophanon latum, 2
Cell-wall structural dynamics, 25–36
Cereolysin, 61
Cheese-making, use of lysozyme in, 122
Chemical biocides, resistance of spores to, 91–102
Chemical resistance, of spores, 54, 55, 91–102
China, botulism in, 105–107
Chitinases, 121
Chloramine-T, 92, 94, 95, 96
Chlorhexidine, 93, 94, 99, 101
Chlorine, 94
Chlorine-releasing agents, 92, 95, 97–99
Chloroform, 92, 93
Cladosporium, 108
Cloning, 11
Clostridium, 2–5, 17, 25, 49, 56
Clostridium
  bifermentans, 92, 94, 96
  botulinum, 70, 72, 74, 76, 77, 82–85, 105–112, 115–125
  bryantii, 3, 4
  lortetii, 3
  perfringens, 32, 33, 41, 64, 84, 122, 123, 124
  sporogenes, 70, 72, 75, 77, 124
  thermosaccharolyticum, 70, 75
Combinations, of antibotulinal factors, 110, 111
Combined treatment, 100, 101
Commercially produced foods, botulism outbreaks from, 107, 108
Commitment reaction, 18, 19, 35
Conditional death, of spores, 41
Continuous flow mixing devices, 73
'Core' enzymes, 32
Cortex electrochemistry, 42
Cortex, functions of, 42, 43
Cortex hydrolysis, 20, 35
Cortex lytic enzymes, 20–22, 49, 53, 54, 122
Crab meat, pasteurized, 124, 125
Cream, UHT, 77
Cyst(e)ine content, 82
Cytosine and thymine residues (CT), 55, 56
Cytosine residues, 55, 56

D-Alanine, 18
D-values, for spores, 57, 74–77, 81, 82, 117, 134
Dehydration, 40–42
  of the spore core, 34
  of spore protoplast, 97
Demineralization, 43, 44
Denaturation, of proteins, 41
Depurination, 41
Desulfotomaculum, 3, 5
Desulfotomaculum
  acetoxidans, 3
  antarcticum, 3
  geothermicum, 3
  guttoideum, 3
  kuznetsovii, 3
  orientis, 3
  ruminis, 3
  sapomandens, 3
  thermobenzoicum, 3

Diarrhoeal type food poisoning, 61–64
Dipicolinate, 40, 42
Dipicolinic acid,
  leakage of, 97, 98
  levels of, 82
Direct heating, 67–69
Direct steam heating, 73
Distribution, of *Cl. botulinum*, 105, 106
DNA base composition, of endospore-forming bacteria, 1
DNA damage, 43
DNA depurination, 57, 58
DNA, spore, 55–58
Dole aseptic canning system, 69, 77
Dormancy, 34
  of spores, 49, 51, 52
Dry heat, 46
Dry heat resistance methods, 74
Dry heat sterilization, resistance to, 67, 69, 74, 75–77
DSC scans, of spores, 43

EDTA, treatment with, 123
End-point method, 117
Electrical heating methods, 68, 69
Emetic syndrome, 61, 64, 65
Endospore cell wall, 27, 28
Endospore-forming bacteria, 1–6
Enterotoxin, from *B. cereus*, 62–64
Environmental factors, effect on radiation resistance, 83–87
Enzymes
  repair of, 82, 83
  with lysozyme activity, 121, 122
Erca Neutral system, 69, 77
Escherichia coli, 132
Ethylene oxide, sensitivity to, 92, 99
Ethylene oxide sterilization, monitors for, 133, 134

Filibacter limicola, 2
Fish, *Cl. botulinum* spores in, 105–107
Flame sterilization, 68
Food poisoning, caused by *B. cereus*, 61–65
Food poisoning organisms, 25
Food preservatives, effect of, on *B. subtilis*, 130, 131
Food products, sterilization of, 67–69
Foods, *Cl. botulinum* spores in, 105–107
Forespore, 40, 49
Forespore core dehydration, 53, 54
Freezing, effects of, on radiation resistance, 83
Fruits, *Cl. botulinum* spores in, 105–107

Genetic approach, to germination, 9–15
Genetic mapping, of germination mutants, 10, 11
Germination-specific lytic enzyme (GSLE), 35, 36
Germinant analogues, 19
Germinant receptor, *see* Germination receptor
Germinants, 9
Germination, of bacterial spores, 9–15, 17–22
Germination genes, 11, 12
Germination mutants, 10, 11
Germination protease, 49, 52, 53
Germination receptor, 9, 10, 17, 18
Glutaraldehyde, 91, 95–98, 99–101

Haemolysins, 61, 62
Haloanaerobium, 3
Halobacteriodes, 3
Halogen-releasing agents (HRAs), 92, 98
Heat killing, 41
Heat resistance, 34, 40–46, 56–58, 81–83
  loss of, 11
  of bioluminescent and non-bioluminescent spores, 130, 131
Heat sensitivity, of irradiated spores, 84, 85
Heat treatment, effect of, 116, 117, 119–121
'High' acid foods, botulism attributed to, 108
Honey, *Cl. botulinum* spores in, 105, 106, 110
Hot-cold incubation, 62
Hydrogen peroxide, 44, 94
  spore resistance to, 54, 55
'Hydrogen-form', 40
Hydrophobicity profiles, of GerAA and GerAC, 13

Inactivation, mechanisms of, 91–102
Incubation temperature, effect of, on survival of *Cl. botulinum*, 119–121
Indirect heating, 67–69
Indirect heating methods, 70–73
Inhibition, of enzyme, 51, 52
Inhibitors, sensitivity of spores to, 123, 124
'Initiation protein', 123
Injury, radiation, of spores, 83–87
Interaction, of sporicides with the spore coat, 94–97
Iodine, 91, 92, 96
Ionizing radiation, 46
Iran, botulism in, 106, 107

Japan, botulism in, 106, 107

Kooiman method, 73

L-Alanine, 18, 19
'Late-blowing', of cheese, 122
Lifecycle, of the bacilli, 25, 26
Localization, of germination proteins, 13, 14
Log survivor–time curves, for *Cl. perfringens*, 123
Low acid aseptic systems, 77
Low temperatures, ability of *Cl. botulinum* to multiply at, 117–119
Luciferase, synthesis of, 130
Lysozyme
  effect of, 116, 117
  properties of, 121, 122, 124

MAFF Predictive Microbiology Programme, 119
Meat, *Cl. botulinum* spores in, 105, 106, 110, 119–121
Meat medium, 119–121
Membrane transport proteins, 12
Membranes, damage to, 43, 44
Metabolism, of germinants, 19

Micrococcus luteus, 121
Milk, food poisoning from *B. cereus*, 65
Milk, UHT, 68
Mineralization, 40, 42, 57, 58
MISTRESS technique, 73
Mixing method, 73
Moist heat, 46
Model food medium, for *Cl. botulinum*, 119–121
Model, for GPR processing and activity, 53
Models, probabilistic, 110, 111
Modified atmosphere, storage of fish in, 110
'Mother cell', 49
Moulds, effect of pH on, 108, 109
Muramic acid δ-lactam synthesis, 29, 35
Mutants, *see* Germination mutants
Mutations, of spores, 57

Neurotoxin, 74, 75
Neomycin, 124
Nitrite, 124
 sensitivity of spores to, 124
Non-proteolytic *Clostridium botulinum*, 115–125
Norway, *B. cereus* food poisoning in, 65

Oscillospira, 3
Octanol, 34
Ohmic system, 69
Oligopeptide transport system, 33
Osmoregulation, 97
Outbreaks, of botulism, 106–108
Oxidative killing, 41, 44–46

Packaging, of food, 108–111
Packaging materials, sterilization of, 44
Packaging, sterilization of, 69, 77
Particle systems, 73
Pasta, food poisoning from *B. cereus* in, 65
Pasteuria, 3
Pasteuria penetrans, 3
Pasteurization, of seafood, 124, 125
Penicillin, 28, 29
Penicillin-binding-proteins, 28
Penicillium, 108
Peptidoglycan hydrolysis, 20
Peptidoglycan, 26–30, 53
 degradation of by enzymes, 121, 122
pH
 effect of on toxin production by *Cl. botulinum*, 108, 109
 of forespore, 51, 52
Phase-darkening, 82
Phosphate, leakage of, 98
Phosphatidylcholine hydrolase, 62
Phosphatidylinositol (PI), 62
Phosphatidylinositol hydrolase (PIH), 62
Phosphoglycerate, 49, 52
Phosphoglycerate mutase, 49, 52
Phospholipases, 61, 62
Pilot scale systems, 72, 73

Planococcus citreus, 2
Plate heat exchangers, 68
Poland, botulism in, 106, 107
Post-germination growth system, 82
Potassium sorbate, effect of, 131
Potatoes, toxin of *Cl. botulinum* production in, 110
Potentiators, 101, 102
Pre-production sterilization, of aseptic processing equipment, 69
Predictive models, 110, 111
Preparation temperature, of spore, 57
Processing, UHT, 67–77
Protein damage, 43
Protein phosphorylation, 21
Proteins, germination, 11–13
'Pseudo-germination', 84

QACs, 91, 99

Radiation injury, 83–87
Radiation resistance, of spores, 55
u.v.-Radiation resistance, of spores, 55, 56
γ-Radiation resistance, of spores, 55
Radiation survival curves, 81–83
Radical formation, 45, 46
Refrigerated processed foods with extended durability, 115–125
Repair, of radiation damage, 82, 83
Receptor, *see* Germination receptor
Regression analysis, 110
Rice, food poisoning from *B. cereus* in, 65

α/β-type SASPs, 57
Salmon, *Cl. botulinum* in, 125
SASPs, *see* small, acid-soluble, spore proteins
Scraped surface heat exchangers, 68
'Sealed structure', 101
'Shoulder', 81–83
Shelf-life, extended, 115–125
Sigma factor, sporulation-specific, 30, 31
Small, acid-soluble, spore proteins, 49, 52–58
Sodium chloride, 85, 86
Sodium nitrite, 86
Sous vide foods, 115
Soviet Union, former, botulism in, 106, 107
Sphingomyelinase, 61
Spo0 genes, 33
Spore coat, role of in biocide resistance, 91, 92, 94, 95
Spore core
 dehydration of, 53, 54
 water content in, 58
Spore cortex, 26–32, 49, 53, 54, 92–97
Spore death, 97
Spore germination, 9–15, 17–22, 30
 inhibition of, 99
Spore heat resistance, determination of, 70–77
Spore photoproduct, 55
Spore protoplast, dehydration of, 97
Spore resistance, 49–59, 67–77
Spores, tolerance of to ionizing radiation, 81–88

Sporicidal agents, 92–102
Sporulation, 31–34
Sporohalobacter, 3, 5
Sporohalobacter
  lortetii, 3, 5
  marismortui, 5
Sporolactobacillus, 2
Sporolactobacillus inulinus, 3
Sporosarcina, 4, 56
Sporosarcina urea, 2, 4
Steam generator, 69
Steam infusion, 68
Steam injection, 68
Sterilization, commercial, 67–77
Stressing agents, 100
Sublethal injury, of spores, 99–101, 122
Substrate/enzyme pairs, within spores, 49, 52
Sulfobacillus, 4–6
'Superdormant' spores, 10
Surface stress theory, 33
Survival, of spores of *B. subtilis*, 49–59
Survivor curves, 72–74, 117
Survivor tails, 72
Syntrophomonas wolfei, 4, 6
Syntrophospora bryantii, 4, 6
Synergistic effects, 84–87

'Tail', 81–83
Targets, of heat damage, 43, 44
Temperature
  effects of on radiation resistance, 83
  optimum, for colony formation, 124
Tetracycline resistance proteins, 12
'Thermal annealment', 83
Thermal death time cans, 72
Thermal death time tubes, 70–72
Thermoactinomyces, 4–6, 56
Thermophiles, 41
'Thermoradiation', 85
'Thermorestoration effect', 83
Thioglycollate, 124
Thymine residues, 55, 56
Thiobacillus ferroxidans, 4–6
Toxins, of *B. cereus*, 61–66
Transpeptidation reactions, 28
Transposon Tn917, 11
Trigger mechanism, 17–24
Trout, *Cl. botulinum* in, 111, 118
Tube systems, 70–72
Tubular heat exchanger, 68
Turgor pressures, of spores, 42
Turkey roll, *Cl. botulinum* in, 110

UDS, 92, 93, 98
UHT food processes, 67–77
Ultra heat treatment processes, 67–77
UME treatment, 92, 93
United States, botulism in, 105–107

Vacuum-packed trout, *Cl. botulinum* in, 111
Vectors, plasmid, 131, 132
Vegetables, *Cl. botulinum* spores in, 105–107
Vegetative cell wall, 26, 27
Vegetative growth, 28–31, 33
Vero cells, 63–64

Water activity, 42
  effect of on growth of *Cl. botulinum*, 109–111
Water content, of the spore's, coat and cortex, 51, 58
Wet heat resistance, 67, 69, 74, 75
Wet heat resistance methods, 70–75

Yeasts, effect of on pH, 108, 109

Zymogen, 52

Printed and bound in the UK by
CPI Antony Rowe, Eastbourne

Printed and bound by CPI Group (UK) Ltd, Croydon, CR0 4YY

27/10/2024

14580391-0005